Zukunftstechnologien für den multifunktionalen Leichtbau

Reihe herausgegeben von
Open Hybrid LabFactory e. V.

Ziel der Buchreihe ist es, zentrale Zukunftsthemen und aktuelle Arbeiten aus dem Umfeld des LeichtbauCampus Open Hybrid LabFactory darzustellen. Es werden neue Denkansätze und Ergebnisse aus der Forschung zu unterschiedlichen Bereichen hybrider und multifunktionaler Bauteile vorgestellt. Dazu zählen Publikationen aus Forschung, Entwicklung, Produktion und Prüfung dieser materialhybriden Komponenten. Die Anwendbarkeit und der Nutzen für den industriellen Einsatz stehen im Vordergrund der Beiträge und sichern nachhaltige Fortschritte in der Fahrzeugentwicklung. Ebenfalls Ergebnisse und Berichte von Forschungsprojekten im Rahmen des durch das Bundesministerium für Bildung und Forschung geförderten ForschungsCampus werden veröffentlicht und Konferenzbände von Fachtagungen im Kontext der Open Hybrid LabFactory werden erscheinen.

Die Bände dieser Reihe richten sich an Wissenschaftler aus den Bereichen der Material-, Produktions- und Mobilitätsforschung und an industrielle Experten der Branchen Technik, Anlangen- und Maschinenbau, Automobil & Fahrzeugbau sowie Werkstoffe & Werkstoffverarbeitung. Der Leser profitiert davon, dass ihm durch diese Buchreihe ein konsolidiertes Angebot der öffentlichen Beiträge zu aktuellen Forschungsprojekten dargeboten wird.

The aim of this new book series is to present key future topics and current work from the LeichtbauCampus Open Hybrid LabFactory. It presents new approaches and research findings from different areas of hybrid and multifunctional components, including publications from the research, development, production and testing of material hybrid components. The contributions focus on applicability and benefits for industrial use and ensure sustainable progress in vehicle development. It also publishes the results and reports from research projects within the framework of the ForschungsCampus scheme funded by the German Federal Ministry of Education and Research, as well as the proceedings of conferences in the context of the Open Hybrid LabFactory.

The volumes in this series address scientists from the fields of materials, production and mobility research and industrial experts in the fields of technology, plant engineering and mechanical engineering, automotive & vehicle construction and materials & materials processing, offering readers a consolidated range of public contributions to current research projects.

Weitere Bände in der Reihe http://www.springer.com/series/16103

Klaus Dröder • Thomas Vietor (Eds.)

Technologies for economical and functional lightweight design

Conference proceedings 2018

Springer Vieweg

Editors
Klaus Dröder
Technische Universität Braunschweig
Braunschweig, Germany

Thomas Vietor
Technische Universität Braunschweig
Braunschweig, Germany

ISSN 2524-4787 ISSN 2524-4795 (electronic)
Zukunftstechnologien
für den multifunktionalen Leichtbau
ISBN 978-3-662-58205-3 ISBN 978-3-662-58206-0 (eBook)
https://doi.org/10.1007/978-3-662-58206-0

Library of Congress Control Number: 2019935981

Springer Vieweg

This Springer Vieweg imprint is published by the registered company Springer-Verlag GmbH, DE
part of Springer Nature
The registered company address is: Heidelberger Platz 3, 14197 Berlin, Germany

Preface

Every two years, Technische Universität Braunschweig in collaboration with ITS Mobility, organises the conference “Faszination hybrider Leichtbau”. The conference brings together experts from research and industry, in order to address leading research priorities and future challenges of the industry in the field of hybrid lightweight Design. The conference is held in the city of Wolfsburg, Germany.

Initiated in 2003, the conference was held for the seventh time in 2018. As in previous years, hybrid and functional lightweight design continued to be the focus of the event. Lightweight design is a key technology in the development towards sustainable and resource-saving mobility. This applies in particular to the automotive industry, as automobile manufacturers are operating in a field of tension between customer requirements, competition and legislation. Hybrid structures combine the advantages of different materials (e.g. fibre-reinforced plastics and metals) and have a high potential for minimizing weight, while simultaneously expanding component functionality. The future efficient use of function-integrated hybrid structures in vehicle design requires major innovations in vehicle development and production technology. This constitutes a great need for new methods and technologies for economical lightweight design in mass production – in particular when taking into account the increasing requirements in terms of variant diversity, safety and quality.

Consequently, the focus of the conference was on new methods and technologies for the development and production of function-integrated, hybrid and high volume ready lightweight solutions in vehicle design, as well as, the exchange of experience between experts from industry and science.

The program included 50 presentations and was framed by top-class keynotes, which broadened the horizon of this conference to digitalization, electro mobility and autonomous driving. A particular highlight was the opening lecture of Johann Jungwirth (Executive Vice President of Mobility Services, Volkswagen Group of America). We would like to thank our speakers and the over 250 participants and are pleased to present these conference proceedings.

Wolfsburg, 23rd November 2018

Prof. Dr.-Ing. Klaus Dröder

Prof. Dr.-Ing. Thomas Vietor

Table of Contents

Projects within the Open Hybrid LabFactory

MULTI MATERIAL DESIGN. A CURRENT OVERVIEW OF THE USED POTENTIAL IN AUTOMOTIVE INDUSTRIES

Benjamin Bader[1], Eiko Türck, Thomas Vietor

[1] Technische Universität Braunschweig, Institut für Konstruktionstechnik, Open Hybrid LabFactory, Hermann-Münch-Straße 2, 38440 Wolfsburg, b.bader@tu-braunschweig.de, www.tu-braunschweig.de/ik

Keywords:

Potentials, Multi Material Design, Automotive Engineering, Component Overview

ABSTRACT

Multi material design represents the current forefront of the lightweight design trend for automotive mass production. The combination of different materials (such as metals and fiber reinforced composites) can be used to improve components in terms of various requirements (component properties, manufacturability, costs, etc.). With the reduction of component weight, low costs and feasibility in mass production, hybrid design approaches are the ideal compromise for the mass automotive lightweight construction of the future. However, there is currently no selection strategy for a specific definition of potential hybrid component areas. The basis for developing a selection strategy for potential vehicle sections in multi material design is a comprehensive overview of current examples. The acquisition and analysis of more than 150 components from the automotive industry and research make it possible to identify the main implementation features and drivers of hybrid design in automotive engineering. This contribution shows that the potential of hybrid design approaches in vehicle technology can usually be limited to a few arguments, such as cost-efficiency or weight reduction. The article shows that the hybrid design is marketed as an innovation feature. A particularly suitable group of components for the application of hybrid design could only be identified to a limited extent. A statement about the “correct” use of multi material design based on the compilation of current examples could not be identify clearly. This confirms the need for a targeted selection strategy for multi material components or an improvement indicator.

1 INTRODUCTION

Today’s drivers of society, such as compliance with regulatory CO_2 limits, comfort requirements of the customer and an acute range problem of electric vehicles, causes the automotive industry to build their vehicles in a lighter way. Additional weight due to larger accumulators, additional comfort features, customization measures and multimedia equipment can be compensated by reducing weight elsewhere in the vehicle. The predominant mass shares of the vehicle are after Schöpf [1] the vehicle body and the drive technology. These provide the best basis for applying lightweight design in the vehicle. For implementing lighter vehicles, the example of automotive body in-in-white is followed by two significant approaches. In the field of high-class vehicles, with low and medium quantity (<200,000 pcs/annum), light metals and fiber reinforcement plastics are increasingly used. Vehicle bodies, such as these of the Mercedes AMG GT or the Audi A8, consist of a high percentage of aluminum. Partially magnesium is used in the area of the outer skin and the attachments as well. Similar extremes are to be noted in bodies with a high proportion of fiber reinforced plastics, as in the VW XL1 or the BMW i3 / i8. The Body of these vehicles is characterized by spaceframe and monocoque design or a hybrid form of it. In the high-volume mass production of automobiles, such as the Volkswagen Golf or the Toyota Auris (> 1,000,000 pcs/annum), lightweight design measures are mainly characterized by the use of

K. Dröder et al (eds), *Technologies for economical and functional lightweight design*, Zukunftstechnologien für den multifunktionalen Leichtbau, https://doi.org/10.1007/978-3-662-58206-0_1

high-tensile steel materials. These can be adjusted in their properties to the respective component requirements and hold, by saving material with the same or improved component properties, a high lightweight design potential. In addition, the use of these steel materials can be integrated without extensive expenses (monetary and structural) in the current production line. The gap between the lightweight steel design in high circulation and the use of light metals and fiber composite plastics in the small and medium series can be assigned to the multi material design.

1.1 Definition of Multi Material Design

Hybrid design is defined in the current literature as the best or targeted combination of different materials to fulfill the technical task of the lightest possible component in an economical way [2, 3]. In addition to the common definition Kleemann et al. [4] divide the multi material design (MMD) into three different levels. The level is fixed to the specific combination of material groups. The lowest level is characterized by a mixture of similar materials, such as the combination of steel alloys in the subgroup ferrous metals. The Level 2 hybrids are characterized by the combination of different material subgroups, such as ferrous metals with non-ferrous metals in the main group of metals. A Level 1 Hybrid is defined as a combination of different major groups, such as the combination of metals with fiber composites or nonmetals. The combination of materials and semi-finished products must always be understood within the limits of a component.

In extension of the plane model of hybrid construction according to Kleemann et al. [4] the boundary of a multi material part for this contribution is defined as follows.

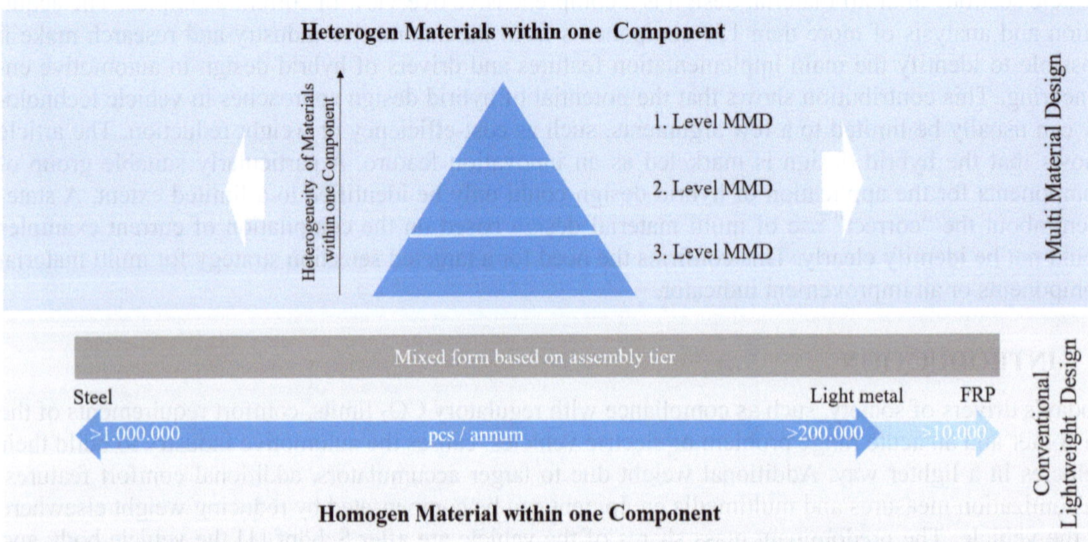

Figure 1: Context multi material design and conventional lightweight design based on Kleemann et al. (2017) [4]

1.2 Challenges and focus of contribution

In the design process of components many higher-level factors play a role in measuring the success of a component. In the case of conventional lightweight solutions, the focus is often on reducing component weight and improving component properties. The evaluation of conventional lightweight design measures can be based on evaluation parameters such as the specific stiffness or the lightweight design grade according to Klein [2]. Well-known lightweight strategies provide additional guidance in the implementation of possible lightweight solutions. Furthermore, there is a wealth of experience in the design of conventional lightweight components made of metals, light metals and fiber reinforced composites. The limits can thus be estimated very well for the developer and the predictive quality through simulation allows a specific statement about the feasibility of developed component concepts.

With regard to multi material design the participation of this knowledge is limited, since the design of hybrid components always requires (according to 1.1) the symbiosis of several materials/semi-finished products in one component. This leaves some questions unanswered which can only be partially estimated using conventional methods and existing experience. What poses new challenges for the product development, are points, such as the combination of different materials, the "correct" use of materials, the high degree of freedom in practicable solutions and usable potential. Since, according to the current state of the art, only limited solutions exist for the evaluation of hybrid components. This contribution is part of the development of a method for evaluating components for implementation in multi material design with optimum potential utilization. As a first overview, the article presents the current state of the art and deals with the specific features of hybrid design methods.

2 METHODOLOGY

For determining and evaluating potentials of multi material design, this article describes the preparation of a market study. This survey deals with the elicitation of the widest possible scope of application of the design in the automotive industry. Therefore, the survey aims to show as an overview a current status which is still in process. In the following, the boundary conditions and the format of the survey are shown. The results are presented in the form of a chronological overview as well as in an evaluation of specific features and potentials of multi material design from research and automotive applications.

2.1 Boundary conditions and format of the survey

In order to be able to depict the current state of multi material design, a more than 2.5-year market research has been carried out. This includes determining specific use cases of the last 30 years. The focus of the study is on structural components made of metals and / or thermoplastic fiber reinforced plastics used in the automotive industry and the associated research with MMD levels (HYL) 1 and 2. The study was created as a collection of components and demonstrators in the form of a table-based database. The sources are information from suppliers, service providers, OEMs and manufacturers of production equipment for the automotive industry or the automotive environment. The cases of application are taken from the usual technical literature, the specific trade press, trade fairs and press releases. The search and selection of possible applications took place in two stages. In the first step what has been selected, were all applications in multi material design and some fiber-reinforced composite components with properties of particular interest. Thereby the considered components must be a structural component or have a reference to the vehicle structure. In a second stage, the component database has been summarized according to the level of maturity and assessable information of the use cases. As a basis for this investigation, 147 use cases remain after the second stage.

2.2 Categorization and structure of the component database

The cataloguing of the use cases takes place in the component database in the form of a tabular list. In addition to the serial number, an illustration, the year of publication and the component name and the use cases are classified in the list according to nine other criteria (see Table 1). As a representative example, this table shows a multi material rear wall of a Mercedes Benz S Class exhibited at the Lanxess booth at K2016 traid fair.

Using the defined criteria, it is possible to create a precise idea of the structure, the materials used, the integration of the individual component in the vehicle and the represented benefits of using the multi material design method. The most important comparative figures are the criteria of the number of pieces and the readiness level. Booth of these allow a realistic statement about the actual reasons for using multi material design. In the case of all publications, it must be assumed that prototype components or concepts can only be implemented to a limited extent and their proclaimed features may be positively exaggerated. Quantity and readiness level are used as weighting in the evaluation of the characteristics and thus reinforce the actually exhausted potentials of hybrid design in addition to the occurrence of a component in the database.

Table 1: Sorting criteria of multi material parts overview

	Possible Content	...	Example
Nr.	1 … 147		1
Figure			
Name	Multi Material Part		Rearwall
HYL	1,2,3		1
Range of application	Structural component Doors, flaps, lids, hoods chassis interior drivetrain vehicle safety holder, brackets, hinges wheels demonstrator		Structural component
Manufacturer	Research Institute OEM Partner		LANXESS Daimler AG
Readyness	Research demonstrator Industrial Demonstrator Serial component (Model, Date)		Serial component, Mercedes Benz S Class (222)
Circulation	Pieces per day		approx. 200
Material information	Material T Material U		Steel (n.n.) Durethan BKV 60 H2.0EF
Composition	Spec. Material V Spec. Material W Internal Material Connection		Metal sheet component Ribs by injection molding Connection by form closure
Connection to Subsystem	By adhesive By welding By joining technology X		Integration in conventional welding and adhesives production line by spot-
Date	2018		2013
Source	Link, Reference Y		[Lanxess, K2016]
Benefit	Weight reduction Benefit Z		Weight reduction Increased stiffness

3 MULTI MATERIAL DESIGN IN AUTOMOTIVE INDUSTRIES AN FURTHER RESEARCH

Multi material design in automotive industry – With the use of a hybrid frontend, a multi material component is being used for the first time in large-scale automobile production. In 1998, Ford installed in the Model Focus (C170) a plastic-coated multi-layered sheet metal component. With this, the manufacturer places the first significant component in multi material design of the last 30 years in a production vehicle [5]. In the following year, Audi equips the A2 light-weight vehicle with a frontend in multi material design as well. In the years following the turn of the millennium, the design is used in countless other hang-on and structural parts. In order to give an overview of the last 30 years in the automotive industry, a summary of the component database is shown in Figure2. In the process, representative components have been selected, through which the milestones of the multi material design can be demonstrated. The examples are arranged according to the year and the approximate number of pieces (estimated from the date of the vehicle or taken from the source). In this case, only one source or the respective vehicle was named, although several companies and research institutions could be involved in the development and implementation of the application examples.

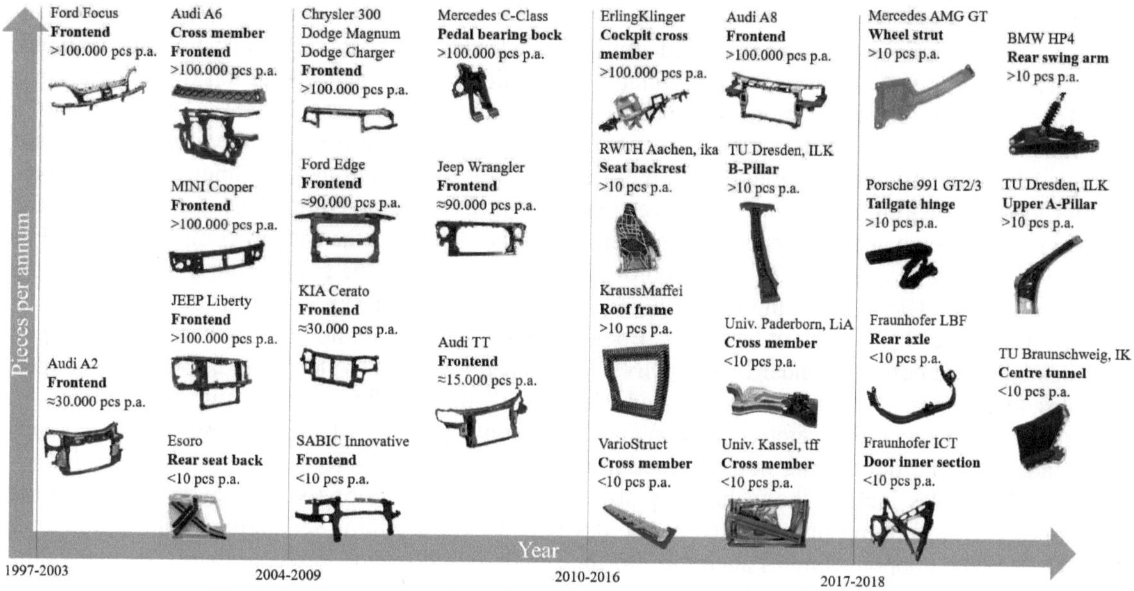

Figure 2: Time course of multi material design in automotive Industries
Additional picture sources: [5, 6, 7, 8, 9,12,13,18,19,20,21,22,23,24,25]

The first milestone in the history of multi material design in the automotive industry is the fundamental combination of polymers and metals in a mass-produced component. In most cases, one or more metal parts, later also fiber reinforced plastics, serve as a supporting structure. To realize complex structures in the connection and integration of add-on parts as well as structural stiffening of the basic structure, for example in the form of ribs, thermoplastics are used in the molten state due to their high flexibility of form. Short and long fiber reinforced thermoplastics are applied by injection molding or extrusion molding. From the beginning, the combination of different materials is the biggest challenge in the focus of the design approach. Possible solutions for the connection problem are form closure, adhesion promoter systems and adhesives. For the first components around the turn of the millennium, the connection via form closure connection is preferred. The expression of form closure is defined in this case by the term “undercut”, as they arise during the injection molding or encapsulation of a perforated sheet metal semi-finished product in the form of buttons or collar. The roof cross member and the frontend of the Audi A6 C5 from 2001 can be cited as representative examples. In the later stages of multi material designs, surface structuring by machining, laser treatment or the welding of pin structures on the sheet metal semi-finished products, extend the classical application. In contrast, technologies such as primer systems and adhesives usually have to be applied to the corresponding semi-finished products in an additional process. That’s why these connection methods where considered as uneconomic in the early phase. From the years around 2010, the philosophy of multi material design changes significantly. The industry’s ambition to use the design as a cost-effective lightweight alternative for mass production implies efficient production with good integrity in existing industrial processes and low cycle time. This characterizes the concept of the multi material component, created in the so-called “one-shot” process [10]. The second milestone in the timeline of multi material design is therefore the industrial use of bonding agent systems. Examples include the cockpit cross member and the lock carrier of ErlingKlinger AG from 2015 [11]. The cross member was the first component in the form of an aluminum based structure in combination with a powder bonding agent system and short fiber reinforced plastic in the automotive mass production (Mercedes Benz). Another difference in the development of the multi material design is expressed around the year 2010. Until 2010, the automotive industry (OEM) and their supply chain (plastics industry, machinery and plant construction, tool manufacturers) involved in the implementation of serial components, as of 2010, more and more research institutions (universities, research societies) engaged in the development of this design approach and the production technologies. In addition, the period after 2010 is characterized by diverse, new technology forms in the multi material design. The combination of metal shells with cast rib structures [12], use of biopolymers and natural fibers in structural components [13] as well as the intensive use of continuous fiber reinforced semi-finished products with thermoplastic matrix (tapes, laminates, organic sheets) in load path equitable production or even as substitution of the metal part [14, 15] show manifold applications of hybrid design in industry.

Multi material Design in Research – Research generally forms the basis for developing new technologies. In terms of multi material design it exists a variety of different test specimens and test setups similar to the industrial use cases. In the investigation of multi material composites, the focus is on the material combination and its connection, the achievable properties of the overall composite as well as the manufacturing processes and production of specimens regarding the application in the automotive mass production. The basis of the research are standard tests already known from material research, such as the tensile shear-, 3 point bending-, 4 point bending test or a torsion test, with classical coupon samples based on defined standards (e.g. DIN 53283 or DIN EN ISO 178). In addition to the coupon samples, test specimens have been established which deal with specific issues of multi material design. The specific test specimens referred to below as technology specimens focus on new connection concepts, adapting and developing new production methods as well as determining optimal process parameters. Examples include the technology specimen of the Cluster of Excellence "Merge" and the developed thermoforming process [16] or the developed hybrid joining technique of the Fraunhofer IWS [17]. Since coupon samples according to the standard and extended technology specimens take into account the geometric characteristics of a possible application only to a limited extent. The experiments are carried out in an approximate form usually with specimens close to the application (e.g. U-shaped or hat-profiles). Probably the best-known example of applied research in the field of multi material construction is the "Erlanger" carrier. Basic questions were answered on the basis of the test specimen in preparation for the later series production of the Ford Focus C170 front end or the roof cross member of the Audi A6 C5 and various other applications. The spectrum of investigations includes possible combinations of materials and semi-finished products, the combination of materials (form fit, material bond), the manufacturing processes (injection molding, extrusion molding) or the structural design (integration of functional elements, maximum performance, minimum weight, etc.). Figure 3 shows the application spectrum of a hat profile with ribbing, the same or similar "Erlanger" carrier, in relation to its component applications.

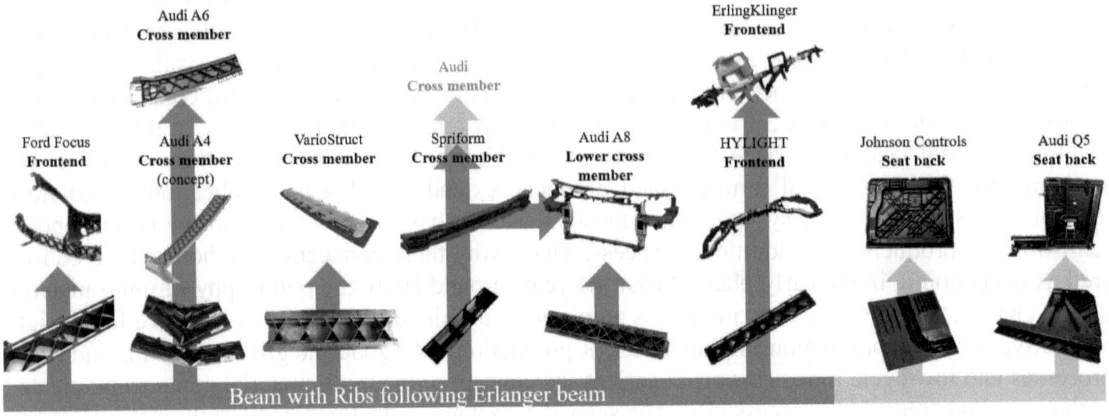

Figure 3: Ribbed profiles as research focus of multi material design in automotive Industries
Additional Picture Sources: [5, 6, 7, 8, 9,12,13,18,19,20,21,22,23,24,25]

4 CHARACTERISTICS OF MULTI MATERIAL DESIGN

To analyze the specific features and potentials of multi material design, the individual components are summarized in groups. The parts are sorted by the subordinate category and frequency of entries in the part database. The main evaluation aspects, the characteristics and the potentials of all the different types of components with the highest frequency, are shown in Table 2. Parts with low quantity in the database are not shown in Table 2. However, they are included in the following overall analysis.

Table 2: Characteristics and released potentials of multi material parts

Component	No. of Parts in Database	Characteristics (geometrical, composition, average HYL)	Released Targets/Potentials
Frontend/ Radiator Support	18	Multi-part sheet metal shells Reinforcement ribs from SFC Partial reinforcement by organic sheet and tape material Functional elements and connections from SFC Ø HYL: 1	Weight reduction Increasing component space Increasing strength Simplified montage of attachment parts Reduction of production complexity Functional integration Economic efficiency (cost neutral) Saving corrosion protection by means of adhesion promoter layer
Seat back/ Seat shell	15	Body of organic sheet and tape material as substitution for sheet metal Reinforcement ribs from SFC Partial reinforcement by sheet metals or profiles Ø HYL: 2	Weight reduction Increasing component space (increase comfort) Increasing strength/rigidity Energy absorption capacity in case of crash Functional integration Economic efficiency (cost neutral) Requirement-appropriate design for various variants
Pillars (a-, b-)	13	Body of sheet metal Full and partial reinforcement by organic sheets, tape material, high strength steel and FRP patches Reinforcement ribs from SFC o. LFC Ø HYL: 1	Weight reduction Part integration Reduction of production complexity Economic efficiency (cost neutral) Requirement-appropriate design for various variants and crash
Cross member (roof, bumper)	12	Body made from sheet metal or organic sheet Reinforcement ribs from SFC Ø HYL: 1	Weight reduction Part integration Reduction of production complexity Reduction of process time Economic efficiency (cost neutral) Requirement appropriate design
…	…	…	…

After summarizing the information from the component database in Table 2, it should be noted that the collected components of multi material design can be subdivided into two basic approaches according to their characteristics. One category includes flat shell designs based on metal sheets or continuous fiber-reinforced semi-finished products (primarily organic sheets). In addition, functional elements and reinforcing structures made from plastics (usually fiber-reinforced) or light metals via casting processes (injection molding, die casting) and press processes (extrusion, patching) are partially applied on the shell construction. On the other hand, there are complex injection-molded components with a high level of freedom of form with load introduction elements and reinforcements (profiles, lock beading sheets) made of metals. Partially, in both approaches additional applications, such as tape materials or fabric fragments, are found for the load path-compliant reinforcement of the component made of continuous fiber-reinforced semi-finished products. Regardless of the approach, the components often occur in the form of a substitution of the materials. The geometry and the internal limits of the part can change. The outer limits of the component usually remain untouched. The integration of the components into the vehicle can thus be ensured by the joining processes established in the series, such as spot welding, screwing, adhesive bonding, etc. With the achieved goals and elevated potentials through by using multi material design, the applications of the component database focus on the economic efficiency and achieved weight reduction with constant or increased component properties. What must be mentioned to increase economic efficiency, are the potentials of component and functional integration as well as a modified (improved) manufacturing process. By integrating individual parts of an assembly into a complex component and by implementing additional functions or functional elements, as well as the production of hybrid components in one production step (one-shot), the economic efficiency can be increased. In particular, eliminating process steps, decreasing process time, saving of components and joining operations as well as tools are argued [5, 26, 27, 28]. Especially multi-layered shell components with many individual parts and interfaces to other attachment or body parts, such as front ends, door panels, A-, B-pillars, etc., are suitable for implementation of potentials named above [29, 20]. The reduction of the component weight matters in terms of multi material design. The saving of CO_2 emissions as well as the mass compensation of heavy accumulators in electric vehicles are the main drivers for multi material lightweight design in vehicle design [6, 30]. As an economical alternative to conventional, more cost-intensive lightweight design, multi material design offers the possibility of saving weight even for high-volume vehicles. The cost effectiveness always depends on a proper material application (correct material in the right place) [31]. In the majority of cases (see Figure 2), the sheet metal component remains to be made from metal and its weight is reduced by optimizing wall thicknesses and using high strength steels or light metals [30]. The additionally applied fiber-reinforced plastics ensure a component performance which is reasonable with respect to the reference structure or mostly improved with less weight. Adhesion promoter systems additionally support the flexible application of the plastic material and allow a more targeted connection to the metal semi-finished product without excess material [7]. Depending on the approach, the metal construction can also be completely replaced by a plastic construction, which is only supported by metals where the loads be applied. Thus, weight savings (e.g., at seat backs) of up to 50 % are possible by using multi material design [32]. The aspects of component integration mentioned in terms of increased cost-effectiveness and the combination of the established technologies of deep-drawing and injection molding into a compact process also contribute to reducing the weight and the required installation space of the component due to the equalized components and joining areas [8].

5 DISCUSSION

Publications from the years 2006 and 2008 represent the multi material design as already established technology in the automotive serial production [5, 8]. The component analysis of these contribution seems to confirm this impression at first glance. A good half of the examples are found in a close-to-production context or even in a vehicle. Multi material design is used especially in add-on parts (frontends, holders, pedals, etc.), in the form of injection-molded components with reinforcements made of metal or organic sheets. However, apart from the components originating in the field of plastics technology, many of the examined components cannot be identified in the high-volume segment. Components such as the roof cross member of the Audi A6 C5 were even replaced by conventional metal parts in the following vehicle generation. Multi-material design methods are often used in high-priced segment vehicles or as sample components that were produced in small quantities in a laboratory process close to mass production. Compared to pure plastic parts or sheet metal parts in steel shell construction of current vehicles, multi material design in the large-scale automotive series are therefore only conditionally considered to be established in 2018. However, Figure 2 shows a trend in which the designs could find more use in the following vehicle generations. The increasingly high number of research projects and the stable use of hybrid design approaches in the high-price segment suggest that this design approach will increase in the next decade. As a cost-effective alternative to conventional lightweight design, the cost-effectiveness for the further development of multi material design will be the decisive factor. In every publication, economics has always been associated with reducing weight and increasing component performance or the improved manufacturing process. In comparison to the determined diversity of potentials by Inkermann et al. [33] and Kleemann et al. [4] only few of the theoretically possible aspects are shown in the publications. On the one hand, it is possible that some of the potentials are subordinated to those mentioned above, or that they have not been communicated in the publications for specific reasons (secrecy, strategy, etc.). On the other hand, the question remains unanswered why only a minimal part of the potentials, reduced to the points of economy, manufacturability in the "one-shot" process and lightweight design for hybrid design, is significant. In the interplay of large-scale production and competition of conventional lightweight design approaches, these are probably the marketable aspects which make multi material design interesting. How precisely the components were selected and their potentials for improvement by multi materials design were estimated, can only be inferred from the sources to a limited extent. In view of the restricted potentials and only marginally published reasons for the application of the design on explicit components, this currently indicates a strategically shaped selection in the interests of boosting the company's image. Multi material design serves as an innovation feature for the products of manufacturers and plant manufacturers. This strengthens the chain of reasoning of marketable aspects at the heart of the development of multi material parts and emphasizes the importance of economics in design.

6 CONCLUSIONS AND FURTHER RESEACH

The study presented in this paper deals with the analysis of characteristic and potentials of components in multi material design. 147 components are described by properties such as component structure, material combination used and advantages of a multi material design. As a result of the analysis, the current state of research and industry as well as the main application reasons are shown. It should be noted in this context that the application of the design method depends on three main features. In addition to the reduction of the component weight and the shortest possible production time / a highly optimized manufacturing process, the economy is the focus of the design. All determined potentials of the design are subordinated to the cost question as a kind K.O. criterion. Furthermore, it has been found that due to the strict cost limit, multi material design can be well marketed as a type of low-cost lightweight design. The targeted use of materials creates the impression of the highest technology and innovation, which can be integrated into large scale production vehicles through low component costs. Whether the proclaimed improvements from the sources can ultimately be implemented on a scale suitable for mass production can only be assessed to a limited extent. The selection of the published applications in multi material design and the identification of potential for improvement could not be answered clearly from the sources of the study. It could only be found a tendency to attachments in injection molding and shell construction, as the most common applications. Since the component analysis is currently based on industrially influenced sources, the survey will be expanded in the further course to scientific studies on the use of multi material design methods. These could show a different perspective on the design and, assess the aspect of cost-effectiveness more neutrally. Furthermore, the influence of the "Erlanger" beam on the design is to be examined more closely. It is important to understand why exactly this geometry is particularly suitable in a large number of examples for the implementation of multi material design. The results derived from this study will, in conjunction with the overall scope of the component analysis, be incorporated into the creation of an assessment strategy of components in multi material design.

REFERENCES

[1] Schöpf, H. J., Vorlesungsskript: Kriterien zukünftiger Kraftfahrzeuge, Daimler AG, 2005.

[2] Klein, B., Leichtbau-Konstruktion, DOI 10.1007/978-3-658-02272-3, Springer Vieweg, Wiesbaden 2013.

[3] Friedrich, H., Leichtbau in der Fahrzeugtechnik, DOI: 10.1007/978-3-8348-2110-2, Springer Vieweg, Heidelberg 2013.

[4] Kleemann, S. et al., "A Semi-Formal Approach to Structure and Access Knowledge for Multi-Material-Design", DOI: 10.24355/dbbs.084-201708301114, ICED2017, Vancouver CA, 2017.

[5] Lanxess, Technische Information, Hybrid-Frontend Ford Focus, Ausgabe 18.06.2008, Leverkusen 2008.

[6] Lanxess, Presseinformation, Das erste Frontend in Aluminium-Hybrid-Technik – Geringerer Kraftstoffverbrauch, besseres Fahrverhalten, Leverkusen 2007.

[7] Evonik Industries, elements48, Quaterly Science Newsletter, Ausgabe 3, 2017, S. 6–11.

[8] Lanxess, Technische Information, Hybridbauteile in der Serienfertigung, Ausgabe 20.09.2006, Leverkusen 2006.

[9] Malnati, P., Composite World, What's new in automotive front-end modules?, Posted on 12/1/2008, URL: https://www.compositesworld.com/articles/what39s-new-in-automotive-front-end-modules (Accessed: 23.04.2018).

[10] Lanxess, Pressemitteilung, Entwicklungsservice auf Topniveau Organoblech-Hybridteile per integrativer Simulation präzise, Leverkusen 2011.

[11] ErlingKlinger, Pressemitteilung, ElringKlinger erhält weiteren großen Serienauftrag für Leichtbauteile, Dettingen/Erms 2018.

[12] v. Kulmitz, G., VarioStruct – Light Weight Structures, Road Show 2012, Aachen 2012.

[13] Kohl, D., Böhm, S., Schlussbericht zum Vorhaben: Holzformteile als Multi-Materialsysteme für den Einsatz im Fahrzeug-Rohbau Akronym: HAMMER, Teilprojekt: Integration von holzbasierten Multimaterialsystemen in Fahrzeugstrukturen durch geeignete Fügetechnologien und in Rohbaufertigungen durch geeignete Prozessketten, Kassel 2016.

[14] Esoro, E-LFT Musterbauteil, http://www.esoro.ch/deutsch/content/kernk/bautent/elft_muster.htm(Accessed: 23.04.2018).

[15] Lanxess, Technische Information, Frontend in Hybridtechnik mit Organoblech, Ausgabe 14.06.2010, Leverkusen 2010.

[16] N.N., Forschungshighlights und Technologiedemonstratoren, TU Chemnitz, https://www.tu-chemnitz.de/ MERGE/ research_highlights.php (Accessed: 18.04.2018).

[17] Zimmermann, F., Fraunofer IWS Jahresbericht 2016, Thermomechanisches Fügen von Metall und Faserkunsstoffverbund, Dresden 2016.

[18] Dynamit Nobel Kunststoff GmbH, Weißenburg.

[19] LKT, Univ. Erlangen, https://www.lkt.tf.fau.de/forschung/forschungsschwerpunkte/leichtbau-und-fvk/kunststoff-metall-hybridtechnik/ (Accessed: 26.04.2018).

[20] Küsters, K., IKV Aachen, Hylight – Innovative Hybrid-Leichtbautechnologie für die Automobilindustrie, Hochschule trifft Mittelstand, Aachen 2011.

[21] N.N., Kunststoff Magazin Online, Evonik und der Leichtbau: Aluminium-Kunststoff-Hybrid geht in die Serienproduktion, https://www.kunststoff-magazin.de/fvk-werkstoffe/aluminium-kunststoff-hybrid-geht-in-die-serienproduktion.htm (Accessed: 26.04.2018).

[22] Pfefferkorn, T. et al., Vom Laminat zum Bauteil, Kunststoffe 12/2013.

[23] N.N., Plastiker-News, LKT: Fachtagung – "Thermoplastische Faserverbundkunststoffe", http://plasticker.de/Kunststoff_News_13560_LKT_Fachtagung___Thermoplastische_Faserverbundkunststoffe?begriff=LKT&firmid=285&submit=true (Accessed: 26.04.2018).

[24] Wacker, M., Oechsler AG, Hochleistungsfaserverbunde in der Großserie, Ansbach 2012.

[25] Jäschke, A. and Dajek, U. (2004): "Dachrahmen in Hybridbauweise", Sonderdruck aus VDI-Tagungsband Nr.4260, VDI Verlag, Düsseldorf, S. 25–45.

[26] Pudenz, K. Automobil+Motoren, 800g leichter: Lanxess zeigt PKW-Sitzschale aus Hochleistungscomposite Tepex, Onlineartikel 16.10.2013.

[27] Fraunhofer (LBF,ILT), Presseinformation, Come together: Gemeinsam zur optimalen Mischabauweise, Aachen, Darmstadt 2018.

[28] N.N., Stahlblech im Alumantel, Automobil Industrie, Ausgabe 6/2010, S. 46–47.

[29] Ahrens, M. Leichtbaugipfel 2015, Leichtbaupotenzial von Leichtmetallguss-Metallhybriden, Würzburg 10.–11.03.2015.

[30] Quitter, D., Konstruktionspraxis, 3D-Hybrid-Bauweise bereit für die Großserie, Posted on 02.03.2018, URL: https://www.konstruktionspraxis.vogel.de/3d-hybrid-bauweise-bereit-fuer-die-grossserie-a-691775/(Accessed: 15.04.2018).

[31] Muhr, T. et al., Wirtschaftliche Leichtbauweise für eine hybride B-Säule, ATZ, Ausgabe 03/2015, 117. Jahrgang, S. 16–20.

[32] Schulte, T., et al., Strukturauslegung einer Multimaterialstruktur für Automobil-anwendungen, lightweight design, Ausgabe 6/2015.

[33] Inkermann, D. et al., "Ein Potentialmodell für die Nutzung neuer Technologien in der Produktentwicklung", SSP2017, Stuttgart, 2017.

DATA MINING APPLICATIONS IN MANUFACTURING OF LIGHTWEIGHT STRUCTURES

Sebastian Gellrich[1,2,3], Marc-André Filz[1,2], Johannes Wölper[1], Christoph Herrmann[1,2], Sebastian Thiede[1,2]

[1] Technische Universität Braunschweig, Chair of Sustainable Manufacturing and Life Cycle Engineering, Institute of Machine Tools and Production Technology (IWF), 38106 Braunschweig, Germany, www.tu-braunschweig.de/iwf/pul
[2] Open Hybrid LabFactory e.V., 38440 Wolfsburg, Germany, www.open-hybrid-labfactory.de/
[3] s.gellrich@tu-braunschweig.de

Keywords:

Data mining, Lightweight structures, Shop floor data acquisition, Machine state recognition

ABSTRACT

Advanced manufacturing of automotive lightweight structures implies the introduction of new process steps into traditional process chains. Due to the combination of materials and their functional integration, those process steps show increased complexity. As a result, manufacturing faces new challenges regarding a constant and high product quality. A widely discussed approach to encounter these new challenges is the analysis of manufacturing process data by applying data mining methods. Benefits of the underlying digitalization approach are found in extensive transparency, product quality assurance, decision support or even in an automated manufacturing control.

Application fields of data mining in manufacturing of lightweight structures, the design of an appropriate context-based data acquisition infrastructure and special aspects of lightweight structures manufacturing influencing the CRISP-DM data mining workflow are discussed. The application of machine state recognition in extrusion of a glass fiber reinforced plastic rib structure exemplifies the proposed aspects.

1 INTRODUCTION

Data mining is a process of analyzing a given dataset in order to derive models and detect interdependencies as well as patterns. The two main targets of data mining are description and prediction. Descriptive data mining produces new and nontrivial information based on a given dataset. Predictive data mining generates models of the system described by the underlying dataset [1]. A widely used standard of industrial applications of data mining is the Cross Industry Standard Process for Data Mining (CRISP-DM) [2]. The methodology consists of six phases with forward and backward dependencies between phases making it nonlinearly. Descriptive and predictive data mining in manufacturing can be used in different ways, such as monitoring and reporting, planning and optimizing, recognizing and reacting as well as predicting and acting. Especially predict and act offer many possible applications to support manufacturing, e.g. causal analysis of quality or manufacturing problems and prediction of failures or disturbances [3].

However, there are several reasons preventing the widespread application of data mining in manufacturing. Programmable logic controllers (PLCs) of manufacturing machines are able to exchange shop floor data in real-time, but cannot provide all relevant data for analytics, e.g. vibrations, by themselves. Consequently, reliable, accurate and efficient sensors need to be selected and installed. In the field of data integration and communication, manufacturing is faced with data from various

K. Dröder et al (eds), *Technologies for economical and functional lightweight design*, Zukunftstechnologien für den multifunktionalen Leichtbau, https://doi.org/10.1007/978-3-662-58206-0_2

sources having varying meaning, wording, units and values. Approaches for standardization like AutomationML and OPC UA try to unify data exchange, but still face considerable implementation efforts. Even the modeling of manufacturing data with data mining methods still faces obstacles because of a lack of effective, accurate and reliable algorithms and analytics methods [4].

In the automotive sector, lightweight structures are an approach to reduce tailpipe emissions during the use phase of a car. A hybrid approach, e.g. multi-material mix of metal and fiber reinforced plastics (FRP), even enhances the lightweight potential under consideration of mechanical requirements. However, in comparison to common single material structures, the manufacturing of hybrid lightweight structures is more complex due to more production steps, making process and quality control substantial challenging. As manufacturing becomes more complex, data mining can be a valuable approach for mastering the complexity. Throughout this paper, the terminology of lightweight structures is interpreted as hybrid structures in the context of the automotive industry.

From this point forward the paper is structured as follows:

Section 2 reviews aspects of the acquisition and analysis of manufacturing data, covering challenges in data acquisition as well as goals, procedures and application fields of data mining in manufacturing. Then, section 3 presents a framework for applying data mining in the context of lightweight structures manufacturing. Within the framework of context-specific application fields, a tailored data acquisition infrastructure and impacts on the CRISP-DM are discussed. Afterwards, section 4 demonstrates the application of the framework in the scope of an extrusion process of glass fiber reinforced plastic rib structure as an exemplary process. Special emphasize is on the data mining workflow and performance evaluation of different data mining methods in order to perform a machine state recognition. Finally, the essential findings of the framework are sketched and necessary research is outlined.

2 ACQUISITION AND ANALYSIS OF MANUFACTURING DATA

2.1 Acquisition of Manufacturing Data

As the underlying data set strongly influences the performance of data mining methods, accurate and reliable manufacturing data needs to be acquired [5]. The variety of data and possible data sources in manufacturing is huge and depends on the individual production process. In this way, a production process can be described by input and output variables, such as raw materials and their material characteristics. The process is designed and controlled by design and control parameters such as holding times and tool temperatures for instance in the case of hot forming. Actual process information such as actual tool positions, occurring forces and pressures is available by means of status data. Performance criteria can be determined at runtime of the process, such as time per part, or downstream in test procedures, such as wall thicknesses. Further data can be acquired through capturing

- logistic-related data,
- internal and external influencing factors such as actual humidity,
- job-related data from ERP systems or
- publicly accessible data such as weather or social media data.

However, not all data sources have to be used for a specific analysis task [3]. The performance of data mining methods, e.g. Support Vector Machine, can even decrease due to too many different features. Based on feature selection methods, it is possible to exclude irrelevant features and thus reduce the complexity and computing time of the model [6]. In the course of data acquisition, several technical challenges have to be solved and thereby pre-processing efforts have to be taken into account. An overview of challenges and examples from industrial praxis is shown in Figure 1. Depending on the application, however, further challenges may arise.

Challenge	Industrial example
Diversity of data sources	Necessity of integrating different data sources, e.g. PLCs, sensors, MES, ERP, analytics
Temporal asynchrony	Seamless procedure of data handling and transfer controlling, e.g. divergent timestamps, by utilizing communication standards like OPC UA
Inconsistent temporal resolution	Integrating inconsistent temporal resolutions, e.g. ms and s, potentially resulting in an loss of information
Lack of information content	Data without relevancy for analytics but being required for program code execution
Data interpretability	Unlabeled labled data preventing interpretability of physical meaning
Missing data	Dealing with missing data, e.g. caused by machine downtimes and maintenance
Volume, Velocity, Variety	Trade-offs between high amount of information and facilitation of a real-time manufacturing control on the one side and economic efficiency concerning data acquisiton and storage costs as well as calculation time and capability of data mining algorithms to handle high-dimensional data on the other side

Figure 1: Challenges in acquisition and preprocessing of manufacturing data [5,7,8]

2.2 Analytics of Manufacturing Data

Data mining describes a process of discovering novel patterns from a large given dataset in which algorithms are applied to extract new information. Moreover, data mining is in some cases used as a synonym for the process of knowledge discovery from data (KDD) [1]. However, the focus of KDD is on the overall process of discovering useful pattern in data, whereas data mining is one process step in this [9]. The KDD can be divided into three major phases: data pre-processing, data mining and data presentation [10].

Data mining, as a general technology, can be applied to various kinds of data as long as the given data is meaningful enough for a specific application. Depending on the goal of the process, data mining methods can be divided into predictive and descriptive (Figure 2). Even if the boundaries between these two categories are not sharp, the differentiation is helpful to understand the overall discovery goal [9].

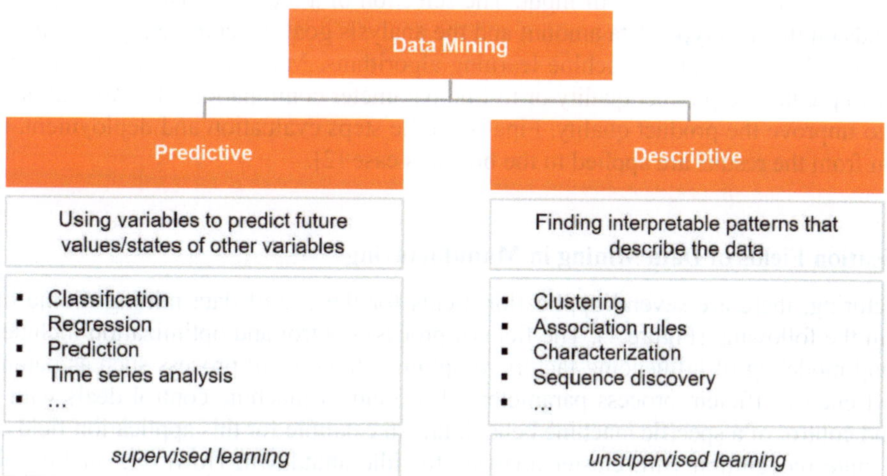

Figure 2: Overview of data mining tasks [11,12]

Prediction is based on the selection of some data or variables to predict unknown or future values of other variables. However, description tries to find patterns which outline the data in an adequate way that can be interpreted by the user [9].

In order to achieve the specific objectives of prediction and description, there are various data mining methods. Classification designates the learning of a function that assigns data elements to one of several predefined classes. Regression is the learning of a function that can predict data elements with a continuous variable. Unlike classification, clustering maps data into several classes (clusters) that are naturally defined by the data through similarity metrics or probability density models [13,14].

An often referenced process model for data mining is the CRISP-DM shown in Figure 3. It is useful for all kind of production processes. This includes the manufacturing of lightweight structures.

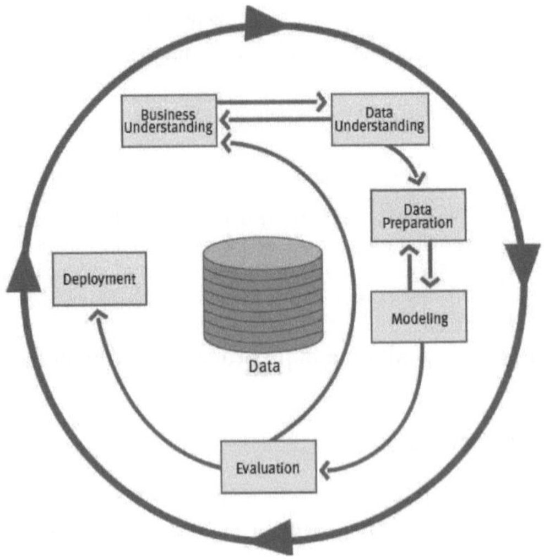

Figure 3: Framework for data mining – CRISP-DM [2]

First step to get into the CRISP-DM cycle is the business understanding. Based on the business objective, this step determines the data mining goal. The next step in CRISP-DM is the data understanding. The challenge within data understanding is to determine and collect all relevant data and get a first idea of possible interactions. Once the data has been collected they must be merged and transformed to fit the requirements for the selected analysis model. This takes place in the data preparation phase. The following path in CRISP-DM towards modeling is not just unidirectional. Differed types of mathematical models need different types of input. The selection of a mathematical model for the modeling phase depends on the data type, data amount and the analysis goal. Its complexity can range from simple linear regression to complex machine learning algorithms. Machine learning algorithms offer the opportunity to predict the product quality or to find parameter combinations that are not obvious at the beginning to improve the product quality. Finally, in the steps evaluation and deployment, the conclusions drawn from the results are applied to the business case [2].

2.3 Application Fields of Data Mining in Manufacturing

In manufacturing, there are several application fields for the use of data mining. Some of them are explained in the following (Figure 4). The field of process control and optimization includes the identification and modeling of influencing factors to optimize the overall process such as parameter models for most energy efficient process parameters. The field of machine control deals with the understanding and control of a specific machine behaviour. An example for this application field could be an automated state recognition with cluster analysis for idle shutdown. However, quality management ensures quality compliance through efficient checks and feedback. Therefore, decision trees for prediction of product quality based on process parameters could be used. The field of maintenance pursues the goal of preventive avoidance of failure for example by predicting failures. Moreover, energy monitoring systems can be used to optimize the energy consumption within a plant. Furthermore, the knowledge about the energy consumption of machines can also be used for predictive maintenance [3,15].

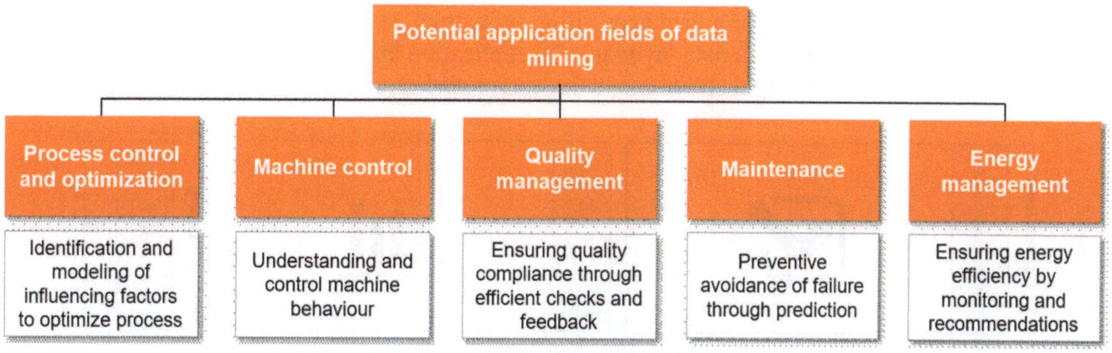

Figure 4: Potential application fields of data mining

3 FRAMEWORK – DATA MINING IN MANUFACTURING OF LIGHTWEIGHT STRUCTURES

3.1 Application Fields of Data Mining in Manufacturing of Lightweight Structures

The manufacturing of lightweight structures consists of several sequential process steps, where different materials are produced and subsequently joined into one multi-material preform. Afterwards process steps such as injection moulding and hybrid pressing take action.

In this production environment, knowledge of the interaction of the process parameters is particularly important in order to be able to manufacture the products in the desired quality. In addition, data mining can support quality management, e.g. through virtual models that enable direct quality control during the production process and a more detailed cause-effect analysis.

There are several application fields where data can be exploited in manufacturing of lightweight structures via different data mining methods. In the following some applications for hybrid multi-material components and the role of machine state recognition is introduced.

With the help of data mining, the complexity of the production of hybrid structures can be reduced. On the one hand, this allows product quality to be optimized for complex multi-stage value-added processes in the hybrid lightweight manufacturing chain. Such an application case can exist, for example, with a large number of possible influencing variables which may be non-linearly connected and which could hardly be controlled by humans themselves. Moreover, with the help of data mining, drape effects for the positioning during hybrid press processes can be analysed and optimized. Hence, the production failure due to wrong positioning can be counteracted.

On the other hand, machine state recognition can serve as an enabler to support process planning. In general, information about the current process step can be supplemented with further data such as temperatures. This is particularly elementary in complex, converging processes, such as the production of hybrid structures, since these must be coordinated to ensure high product quality. Furthermore, standard behaviour can be determined for different process states. When deviations occur, quality assurance checks can be triggered automatically. Moreover, data mining can be used for predictive maintenance. If specific process data drift, this can indicate a defect in individual machine components (e.g. higher pressures of the hydraulic pump).

3.2 Framework for the Design of a Data Acquisition Infrastructure

The data acquisition and processing challenges described in section 2 affect the design of a data collection infrastructure. For example, an appropriate database concept can be determined by evaluating the volume, velocity and variety of data to be managed. The design of a data acquisition infrastructure also strongly depends on the respective application (e.g. cross-site analyses), the integration of existing IT systems as well as the analysis goal. For example, if process parameter correlations are to be revealed once, there is no need for a cost-intensive IT infrastructure; manual data collection over a short period of time can serve the purpose.

The framework for the design of a data acquisition infrastructure consists of three layers, as shown in Figure 5: (1) shop floor layer, (2) data layer and (3) application layer. The framework is exemplified by the manufacturing of lightweight structures in the automotive industry.

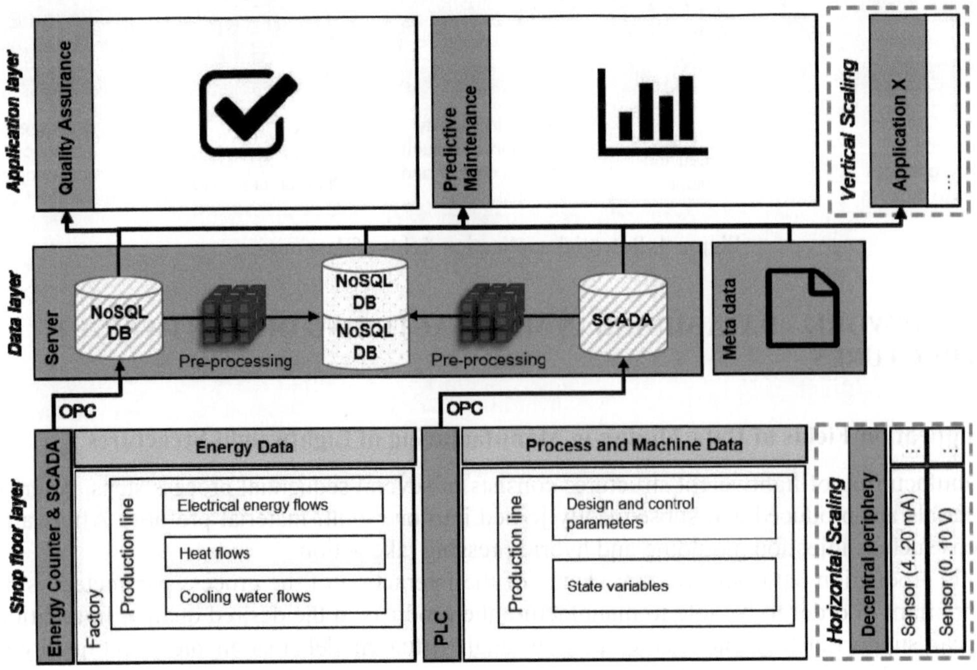

Figure 5: Technical framework for data acquisition and analytics

Influences on the application layer result out of a variety of disciplines that are involved in research, development, manufacturing and testing of lightweight structures – e.g. design, production technology, materials science and life cycle engineering. The availability of live or historical data can add value within each discipline and to an interdisciplinary collaboration. In the field of life cycle engineering, for example, life cycle assessments can be performed in real time [16]. For this reason, vertical scalability across several applications and user groups accessing the central historical database as well as the ability to evaluate production data in real time must be taken into account when designing the data infrastructure. Consequently, the data layer has to serve the requirements of real time capability and historical data access. One approach to satisfy the requirements is the lambda architecture [17]. A speed layer for real time handling of new incoming raw data stores data temporarily and parameterizes precomputed models. In Figure 5 the speed layer is represented by the diagonal lined databases and their direct links to the application layer. The historical data access is made possible by a batch and serving layer architecture, represented by the loosely dotted databases in Figure 5. Both layers are characterized by a high latency, whereby pre-processed data, e.g. production machine is turned on, is stored persistently in the batch layer and application-tailored calculations on the historical data are stored in the serving layer. The NoSQL database format allows flexibility in data storage due to the highly inhomogeneous manufacturing processes involved in the value-added chain for hybrid components. The lambda architecture and data pre-processing, e.g. filtering of irrelevant raw data, ensures a high performance and cost-efficient data layer. The manufacturing data can be enriched via metadata such as layouts and test results or data out of ERP systems. On the shop floor layer, the manufacturing data is acquired via energy meters and PLCs. Depending on the data requirements of the application, further data can be acquired through horizontal scaling with additional sensors. For instance, ambient data can be made accessible through low-cost sensors that are capable of modern communication protocols like MQTT [18].

3.3 Framework for the Application of Data Mining in Manufacturing of Lightweight Structures

The general CRISP-DM cycle is presented in section 2.2. To describe why data mining is especially interesting for hybrid structures, the focus will be on the cycle steps, data understanding and data preparation. Figure 6 shows the comparison between a hybrid and a metal structure for these two steps. The chosen example is quite generic. However, one possible goal of this data mining example could be product quality. For example, answering the question which parameter or parameter combination leads to the most stable quality level. The three groups of parameter considered in this example are:

- Quality parameter for each raw material (e.g. yield strength) – (Qp)
- Process parameter for transforming the raw material into a part (e.g. cutting speed) – (Pp)
- Interacting parameter for two materials in a hybrid part (e.g. surface quality for bonding) – (Ip)

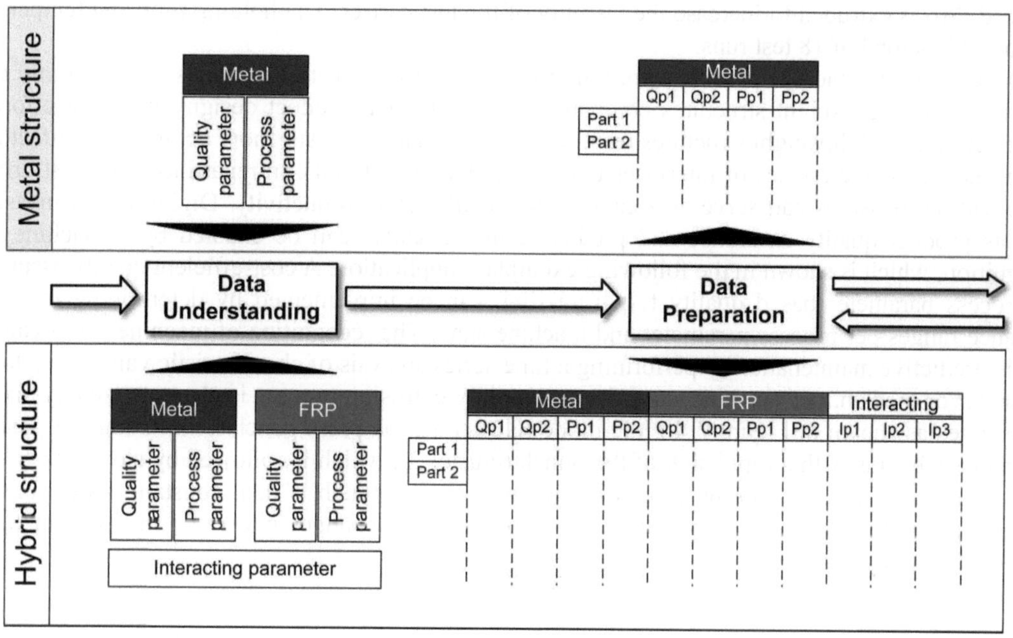

Figure 6: Comparison of data understanding and data preparation for metal and hybrid structure

These three groups represent the input for the data understanding step. The data must be collected for each material or material combination. A part of the data understanding also is a first investigation on global interrelationship. Figure 6 shows the influence on the data understanding step if just one further material is added for the hybrid structure. This more than doubles the number of relevant parameter groups.

The strength values, as a quality characteristic for a material, is a good example to show what this means in detail. For a metal, the relevant strength value is the yield strength. For an FRP, on the other hand, several strength values for fibers and matrix are required, which, in connection with the stacking, result in several direction-dependent strength values. It is extremely difficult to identify interrelationships in such data without pre-processing.

Therefore, the data preparation phase must start early in order to calculate the previously described relationships and bring them in a specific format. The target data format of the data preparation step depends on the selected model (e.g. machine learning algorithms). Figure 6 shows an example of how a second material in a hybrid structure significantly increases the number of parameters in the structured table. In the example, FRP and metal have the same quality and the same amount of process parameters. However, as described above, this is a clear underestimation.

In summary, the number of parameters determines the use of a structured data mining approach. For hybrid structures, however, this is not the only reason to use data mining. The combination of different materials and manufacturing processes leads to many possible interactions which are not obvious at first glance but which could influence the quality of the final structure.

4 EXEMPLARY APPLICATION – MACHINE STATE RECOGNITION IN EXTRUSION OF GFRP RIB STRUCTURES

To demonstrate the application of data mining in the light of manufacturing of lightweight structures a machine state recognition of an extrusion process at the Open Hybrid LabFactory is performed. The Open Hybrid LabFactory, located in Wolfsburg, Germany, offers excellent possibilities and resources for the research, development, production and testing of hybrid components for the future topics of the automotive industry, digitalization and resource efficiency. The extrusion is carried out on a 2,500t hybrid metal forming press. The multifunctional press is capable of different process variants like cold forming, warm forming, sheet moulding compound (SMC) and resin transfer moulding (RTM) and is characterized by a combination of high pressing force and high speed. On a hybridized demonstrator (u-profile consisting of high-strength steel and organosheet) a rib structure made of glass-fiber-reinforced PA6 is extruded to increase the stability of the test carrier. The holding time and temperature are varied in a total of 18 test runs.

The exemplary machine state recognition is motivated by industrial demands for economic competitiveness of lightweight structures compared to conventional product design. Increased efforts in manufacturing of lightweight structures result due to high material costs, more complex manufacturing processes and characteristics of interfaces between different materials. In regard to these cost drivers, digitalization measures can serve as a catalyst for quality and productivity. Digitalization measures, such as product quality assurance and predictive maintenance, can be enabled by a machine state recognition, which is shown in the following exemplary application. A cost-efficient quality assurance, by process parameter based quality test triggering, can be implemented by determining individual tolerance ranges per process parameter and machine state. The recognition of machine states can also enable predictive maintenance by performing a time series analysis of characteristic values, e.g. lowest parameter or median, per machine state. The exemplary extrusion process is characterized by six machine states as shown in Figure 7. In idle mode (cluster 1), the press punch is in its highest position. The process begins with a rapid sink of the punch (cluster 2), which is followed by slow sink (cluster 3) when entering the die. During processing (cluster 4), the punch is held constant. After finishing processing, the punch rises slowly (cluster 5) at first, until it changes into rapid rise (cluster 6) and finally idle mode again.

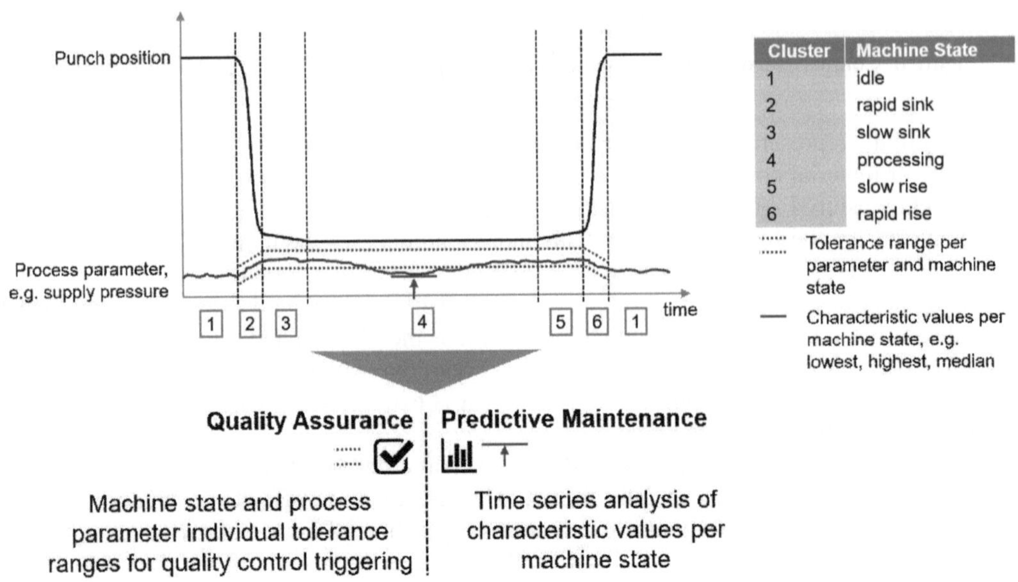

Figure 7: Machine state recognition as enabler for quality assurance and predictive maintenance

According to CRISP-DM the goal of the data mining measure is defined at first. The goal of the exemplary application is to generate a reliable machine state recognition through testing a selection of supervised learning algorithms against an optimal machine state dataset, generated in an initial clustering phase. The complete data mining workflow is shown in Figure 8. Business understanding is followed by data understanding. The data is collected with a sampling rate of 50 ms, comprises 48 variables, like timestamp and data of each of the four press cylinders, e.g. actual position, velocity and supply pressure adding up to 1,6 mio. data points. In order to explore the data, a linear correlation analysis of all variables is performed, utilizing this as knowledge for data preparation. Four independent variables (actual press force [kN], actual velocity [mm/s], actual supply pressure [bar] and actual cylinder force [kN] of cylinder 1) where selected for k-means clustering.

As stated above, the exemplified extrusion process is characterized by six obvious machine states (idle, rapid/slow sink, processing and rapid/slow rise), which should be identified by the algorithm. The k-means clustering results of different choices of k are visualized in Figure 8, where the actual velocity (positive and negative values for punch rise and sink) is plotted on the x-axis and the actual punch position on the y-axis. An analysis over different choices of k showed that k = 8 delivers the best fit, capturing all six target states (Figure 8, III). The reason for this is to be found in the idle state, which has several characteristics (e.g. supply pressure on/off), that are more clearly detected by the algorithm, than the switch between fast and slow sink. For instance, k = 6 (Figure 8, I) delivers two different idle clusters and k = 7 a third idle cluster (Figure 8, II). Moving backwards to the data preparation phase of CRISP-DM, the three clusters all describing idle modes are pooled and mislabeled data points were manually clustered (Figure 8, IV).

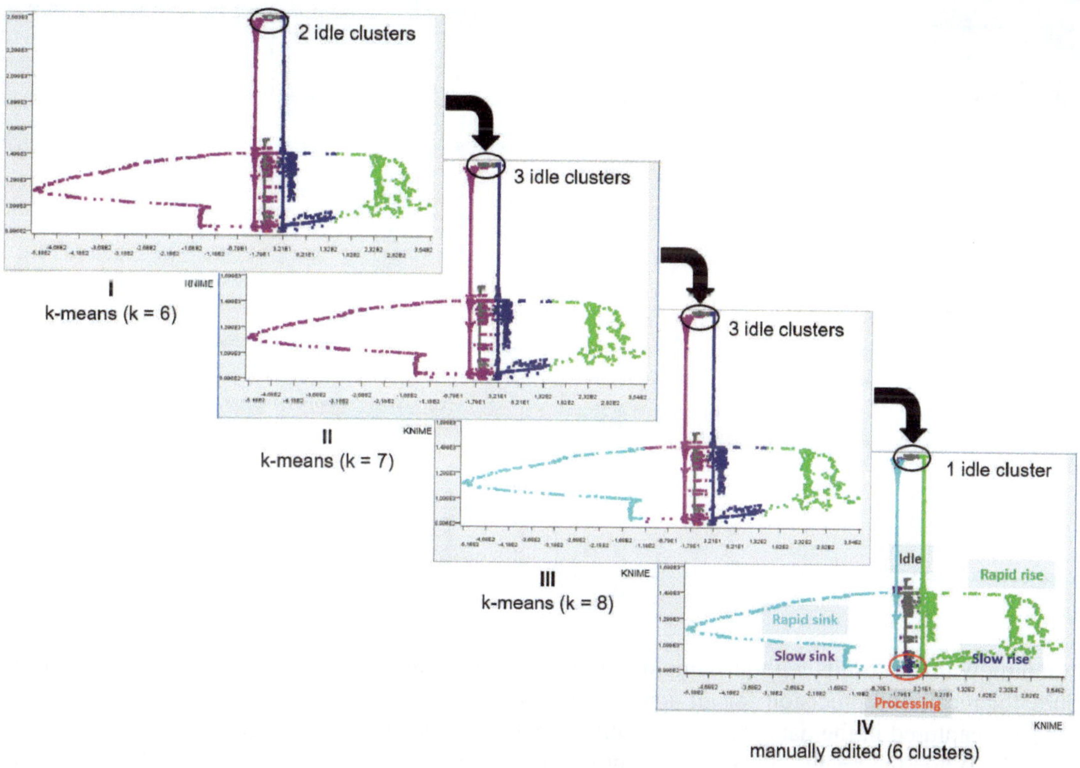

Figure 8: Generation of ideally labeled datasets via iterations of k-means clustering

After dividing the process data into meaningful machine states by applying a k-means clustering, the focus of the further analysis is on classification to assign given process data to predefined machine states. Within the scope of this paper, methods of probabilistic neural network (PNN), decision tree, random forest and support vector machine (SVM) are taken into account. For each method, one data mining model was developed for the given data and implemented in the software KNIME. For the purpose of modeling different machine states, 785.650 data sets were included. The input data was split into two partitions, one for training and one for test data. The share of the training data set is 80 %, while the test data set comprises 20 %.

Table 1 graphically shows the results of the analysis of the different data mining methods considering the different machine states. To ensure comparability between the results of the methods, recall as well as precision are considered. Using recall, the proportion of objects correctly classified as positive in the total number of objects that are actually positive can be determined. However, precision can make a statement about the proportion of objects correctly classified as positive in the total number of positive classified objects [19]. In order to display the results properly, the method of heat mapping is used to make the large amount of data intuitively understandable for the observer. Therefore, the heat map is separately calculated for recall and precision over all machine states. In this context, a dark red hue means a low value, a yellow hue a medium value and a green hue a high value of the respective criteria. Furthermore, all data mining methods were performed with and without sampling. The aim of sampling is to be able to examine and compare possible data sets for the individual clusters.

Table 1: Results of recall and precision for several data mining methods

		Decision Tree						Random Forest					
Machine State		Idle	Rapid sink	Slow sink	Processing	Rapid rise	Slow rise	Idle	Rapid sink	Slow sink	Processing	Rapid rise	Slow rise
Without Sampling	Recall	1,00	0,98	0,98	1,00	0,99	1,00	1,00	1,00	1,00	1,00	1,00	1,00
	Precision	1,00	0,99	0,99	1,00	1,00	1,00	1,00	1,00	1,00	1,00	1,00	1,00
With Sampling	Recall	0,99	1,00	0,96	1,00	1,00	1,00	1,00	1,00	0,99	1,00	1,00	1,00
	Precision	1,00	0,58	0,55	0,95	0,80	0,82	1,00	0,92	0,87	1,00	0,99	0,92

		Idle	Rapid sink	Slow sink	Processing	Rapid rise	Slow rise	Idle	Rapid sink	Slow sink	Processing	Rapid rise	Slow rise
Without Sampling	Recall	1,00	0,34	0,00	0,69	0,22	0,81	1,00	0,80	0,96	1,00	0,86	0,97
	Precision	0,98	0,89	0,00	1,00	0,96	1,00	1,00	1,00	1,00	1,00	1,00	1,00
With Sampling	Recall	0,72	0,99	0,46	0,43	1,00	0,92	0,97	1,00	0,98	0,98	0,91	0,95
	Precision	1,00	0,24	0,83	1,00	0,02	0,70	1,00	0,08	0,78	1,00	1,00	0,95
Machine State		Idle	Rapid sink	Slow sink	Processing	Rapid rise	Slow rise	Idle	Rapid sink	Slow sink	Processing	Rapid rise	Slow rise
		PNN						SVM					

Beginning with the results without sampling, the values of predicting the idle machine state show high values for recall and precision for all applied methods. This can be traced back to a high number of idle states captured in the data set, which facilitated all methods to learn the characteristics of the idle state well. While the results of rapid sink show for precision values of almost 1 for all methods, significant fluctuations occur for recall. The worst result is achieved by the PNN with a value of 0.34 followed by SVM with 0.80. The methods of decision tree and random forest reach a value of 0.98 and 1. With respect to the slow sink, the PNN achieves a recall of nearly 0, while the results of the other methods are between 0.97 and 0.996. For precision, comparable results appear. Regarding the recall of processing, PNN performs with the lowest value of 0.69 while the methods of decision tree, random forest and SVM have results of almost 1. In contrast to this, all methods show a strong performance for precision with values above 0.95. The subsequent machine state of rapid rise has a recall of 0.22 for PNN, 0.86 for SVM and values of 1 for decision tree and random forest. Precision shows similar results except for SVM which achieves 1. The slow rise can improve its recall result for PNN and SVM to 0.81 and 0.97. For precision, all methods achieve a value of 1.

Compared to this, the results with sampling show comparable results. Overall, improvements in recall can be observed for previously underrepresented clusters like rapid sink or rapid rise. This becomes particularly clear with the example of the data mining methods PNN and SVM. In addition, it can be stated that the overall recall improves, while a deterioration occurs in precision when using sampling. This can be related to the fact that when calculating the precision value, data is used that is not subject to sampling. As a result, the distribution of data between the individual clusters is inhomogeneous.

In the course of benchmarking different data mining methods, the execution time of the models is of particular importance. Table 2 shows the execution time of every method with and without sampling.

Table 2: Execution time of data mining methods

Method	Without sampling [s]	With sampling [s]	Difference [%]
Decision Tree	688.8	330.9	–52
Random Forest	205.9	370.2	+79.8
PNN	1168.3	340.7	–70.8
SVM	20624.2	1248.3	–93.9

Based on the data without sampling, random forest shows the shortest execution time with 205.9 seconds, while SVR with 20624.2 seconds shows the longest time. Within these limits, decision tree has an execution time of 688.8 seconds and PNN of 1168.3 seconds. Compared to this, using sampling can reduce the execution time up to 93.9 percent for SVM. The comparably high execution time of the random forest with sampling differs widely from the results of the other methods. This should reduce the execution time during sampling. The fact that this did not occur in this case can be attributed, if necessary, to high calculation requirements when executing the models in parallel. Therefore, further tests must be carried out with a higher computing power to check whether the results can be reproduced.

Overall, the analysis of the results of all machine states shows that PNN is less suitable for the classification of machine states compared to the other methods examined, even considering the sampling. In a direct comparison of all machine states, random forest and decision tree show the best results for the evaluation criteria recall and precision. The results of both methods are comparably good and are characterized by a short execution time. Furthermore, the analysis of the results concludes that SVM is less suitable for rapid changes in process parameters. This becomes particularly clear by comparing the results for recall when changing from rapid to slow sink and rise machine state. Thus, it can be concluded that data points that are located well apart are less suitable for classifications with SVM. Moreover, the method requires a comparably long processing time.

5 SUMMARY AND OUTLOOK

In this paper, the application of data mining in the context of lightweight structures manufacturing is covered. As the manufacturing of lightweight structures often draws on emerging production technologies and combines different process routes (e.g. prepregs and metal inserts being combined in hybrid injection moulding), quality and process control becomes more challenging. In order to enhance competitiveness, data mining-based information can be utilized in a wide selection of potential application fields. Prerequisite for the application of data mining is the collection of relevant and reliable manufacturing data. In anticipating challenges in data acquisition, a cost-efficient and goal-oriented data infrastructure can be designed to boost the efficiency of the digitalization measure. The CRISP-DM can serve as a methodological framework for conducting a data mining approach. As highlighted, the application field of lightweight structures influences mainly two phases of the framework – data understanding and data preparation in terms of higher efforts for selecting and pre-processing relevant data.

The exemplary application of machine state recognition demonstrates one application field of data mining in manufacturing to enable extensive applications like quality assurance and predictive maintenance. By evaluating precision and recall, an excellent performance of the methods decision tree and random forest has been revealed. However, the performance of a data mining method strongly depends on parameterization and data set. Those effects are sketched out by the performance improvement of SVM and PNN when applying sampling. Furthermore, it would be worth investigating to what extent the performance of SVM can be increased by previous feature extraction. Additional studies should be conducted regarding model robustness. In the scope of the exemplary application, no statements concerning transferability are yet possible for different process variants run on the same machine, e.g. hot forming, cold forming, or in an extreme manufacturing of lot size one. The expansion of the outlined approach from data mining to cyber-physical production systems could further increase the economic potential of the digitalization measure.

ACKNOWLEDGEMENTS

The results published in this paper are based on the project MultiMaK2 aiming at developing design and evaluation tools to ecologically optimized multi-material vehicle components.

This research and development project is funded by the German Federal Ministry of Education and Research (BMBF) within the ForschungsCampus Open Hybrid LabFactory and managed by the Project Management Agency Karlsruhe (PTKA). The authors are responsible for the contents of this publication.

REFERENCES

[1] M. Kantardzic, Data mining: concepts, models, methods, and algorithms, Vol. 2, John Wiley & Sons, Inc., Hoboken, New Jersey, 2011.

[2] P. Chapman et al., CRISP-DM 1.0: Step-by-step data mining guide, 1999.

[3] P. Gölzer, Big Data in Industrie 4.0 – Eine strukturierte Aufarbeitung von Anforderungen, Anwendungsfällen und deren Umsetzung, 2017.

[4] C. Liu, X. Xu, Cyber-physical Machine Tool – The Era of Machine Tool 4.0, Procedia CIRP, 63, 2017, pp. 70–75.

[5] T. Wuest et al., Machine learning in manufacturing: advantages, challenges, and applications, Production & Manufacturing Research, 4, 2016, pp. 23–45.

[6] I. Guyon et al., An Introduction to Variable and Feature Selection, Journal of Machine Learning Research, 3, 2003, pp. 1157–1182.

[7] J. Lee et al., A Cyber-Physical Systems architecture for Industry 4.0-based manufacturing systems, Manufacturing Letters, 3, 2015, pp. 18–23.

[8] M. Chen et al., Big Data: A Survey, Mobile Netw Appl, 19, 2014, pp. 171–209.

[9] U. Fayyad et al., From Data Mining to Knowledge Discovery in Databases, AI Magazine, 17, 1996, pp. 37–54.

[10] F. Chen et al., Data mining for the internet of things: Literature review and challenges, International Journal of Distributed Sensor Networks, 2015.

[11] P. Ponniah, Data Warehousing Fundamentals: A Comprehensive Guide for IT Professionals, John Wiley & Sons, Inc., 2001.

[12] P.-N. Tan et al., Introduction to Data Mining: Instructor's Solution Manual, Pearson Addison-Wesley, 2006.

[13] U. Fayyad et al., The KDD process for extracting useful knowledge from volumes of data, Communications of the ACM, 39, 1996, pp. 27–34.

[14] J. Lee et al., Recent Advances and Trends of Cyber-Physical Systems and Big Data Analytics in Industrial Informatics, Proceeding of Int. Conference on Industrial Informatics (INDIN), Porto Alegre, Brazil, 2014.

[15] S.G. Pease et al., An intelligent real-time cyber-physical toolset for energy and process prediction and optimisation in the future industrial Internet of Things, Future Generation Computer Systems, 79, 2018, pp. 815–829.

[16] F. Cerdas et al., Shop-floor Life Cycle Assessment, Procedia CIRP, 2017, pp. 393–398.

[17] N. Marz, J. Warren, Big Data: Principles and best practices of scalable real-time data systems, Manning Publications, 2015.

[18] XDK Cross Domain Development Kit, https://xdk.bosch-connectivity.com/de/overview, accessed March 23, 2018.

[19] J. Davis, M. Goadrich, The relationship between Precision-Recall and ROC curves, Proceedings of the 23rd International Conference on Machine Learning – ICML'06, Pittsburgh, PA, 2006, pp. 233–240.

DEVELOPMENT AND NUMERICAL VALIDATION OF COMBINED FORMING PROCESSES FOR PRODUCTION OF HYBRID PARTS

Bernd-Arno Behrens, Sven Hübner, Alexander Chugreev, André Neumann, Nenad Grbic, Henrik Schulze, Ralf Lorenz, Moritz Micke, Florian Bohne[1]

[1] Institut für Umformtechnik und Umformmaschinen, An der Universität 2, 30823 Garbsen, bohne@ifum.uni-hannover.de, www.ifum.uni-hannover.de

Keywords:

Hybrid forming, Process simulation, Interface analysis

ABSTRACT

This paper gives an overview about different one-shot processes of hybrid forming and joining of semi-finished products using conventional forming presses. An example of forming with a variothermal tool is given by forming a sandwich structure. Different aspects of the forming process are addressed, including the interlayer formation during the process. Here, the objectives are the complete impregnation of the fibres, the consolidation and a firm bond. It is shown, that in contrast to that, the objective when forming a sandwich sheet is to prevent delamination. For forming a hybrid battery tray, a process was developed that combines the advantages of organic sheet forming and compression molding. In order to predict the quality of the bonding of the glass mat reinforced thermoplastic to the organic sheet the temperature evolution is numerically analysed.

1 INTRODUCTION

The importance of fiber reinforced plastics (FRP) in automotive lightweight construction as well as the associated substitution of metallic materials is steadily increasing. In addition to the traditional applications in the interior, a rising number of structural parts consist of fiber-reinforced plastics or a combination of different materials (hybridization). This shift from the application of mono-materials to multi-material structures causes considerable research needs in different fields of production engineering, ranging from the basic materials and the associated processing techniques to the applied analysis methods. The overall objective is to enable affordable lightweight construction.

At the Institute of Forming Technology and Machines (IFUM), the main focus is set on the development of hybrid forming processes, which can be integrated in conventional pressing technology such as that used in the pressing plants of the forming industry. A further focus is put on the development of reliable analytical methods in order to support the process design. Hereby the high-production capabilities and analysis know-how of the already existing forming technologies is exploited, reducing the capital investment for the production of hybrid structures to a minimum. The evolution of analysis methods is based on the development of simulation approaches using commercial software of the respective field. The objective is to properly predict the influence of process factors on the forming process and reducing the number of expensive experimental investigations to a minimum.

In case of the production of hybrid materials the interlayer formation can be identified as a main process factor, which must be considered in the process analysis. Depending on the semi-finished products either a bonding must be achieved or a debonding must be prevented. By combining the individual materials, the advantages of the individual material groups can be combined with each other, taking advantage of synergy effects and optimally exploiting lightweight construction potential.

K. Dröder et al (eds), *Technologies for economical and functional lightweight design*, Zukunftstechnologien für den multifunktionalen Leichtbau, https://doi.org/10.1007/978-3-662-58206-0_3

1.1 Hybrid forming processes

There are two principal approaches regarding the forming of metal-FRP-hybrid components. On one hand, the composite partners can be joined before the forming process or after two separate forming processes [1, 2]. On the other hand the joining and forming can be realized in one single process step. Sandwich composites with a pre-impregnated carbon fiber reinforced polymer (CFRP) core and steel cover sheets as well as one-sided pre-impregnated reinforced CFRP steel composites can be formed together and cured later. The epoxy resin of the prepreg adheres to the steel surfaces and produces a firm adhesive bond after curing [3, 4]. While predominantly thermoset plastics have been used in industry so far, FRP with a thermoplastic matrix are increasingly being used. The main reasons for this are the shorter processing time, better recyclability and the applicability of thermal joining processes [5].

A series of research projects for the production of multilayer composite components from thermoplastic hybrid fabrics and metal sheets in combined one-step forming processes are carried out at the Dresden University of Technology [6, 7]. The combination of metallic sheets as well as flat FRP semi-finished products with 3D-structures formed by injection and compression molding enables the introduction of a further dimension in the production of components. For example, car B pillars can be reinforced with a three-dimensional rib structure [8]. Furthermore, assembly groups can be combined in one component due to the increased freedom of design.

Alternative methods to pure forming technologies are hybrid injection molding processes, which integrate the forming of sheet metal semi-finished products when closing the injection mold [9, 10].

1.2 Simulation and modelling of hybrid forming processes

The development of new forming processes for hybrid components involves the development of numerical methods, which support the process analysis. The applied simulation software must provide different modelling approaches for the respective geometries, the respective material models and contact options in order to model the variety of different contact interactions.

For modelling metallic components in a forming process a variety of different material model formulations exists, which have been implemented and tested extensively over the past decades, driven by the wide field of application. Standardly, elastic-plastic material models are applied. In case of isotropic material properties the von Mises yield criterion can be used. For taking into account anisotropy approaches such as the Hill yield criterion are available [11].

In case of fibre reinforced thermoplastic sheets the current commercial finite element software packages offer previously homogenized material formulations. Here, due to the lower computational effort, preferably models are used that describe fibres and plastic as a smeared area continuum. In case of fibre reinforced sheet modelling the geometry is discretized with shell elements, whereby the different material behaviours of the fiber and the matrix are modelled simultaneously. Two different approaches can be used: the coupled modelling approach and the homogenized FE modelling approach. The embedded modelling approach is based on a particular discretization, overlaying two separate elements and thereby its respective material models (see Figure 1a), whereas in the homogenized FE modelling approach the different material models are superposed in one material model (see Figure 1b), requiring just a conventional mesh. The forming of a fibre reinforced thermoplastic sheet, which is fully consolidated before the forming process is called draping. The decisive factor in the simulation of a draping process is the determination of the fibre orientation. By calculating the deformation gradient and the associated shear angle, the fibre orientation in the corresponding forming stage can be determined numerically. A prediction of the fibre direction after the forming process forms the basis of an improved structural analysis. The stiffness of the thermoplastic matrix is strongly temperature dependent and has a strong influence on the wrinkling behaviour during the forming phase [12]. Therefore the thermodynamic properties of the composite must be precisely modelled [13].

The modelling of the interface between different materials takes place regarding the material combination and the respective process conditions. In case of forming sandwich sheets, the failure modes of the interlayer are of interest. However, in case of simulating one shot processes, the focus is set on the bond formation in the interlayer of the semi-finished products during the process, predicting its respective influence on the structural stiffness and strength.

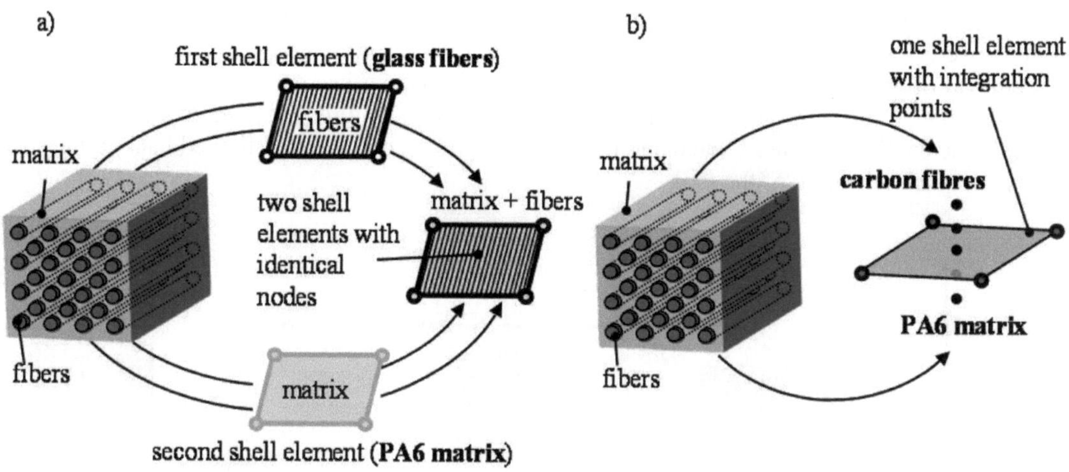

Figure 1: Modelling approaches for the simulation of fibre reinforced thermoplastic: a) coupled modelling approach b) homogenized FE modelling approach [14]

2 PROCESS AND SIMULATION DEVELOPMENT

In this chapter the development of two process strategies for forming hybrid components are presented. Numerical and experimental analyses have been conducted and the results are presented.

In the first section the forming process of a sandwich structure is described. The second process comprises of forming a scaled battery tray consisting of an organic sheet and glass mat reinforced thermoplastic (GMT). The focus in both processes is set on the joining zones of the respective components.

2.1 Stamping of sandwich structures

Commercially available semi-finished products are either fully impregnated and consolidated FRP products or sandwich sheets without any fibers in the core layer. Current process routes focus on processing these kinds of materials.

In order to process partially impregnated, unidirectional CFRP tapes stacked with steel cover sheets a process route was developed, in which the stack is heated, formed, impregnated and consolidated in a single-stage forming tool. The fibers can be orientated to the load direction depending on the component and load requirements. The cover sheets add a ductile component that protects the CFRP core against external loads (impact, chemically aggressive media). Additional Polyamid 6 layers are used between the FRP and the steel sheets to improve the bonding to the steel sheet (see Figure 2). Other adhesives have not been applied. The composite is created solely by thermal direct joining of the Polyamid 6 onto the metal surface.

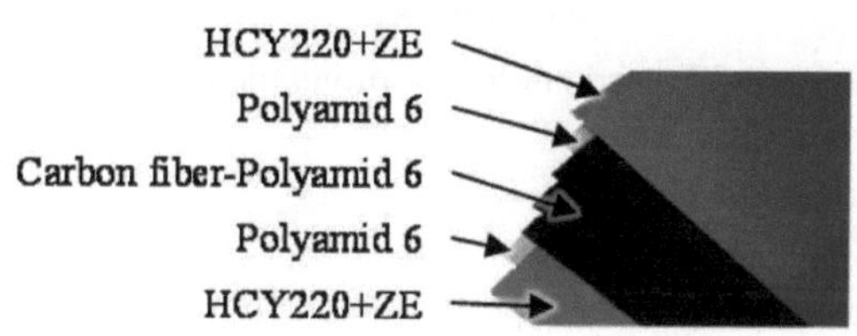

Figure 2: Used semi-finished products for production [15]

2.1.1 Conception of the heating-cooling strategy

For heating, impregnation and consolidation within one forming tool a heating concept according to the temperature-time curve in Figure 3 was developed. The components of the forming tool were heated above the melting temperature of the thermoplastic matrix. Subsequently, the semi-finished products were formed and the fibers were impregnated under prolonged pressure. For consolidation and removal of the component, only the active area along the surface contour of the tool was cooled, while the core of the forming tool remained at elevated temperature. After the component had been removed, the cooling was deactivated and the heat energy stored in the tool flowed back to the surface, allowing a quick reheating of the tool.

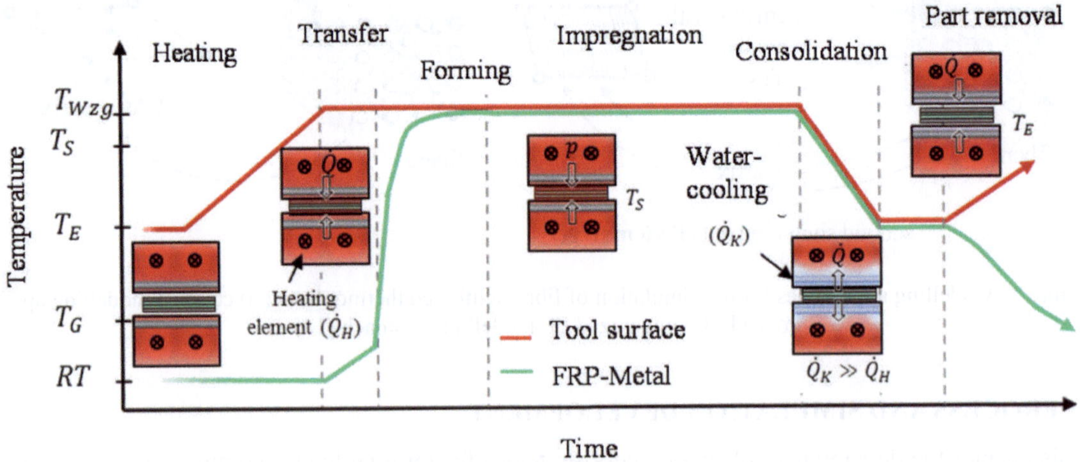

Figure 3: Conception of the temperature-time course of the forming process [15]

Figure 4 shows the thermographic images of the forming die tool during the cooling phase. It can be observed that a few seconds after activating the cooling, a clear temperature gradient was established in the mold. After about 45 s, the removal surface temperature of about 160 °C was reached, whereas the core temperature remained at 250 °C. The reheating of the die surface to the forming temperature of 250 °C was carried out during these experiments within 5–6 min and thus about in a quarter of the time, which would have been required for reheating the whole forming die.

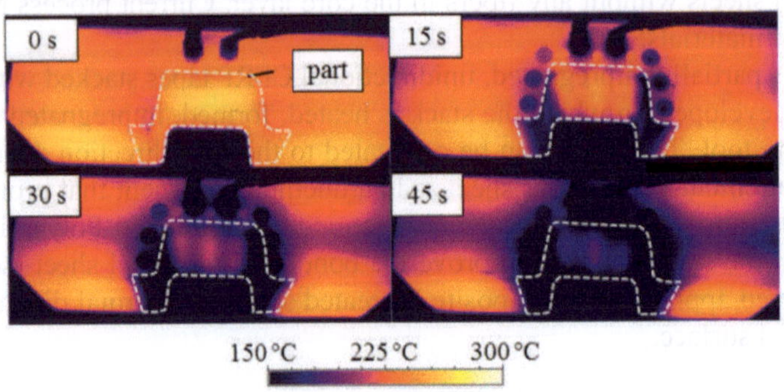

Figure 4: Thermographic image of the variothermal forming die with tool cooling close to the surface [15]

2.1.2 Production of sample components

With the help of the designed tool system, experimental tests were carried out for the production of hat profile components. Such a component is shown in Figure 5. Both, an adhesive bond and a complete impregnation were successfully achieved in the whole part. The highest impregnation and consolidation qualities were obtained at mold temperatures of 245–260 °C.

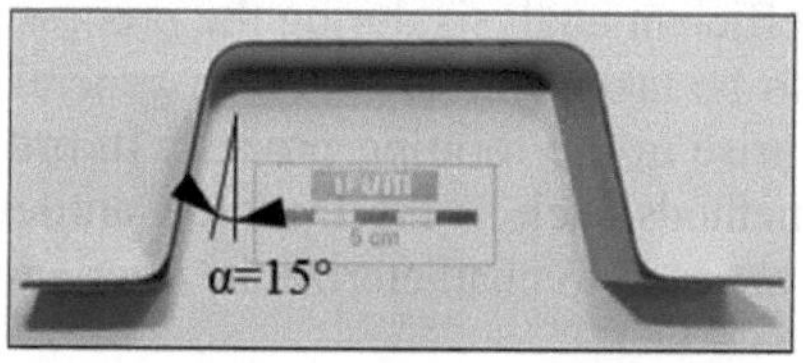

Figure 5: Hybrid steel-CFRP-steel component [15]

Figure 6 depicts different cross sections of the bottom area of the hat profile to show the influence of tool temperature on the consolidation results. It can be observed, that at a tool temperature of 230 °C the impregnation of the fibers and the consolidation of the part were insufficient. Ideal results without any defects were obtained with a tool temperature of 250 °C. By raising the tool temperature to 280 °C, thermal degradation took place over a dwell time of 30 s.

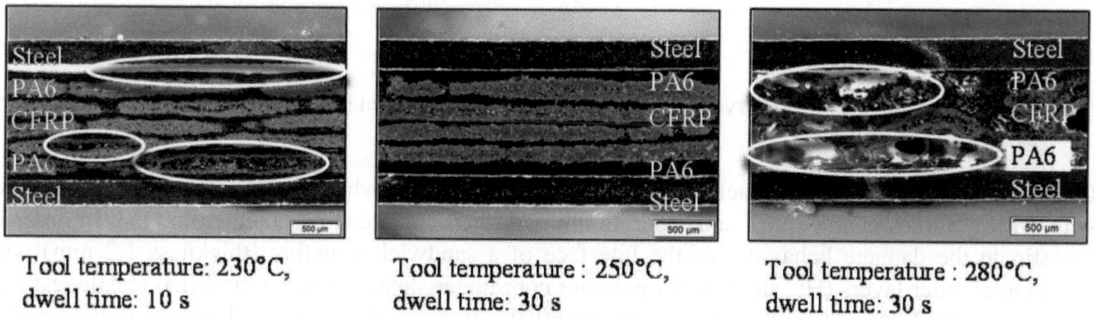

Figure 6: Effect of tool temperature on impregnation and consolidation quality [15]

Due to being the limiting factor of the overall component strength, the objective had been to obtain a firm and stable adhesive bond with the metallic surface within the forming process. This topic is further pursued in the OHLF project Mobilise, in which adhesive and mechanical model approaches are analyzed and evaluated by taking into account the process parameters of pressure and temperature of the interlayer as shown in Figure 7 to create a simulation model for analysing a one-shot process with respect to the interface correlation in order to predict the strength of hybrid forming components.

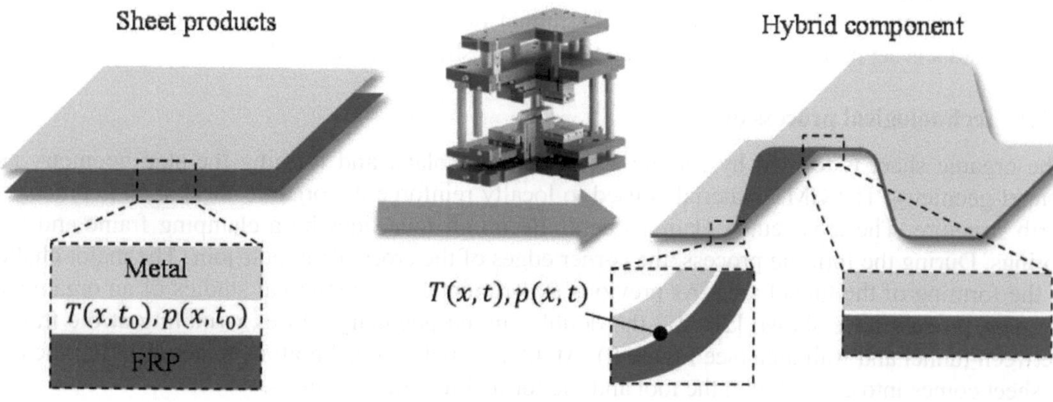

Figure 7: Parameters of the contact interface in a hybrid forming process

2.1.3 Investigations for the interface of sandwich structures

The formation of a firm adhesive interlayer between the hybrid components is important, in case of forming and joining semi-finished components in one forming process step. In case of using pre-bonded materials, the focus is shifted from creating a bond to maintain a stable bond in the forming process. Delamination during the forming phase is an important failure mechanism for sandwich structures (see Figure 8). For an efficient analysis during the process design phase of composite materials the layer interactions have to be taken into account. In general, delamination results from combined loading conditions, which arise in the forming process. In order to measure the strength of the interface, various experimental methods such as the Double Cantilever Beam test [16] or T-Peel test [17] have been developed. These are used to characterize the interlaminar fracture toughness G_C.

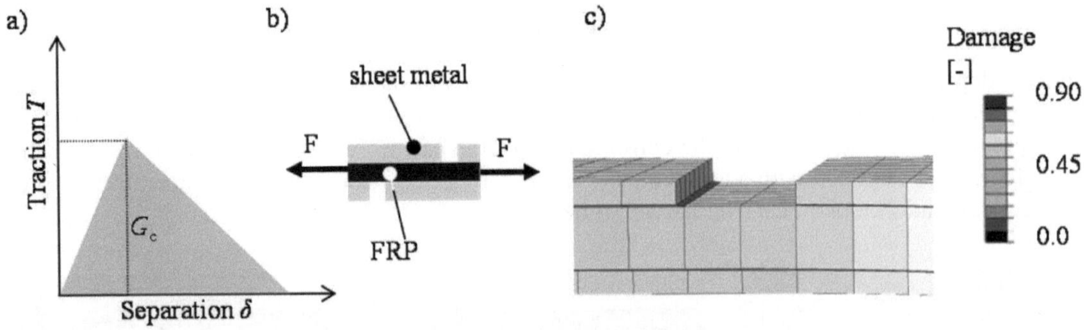

Figure 8: a) Traction-separation curve acc. to [16], b) Sketch for modified tensile test, c) Damage failure value from FE simulation [17]

On the left side of Figure 8a a traction separation curve is shown, which represents an interface failure with critical values of traction T and separation δ_{crit}. More details can be taken from [18]. In order to investigate the damage behaviour of the interface of a sandwich structure (thickness 1.2 mm) with sheet metal outer layers (HC steel) and a polymer core under shear loading a modified tensile test was carried out (see Figure 8b). The interface characteristics were implemented with regard to the traction separation approach and the numerically calculated damage value, presented in Figure 8c. For the considered layer configuration and loading state, no significant failure were detected [19].

2.2 Development of a combined compression molding and draping process

As part of the BMBF project ProVor^{Plus}, the IFUM developed a combined process consisting of organic sheet forming and compression molding process, which enables the production of a scaled battery tray. The battery tray consists of a shell, featuring a step-geometry as well as a tunnel-geometry. The tunnel is reinforced with help of a stiffening rib. For the shell geometry an organic sheet material and for the rib structure a glass mat reinforced thermoplastic was chosen (GMT). A focus in the evaluation of the results is set on the bonding between the organic sheet and the GMT.

2.2.1 Technological process design

The organic sheet is formed by folding a cross-shaped blank and forming the step geometry and a tunnel-geometry. The GMT material is used to locally reinforce the organic sheet shell geometry with a rib structure. The cross cut is clamped centrally on all four sides by a clamping frame and fed by springs. During the forming process, the corner edges of the cross blank butt-join. The major challenge is the forming of the tunnel area. As previous experimental and numerical studies of an organic sheet forming process have shown [20, 21] the double-curved geometry causes wrinkling at the transition between tunnel and wall area (see Figure 9). At 4.5 mm before dead bottom center (BDBC) the organic sheet comes into contact with the tool and the formed fold gets crimped.

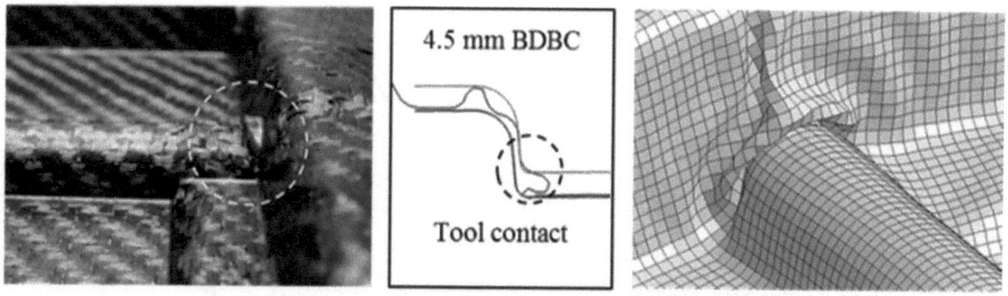

Figure 9: Wrinkling, experimental and numerical results [20]

In order to prevent fiber breakage when compressing the fold, local fiber shearing can be induced by reducing the clamping width in the tunnel area and virtually prevent wrinkling (Figure 9).

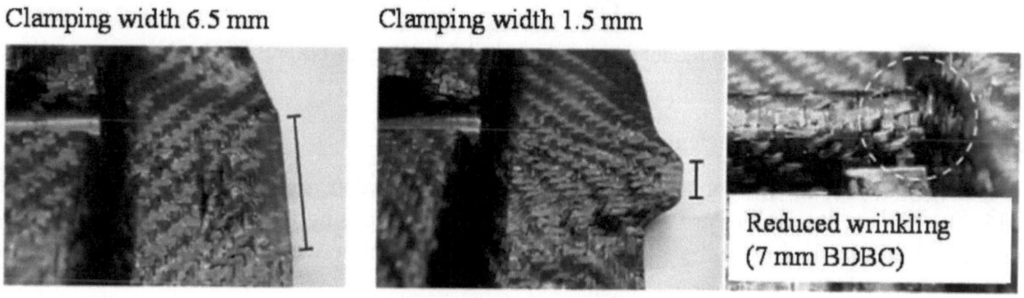

Figure 10: Prevention of wrinkling [20]

In order to develop the organic sheet and compression molding process the organic sheet forming process described in [20, 13] was supplemented by a GMT compression molding process. For this purpose, cavities and an additional sheet gap of 1 mm in the tunnel area were provided for the compression molding process. A main objective is to limit the distribution of the GMT material to the tunnel geometry. The sealing is achieved by a defined contact between organic sheet and forming tool at the front face and flanks of the tunnel.

The punch is used as a lower tool. The heated GMT mass is placed in the gap of the tunnel geometry, followed by positioning the heated organic sheet blank over the punch. Subsequently, the tool is closed.

2.2.2 Numerical process model

The simulation of the combined draping and compression molding process involves the challenge of combining the fundamentally different model approaches of the individual processes. The change in shape of the organic sheet takes place primarily by shear. As previous studies show the thermoplastic matrix cools down strongly during the forming process, which leads to an increase of the bending stiffness of the composite [13]. Since the forming takes place in a temperature range from 150 °C up to 270 °C, the material properties of the organic sheet were measured at elevated temperatures.

The simulation of the combined processes was carried out in LS-Dyna, as it allows the combination of the draping process and the compression molding process in one simulation model. The setup of this model is an extension of the organic sheet forming process model presented in [13]. For modeling the organic sheet the homogenization approach, shown in Figure 1b is chosen. The selected material model combines both, the influence of the fibres and that of the thermoplastic matrix.

In order to determine the evolution of the temperature distribution during the process, it is assumed that the organic sheet has a direction-independent thermal conductivity. Due to the purpose of reinforcing the overall structural strength and stiffness, the rib structure must be properly bonded to the tunnel region. Therefore the temperature of the interlayer between the GMT and the organic sheet must exceed the melting temperature at least on one side of the bond. This thermal process is implemented in the simulation model by modelling the heat transfer between the organic sheet and GMT.

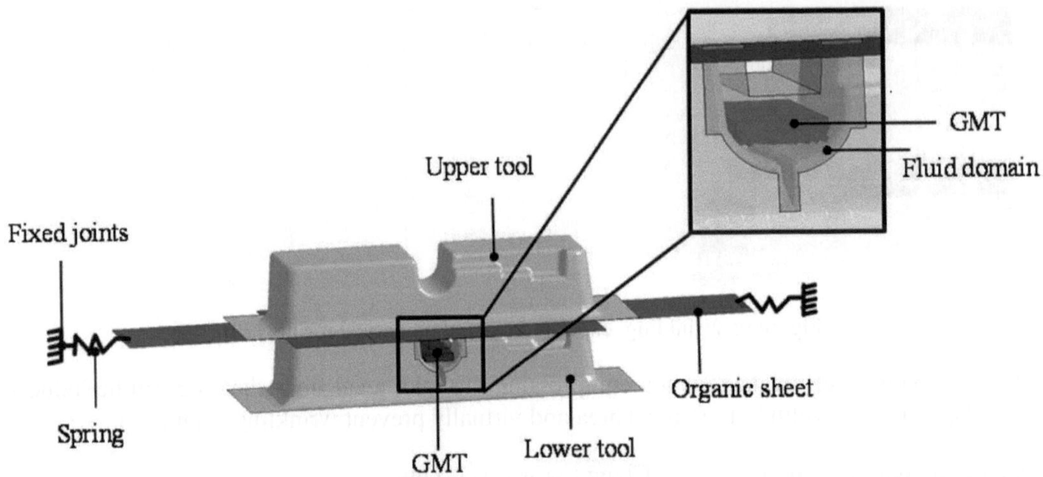

Figure 11: Set up of the simulation model for the combined draping and compression molding process [21,22]

2.2.3 Experimental and numerical results

The experimental results show that at the time of contact between organic sheet and extrusion compound in the tunnel, the cavities are already sealed at the end faces by the unfolded organic sheet. Furthermore, it comes to sealing on the 5 ° steep flanks of the tunnel between organic sheet and the stamp.

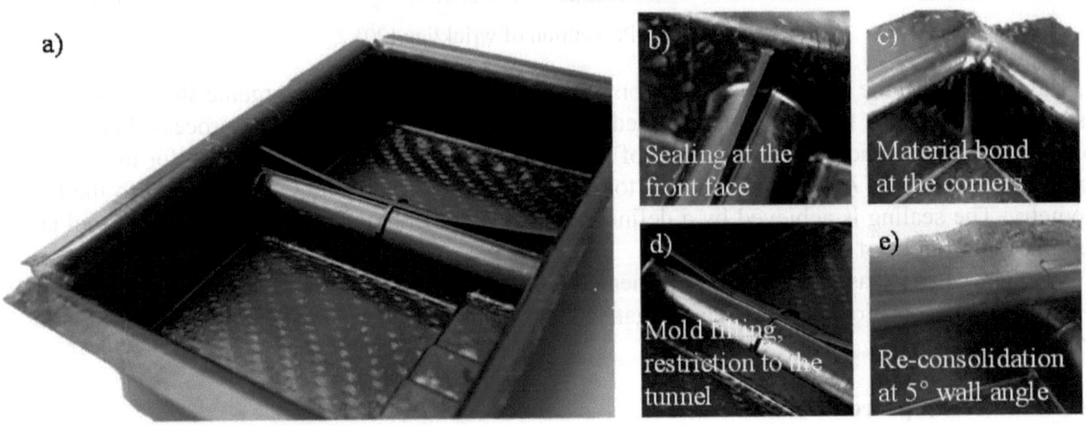

Figure 12: Battery carrier made of organic sheet (fiberglass PA6) and GMT (PA6) [21]

The forming result of the combined process is shown in Figure 12. It comes to a full mold filling of the cavities. Between the GMT and the organic sheet a cohesive bond is formed. The compression molding process was successfully limited to both flanks and to the front face of the tunnel area.

A good overall consolidation in the bottom (Figure 13a) and wall area (Figure 13c) were achieved. A sharp radius and high tensile loads leads to matrix migration, in which fibers accumulate at the inner radius (Figure 13b) and matrix at the outer radius. During the joining at the edges the fiber reoriantate perpendicular to the cutting plan. No clear joining zone and a mixing of the loose ends of the fiber bundels is observed. Furthermore no voids has been detected in the the joining zone. In order to examine the bonding between the organic sheet and the GMT the tunnel area was cut and the intersection prepared for microscopic examination (see Figure 14). The surface of the organic sheet properly bonds to the GMT. A seamless transition from the organic sheet to the GMT is realized.

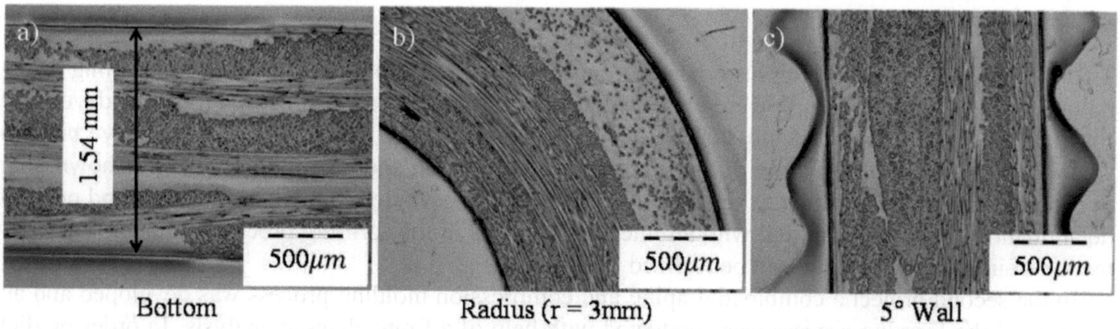

Figure 13: Consolidation quality and fiber orientation in the bottom, radius and wall area of the battery tray

Figure 14: Cross section of a joined edge of the battery tray

The simulative results show that the wrinkling observed in [20] arises as well in the combined process. Thus the GMT component does not prevent wrinkle formation. Furthermore it can be observed that the temperature of the flank areas starts to decrease early. Due to the temperature decrease, the stiffness of the matrix material increases strongly, restricting further forming of the fabric. Later in the process the tunnel areas come into contact with the tool and the temperature starts to decrease until a contact with the GMT is realized. The higher temperature of the GMT leads to a temperature rise in the organic sheet at the end of the forming phase, exceeding the melting temperature. The predicted remelting of the surfaces properly explains the observed bond formation in the experimental process.

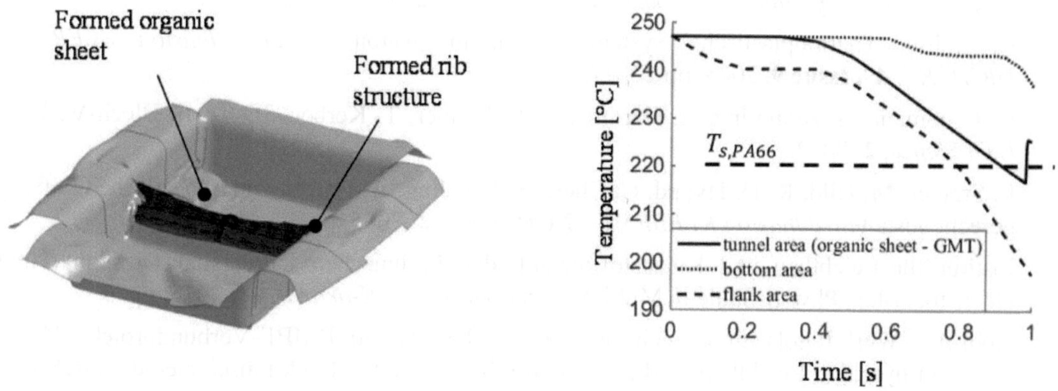

Figure 15: Simulation results of the formed battery tray with rib structure and temperature-time curves for selected regions [21]

3 CONCLUSIONS

Two different processes, which have been in the centre of research at the Institute of Forming Technology and Forming Machines, were presented. In the first process a variothermal tool was developed for the forming of a sandwich structure in one single forming step. The main objectives have been to achieve a complete impregnation of the fibres, the consolidation and a firm bond. It were shown, that in contrast to that, the objective when forming a sandwich sheet is to maintain a firm bond and prevent delamination. It was shown, that with an interface analysis in the forming process simulation, the detection of interlaminar failure can be realized beforehand.

In the second project a combined draping and compression molding process was developed and an analysis of the forming process was conducted with help of a finite element analysis. In order predict the interlayer formation the heat transfer between both constituents was included in the simulation model. The experimental and simulative results were compared and showed a good agreement.

ACKNOWLEDGEMENTS

The research results, presented in this paper, originated from different projects. The project Provor^{Plus} is funded by the German Federal Ministry of Education and Research (BMBF) within the Forschungscampus "Open Hybrid LabFactory" and managed by the Project Management Agency Karlsruhe (PTKA). The subproject TP3 (BE1691/163-1): "Material characterisation and material modelling of composite metal sheets" of the DFG-AiF cluster program "Mass production suitable process chain for high integrated parts made of fiber reinforced-plastic/metal composites" was supported by the German Research Foundation (DFG). The authors thank the respective funders.

REFERENCES

[1] H. Palkowski, P. Giese, V. Wesling, G. Lange, S. Spieler, J. Göllner, Neuartige Sandwichverbunde – Herstellung, Umformverhalten, Fügen und Korrosionsverhalten, *Material Science and Engineering Technology*, **37, Issue** 7, 2006, pp. 605–612.

[2] G. Lange, Beitrag zum Umformverhalten von dreischichtigen austenitischen Sandwichverbunden mit polymerer Kernschicht, Dissertation, Technische Universität Clausthal, 2005.

[3] H. C. Schmidt, C. Lauter, W. Homberg, Stahl-CFK-Strukturen, *ILH insight* – Forschung im Instititut für Leichtbau mit Hybridsystemen, 2013.

[4] M.-S. Lee, S.-J. Kim, O.-D. Lim, C.-G. Kang, The effect of process parameters on epoxy flow behavior and formability with CR340/CFRP composites by different laminating in deep drawing process, *Journal of Materials Processing Technology*, **229**, 2014, pp. 275–285.

[5] H. Lohse, Thermoplastische Systemlösungen im Automobilbau, *adhäsion KLEBEN & DICHTEN*, **49, Issue 9**, 2005, pp 22–27.

[6] K. Großmann, S. Bräunling, A. Hardtmann, E. Staiger, T. Kerber, 3D-Textil-Blech-Verbunde, *UTF Sience*, **2**, 2012.

[7] E. Staiger, M. Hild, R.-D. Hyund, C. Cherif S. Bräuling, A. Hardtmann, Oberflächen haftungsgerecht gestalten, *adhäsion KLEBEN & DICHTEN*, **1-2**, 2005, pp 32–35.

[8] Institut für Leichtbau und Kunststofftechnik der Technischen Universität Dresden (Hrsg.), Höchstbelastete Pkw-B-Säule in Multi-Material-Design, *ILK-aktuell*, **01**, 2015, pp. 3.

[9] Audi AG, Werk Ingolstadt, Gemeinsamer Schlussbericht zum BMBF-Verbundprojekt „Herstellung von hybriden Leichtbau-Verbundrohren mit integrierten Funktionselementen durch Fluidinjektionstechnik ‚FIT-Hybrid'", 2011.

[10] J. Mitzler, J. Renkl, M. Würtele, Hoch beanspruchte Bauteile in Serie, *Kunststoffe*, **3**, 2011, pp. 36–40.

[11] D. Banabic, Sheet Metal Forming Process – Constitutive Modelling and Numerical Simulation, 1, Springer-Verlag Berlin Heidelberg, 2010.

[12] B. Liang, N. Hamila, M. Peillon, P. Boisse, Analysis of thermoplastic prepreg bending stiffness during, *Composites*, **Part A**, , 2014, pp. 111–112.

[13] F. Bohne, M. Micke-Camuz, M. Weinmann, C. Bonk, A. Bouguecha, B.-A. Behrens, Simulation of a Stamp Forming Process of an Organic Sheet and its Experimental Validation, *7. WGP-Jahreskongress Aachen*, 2018, R. H. Schmitt, 2, pp. 11–19.

[14] B.-A. Behrens, M. Vucetic, A. Neumann, T. Osiecki, N. Grbic, Experimental test and FEA of a sheet metal forming process of composite material and steel foil in sandwich design using LS-DYNA, *Key Eng. Materials*, **651-653**, 2015, pp. 439–445.

[15] B.-A. Behrens, S. Hübner, N. Grbic, M. Micke-Camuz, T. Wehrhane, A. Neumann, Forming and Joining of Carbon-Fibre-Reinforced-Thermoplastics and Sheet Metal in One Step, *Procedia Engineering*, **183**, 2017, pp. 227–232.

[16] ASTM, ASTM D5528: Standard Test Method for Mode I Interlaminar Fracture Toughness of Unidirectional Fiber-Reinforced Polymer Matrix Composites, *ASTM International*.

[17] ASTM, ASTM D1876: Standard Test Method for Peel Resistance of Adhesives (T-Peel Test), *ASTM International*.

[18] S. R. Hallet, P. W. Harper: Modelling delamination with cohesive elements, Numerical Modelling of Failure in Advanced Composite Materials, *Woodhead Publishing Series in Composite Science and Engineering*, 2015, pp. 55–72.

[19] B.-A. Behrens, A. Bouguecha, C. Bonk, H. Schulze: Finite element analysis regarding the forming behaviour of symmetric hybrid structures consisting of two sheet metal outer layers and a thermoplastic core, *IOP Conf. Series: Materials Science and Engineering* **179**, 2017.

[20] B.-A. Behrens, A. Raatz, S. Hübner, C. Bonk, F. Bohne, C. Bruns, M. Micke-Camuz, Automated Stamp Forming of Continuous Fiber Reinforced Thermoplastics for Complex Shell Geometries, *Procedia CIRP*, **66**, 2017, pp 113–118.

[21] B.-A. Behrens, S. Hübner, C. Bonk, F. Bohne, M. Micke-Camuz, Development of a Combined Process of Organic Sheet forming and GMT Compression Molding, *Procedia Engineering*, **207**, 2017, pp. 101–106.

[22] B.-A. Behrens, A. Bouguecha, S. Hübner, C. Bonk, A. Neumann, N. Grbic, F. Bohne, M. Micke-Camuz, Umformtechnische Herstellung von hybriden Karosseriebauteilen aus Metallen und faserverstärkten Kunststoffen, 36. Verformungskundliches Kolloquium, Zauchensee, 2017.

Functional Components

HYBRID COMPONENTS WITH FUNCTION INTEGRATION FOR CRASH RELATED APPLICATIONS IN ELECTROMOBILITY

Sebastian Iwan, Frank Schettler, Wolfgang Nendel, Lothar Kroll

Chemnitz University of Technology
Faculty of Mechanical Engineering
Department of Lightweight Structures and Polymer Technology
D-09126 Chemnitz, Germany,
slk@mb.tu-chemnitz.de,
www.strukturleichtbau.net

Keywords:

Lightweight design, Fiber reinforced plastics, Simulation, Electromobility, Crash related

ABSTRACT

In times of major changes in the field of automotive drive concepts, lightweight construction as a key technology is particularly important for the reduction of vehicle mass and thus for sustainable mobility. The most significant synergy effect can only be achieved considering all three principles of lightweight construction: material-, structural- and system-lightweight construction.

As a part of the BMBF-funded project thermoPre®, by pursuing this approach the substitution of a motor carrier in die-cast aluminum construction with steel attachments was investigated. For this purpose, the entire potential of the lightweight construction on the reference component was determined and maximized including a cost-efficient production. The result is a continuous glass fiber reinforced thermoplastic (gfrp) component, which shows the high potential for weight and cost minimization by integrating existing attachments and inserts into the component manufacturing process.

Based on the aluminum component, a material appropriate design with subsequent topology optimization and final component design was carried out. Using the finite element method simulations, a load-capable component design was determined. The final design of the engine carrier and the necessary material blanks came from a series of optimization loops, which ultimately led to the production of prototypes. These were tested according to the requirements of the OEM, inter alia crash tests, whereby an equal and partly even improved performance was proven.

Confirmed by the good results from this research project, the implementation of a mass production of such components is planned.

1 INTRODUCTION

In the field of individual mobility not only new drive concepts but also lightweight constructions are given a major role to reduce greenhouse gas emissions. At the time of combustion engines, lightweight construction is necessary to save fuel. However, in the context of electric mobility the lightweight construction is also highly relevant to balance the high masses of the lithium-ion batteries in order to minimize energy consumption [1]. By reducing the weight of an electric vehicle by using lightweight construction also the mass of the batteries can be reduced, whereby a an iterative reduction of the total mass of the vehicles is initiated [2]. Especially with the already expensive technology of electric vehicles, it is important to focus on cost-effective lightweight technologies that are suitable for a large-scale production.

Within a funded project by the German Federal Ministry for Education and Research (BMBF), the substitutability of an existing engine carrier in die-cast aluminum design with attachments made of steel by a newly developed carrier made of glass fiber reinforced thermoplastic was investigated. The

K. Dröder et al (eds), *Technologies for economical and functional lightweight design*,
Zukunftstechnologien für den multifunktionalen Leichtbau, https://doi.org/10.1007/978-3-662-58206-0_4

major aim was the minimization of component weight and costs while maintaining the mechanical performance of the metal part. For this, the complete development process from component design, component calculation to prototype manufacturing and its testing was realized within the scope of the project. In the following, the results of these single work packages are presented.

2 DESIGN OF COMPONENT

The starting point for the development of an engine carrier made of fiber-reinforced thermoplastic is the standard engine carrier from the VW e-Golf, which is made of an aluminum cast. To mount additional electrical equipment, e.g. power electronics and charging unit, separate steel sheet components are used at the standard engine carrier. These steel sheets are complex stamping and forming components and have to be assembled in an additional manufacturing step. The individual components of the engine carrier are shown in Figure 1.

The engine carrier for the VW e-Golf has been completely redeveloped with the aim to receive a lighter and cheaper component. Thus, various optimizations of the structural design and the manufacturing process were necessary. Due to the low material stiffness of gfrp compared to aluminum, the structural stiffness of the new component had to be improved by using the fully available design space. Furthermore, the component price is also directly affected by the components mass. In addition, especially for the comparatively new materials such as gfrp, the material prices are still significantly higher than those of classic materials such as steel. Therefore, there is still a high motivation for lightweight construction in automotive engineering.

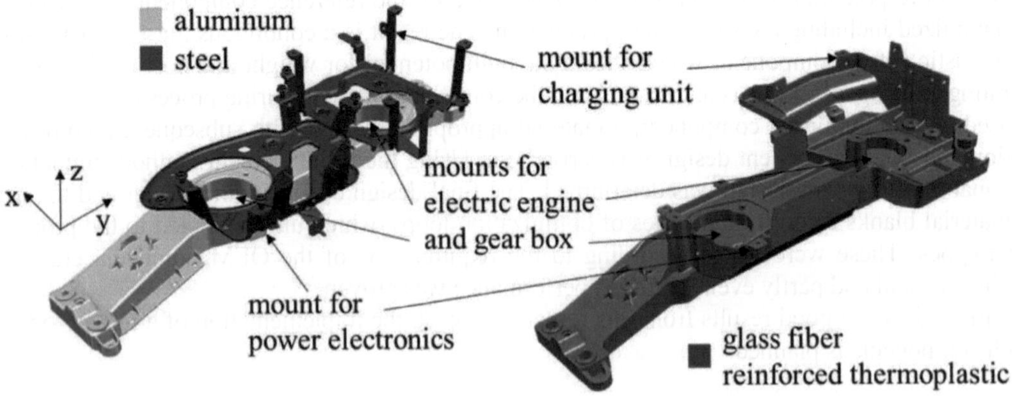

Figure 1: Standard engine carrier made of aluminum and steel sheets (left); new developed engine carrier made of gfrp (right)

The new engine carrier is made of thermoPre® prepregs and glass mat thermoplastic (GMT) material. The unidirectional thermoPre® material with a $[0/0/90/0/0]_n$-lay-up was used as base layer to achieve high strength and energy absorption during crash with no abrupt failure. The preferred direction of this 4:1-lay-up is parallel to global y-direction. To increase the structural stiffness of the component a rib structure made of GMT was used. The component was made in a compression molding process. During the entire development process, attention was paid to the implementation of a cost-effective component. This means that all production steps have to be finished during the compression molding process. For this purpose, through holes as well as metal inserts were inserted directly into the component during the compression molding process. Firstly, this obviates the subsequent milling process and secondly, fibers are forming around the inserts during the process, which increases the load carrying capacity of the load application section. In order to avoid subsequent processing steps like the application of face sheets, the component was designed as an open rib structure, even if closed profiles can achieve higher structural bending stiffness. However, the biggest costs and mass savings are achieved by the function integration. By integrating the mounts of the power electronics and charging unit into the plastic component, the complex production of steel brackets can be omitted. In addition to the weight reduction and the gain in stiffness, the reduction of assembly time is a major advantage, too. Figure 1 shows the comparison of the metal and the plastic design. It can be clearly seen, that the integration of the mounts results in a more complex design. However, the component can be produced in one production step due to the good flow properties and fulfills the same requirements as the standard metal part.

In summary, it was possible to develop a component design that is both lighter (up to 23 %) and has the potential to save costs (up to 47 %) by integration many of the several components into one structure and its production process.

3 COMPUTATION AND SIMULATION

Load cases for the computation and simulation of the new developed engine carrier base on the mechanical requirements of the standard aluminum cast component. For this, numerous static load cases were calculated with the relevant load directions x- and z-direction, according to the vehicle coordinate system.

For an ideal design of the rib structure, a topology optimization was carried out. The completely available design space was determined. Therein, the density and thus the mechanical characteristics were varied over the component in local sections. Taking into account the restrictions of the stiffness and the production technology, a lightweight and production-ready rib design with the same stiffness like the standard part was determined. In this case, manufacturing restrictions such as minimum and maximum rib width, spacing between ribs and the direction of demolding of the manufacturing process were considered during the optimization. The result of the optimization is shown in Figure 2 a). This result forms the basis for the design of the new developed component. Figure 2 b) shows the final rib design.

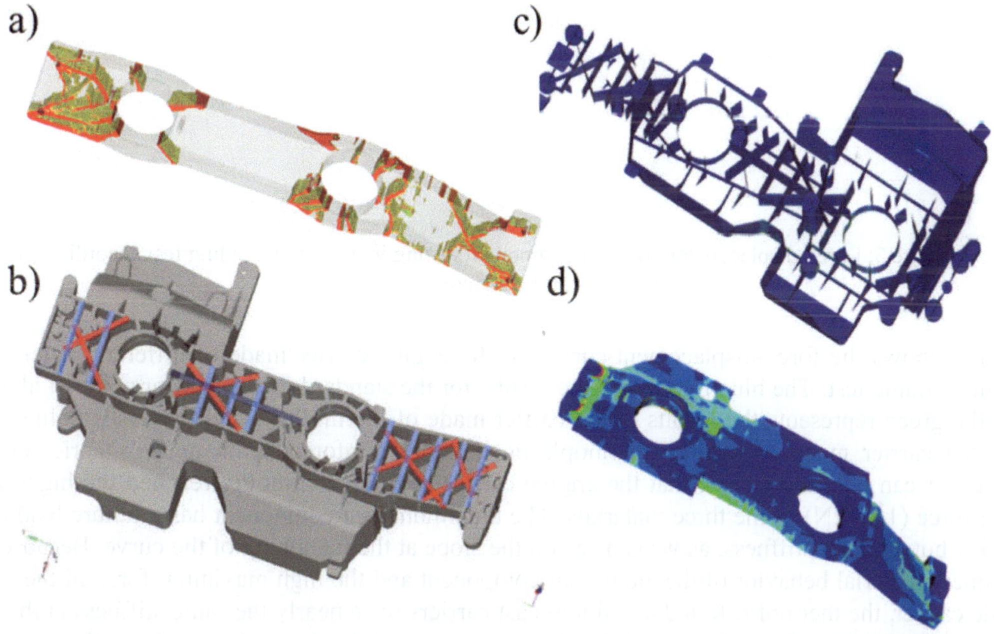

Figure 2: Results of the topology optimization a) and b) and the material specific analysis of component failure c) and d)

The assessment of the component strength has to be material specific. Due to the mainly isotropic material behavior of the GMT, the equivalent stress according to "von Mises" is used to evaluate the strength and the failure of the ribs (see Figure 2 c). The continuous fiber reinforced material thermoPre® has an anisotropic material, more precisely a transversal isotropic 23 behavior. For this unidirectional material behavior, special physically based failure criteria have to be used. For computation of the engine carrier the Failure Mode Concept (FMC) according to Cuntze was used. This failure criterion differentiates and evaluates five possible failure modes of an unidirectional material: fiber failure due to compression and tension loads as well as matrix failure due to compression, tension and shear loads [3]. Figure 2 d) side shows the inverse reserve factor according to the FMC for a chosen load case.

As a result of the simulation, the new developed engine carrier fulfills the same strength requirements as the aluminum cast carrier. Due to the small available design space, the stiffness requirements were only partially fulfilled.

4 TESTING RESULTS

The basis for comparing the performance between the standard and the new developed engine carrier was a static 3-point bending test, wherein a load was applied in the z-direction to the mount of the gearbox.

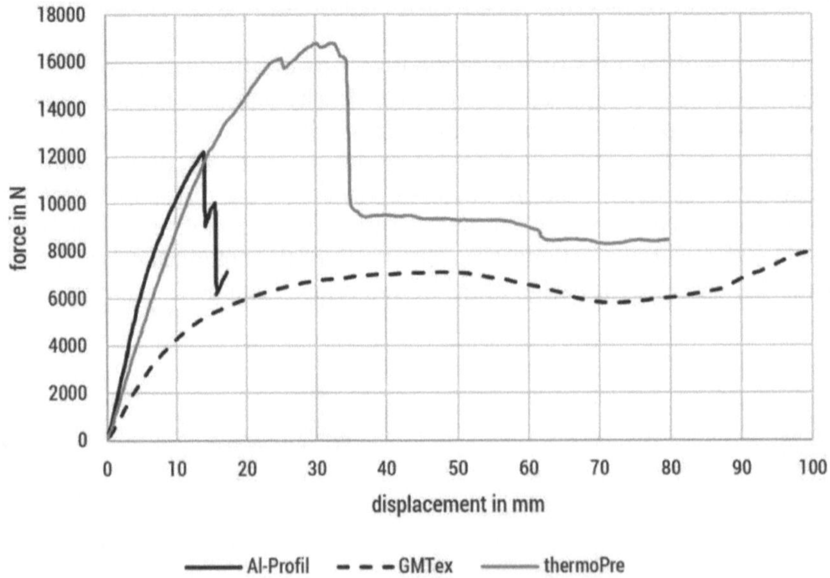

Figure 3: Force-displacement-curve of component testing in a 3-point bending test according to z-direction of the car coordinate system

Figure 3 shows the force-displacement-curves of the engine carrier made of different materials in a 3-point bending test. The blue line shows the results for the standard engine carrier made of aluminum cast, the green represents the results of the carrier made of thermoPre® and the dashed blue line depicts the carrier made of another thermoplastic material reinforced with an 4:1 fabric, Quadrant GMTex. It can be clearly seen, that the engine carrier made of thermoPre reached the highest maximum force (16.8 kN) of the three materials. The aluminum cast component has a failure load of only 12.1 kN but a higher stiffness, as we can see on the slope at the beginning of the curve. Because of the non-linear material behavior of the aluminum component and the high maximum force of the thermoplastic carrier, the thermoPre® and aluminum cast carriers have nearly the same stiffness at the breaking point of 12.1 kN. The deflection of the aluminum cast and the thermoPre carrier at the load of 12.1 kN are 14.0 mm and 14.5 mm. Therefore, the stiffness deviates only about 3.4 % between these two carriers. In addition, the brittle material behavior of the aluminum cast carrier is clearly visible. Achieving the maximum force results in a sudden reduction of the force and a component failure. The thermoPre® carrier has a much more benign failure behavior. After reaching the breaking load, there is also a sudden reduction of force, but the component does not fail totally. After the breaking jkload, the thermoplastic carrier shows a ductile behavior with a high residual strength over a displacement length of at least 30 mm. The GMTex carrier, although made in the same tool like the thermoPre® carrier, shows significantly lower stiffness and strengths but a more ductile behavior than the thermoPre® carrier.

Table 1 depicts the results of the 3-point bending test. In addition to the maximum forces, the stiffness as a function of force and displacement, the energy and the mass of the three tested carriers are also listed. The thermoPre® achieves a 40 % higher maximum force and a six times higher energy absorption than the standard aluminum cast carrier.

Table 1: Results of the 3-point bending test according to z-direction of the car coordinate system

	Al-Profil	thermoPre	GMTex
maximum force in kN	12.1	16.8	7.1
maximum force in %	100.0	138.8	58.7
stiffness of the carrier in N/mm (slope of the f-u-curve)	1289.8	947.8	477.0
stiffness of the carrier in %	100.0	73.5	36.9
energy in J	129.9	808.7	608.5
energy in %	100.0	622.6	468.4
mass in g	7645	5900	4800
mass in %	100.0	77.2	62.8

In addition to the static tests, dynamic tests were performed on a drop tower, too. Because the drop tower had no measuring instruments, only a comparative testing was carried out. Meaning that both components (standard aluminum cast and thermoPre® carrier) were loaded by a punch with a diameter of 256 mm, an initial velocity of 33,8 km/h and an impact mass of 45 kg. In total, both components were loaded with about 2 kJ of impact energy. The crash behavior was evaluated during the test by means of a high-speed camera.

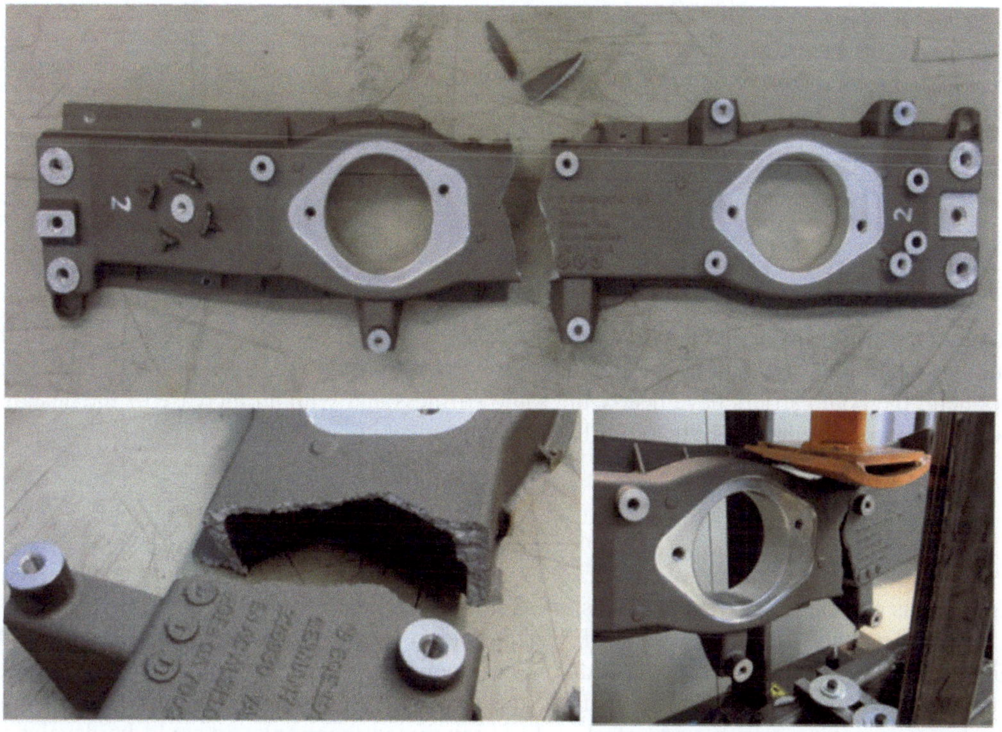

Figure 4: Failure behavior of engine carrier made of aluminum after an impact test

Figure 4 shows the standard carrier after the crash test. As already seen in the static test, also in dynamic testing the aluminum cast carrier shows a very brittle material behavior. The carrier splits into at least four fully divided parts with no residual carrying capacity. High-speed videos from the testing also confirm the low absorption of energy.

Figure 5 depicts the new developed engine carrier made of thermoPre® after the impact test. It can be seen, that there are some damages around the mount of the gearbox as well as the load introduction points. However, the whole component is still attached so that the component has a residual carrying capacity.

Figure 5: Failure behavior of engine carrier made of gfrp after an impact test

In summary, it can be said, that the new developed engine carrier exceeds the requirements of the aluminum cast carrier in terms of strength and energy absorption. At the breaking point of the standard component, the both carriers have nearly the same stiffness and with an increase of the design space, the stiffness requirements can be also fulfilled at lower forces.

5 CONCLUSION

The aim of the research project was the development of an engine carrier including design, technological as well as economic restrictions. The starting point was the standard engine carrier, which is made of aluminum cast and has got a weight of 7.9 kg. Due to the development of the new, cost-effective semi-finished product thermoPre®, the reduction of manufacturing steps through their integration into the compression molding process as well as the usage of effective and high-precision simulation methods a new engine carrier made of glass fiber reinforced thermoplastic could be developed. The new engine carrier made of thermoPre® has a weight of 5,9 kg, which means a mass reduction of about 23 %. In addition to the fulfillment of the strength and energy absorption requirements, the price of the component can be reduced as well. Due to the integration of the mounts of the power electronics and charging unit, a reduction of tooling and assembly costs can be achieved in addition. Thus the total minimal price of the new developed carrier, depending to the production volume, is about 48 % less than the price of the standard metal part. Due to the very good material damping behavior of thermoplastic materials, the new engine carrier also offers many advantages to reduce noise, vibration and harshness levels [4].

The newly developed technologies were presented using the example of the engine carrier. In several other projects it has already been shown that the results of this project can be transferred to other components. Part-related positive results can be achieved there, too.

REFERENCES

[1] Stegelmann M.: Elektromobilität und Leichtbau sind große Zukunftsthemen für Deutschland. Werkstoffe für die Zukunft, DresdnerTransferbrief, 23. Jahrgang (2015), 26–27.

[2] Eckstein L., Schmitt F., Hartmann B.: Leichtbau bei Elektrofahrzeugen. ATZ-Automobiltechnische Zeitschrift, volume 112 (2010), 788–795.

[3] Cuntze R.G.: Efficient 3D and 2D failure conditions for UD laminae and their application within the verification of the laminate design. Composites Science and Technology, volume 66 (2006), 1081–1096.

[4] Klaerner M., Wuehrl M., Kroll L. and Marburg S.: Modelling and FEA-Simulation of the Anisotropic Damping of Thermoplastic Composites. Advances in Aircraft and Spacecraft Science, volume 3 (2016), 331–349.

COMPOSITE ENGINE BLOCK – CHALLENGES FOR DESIGN AND MATERIAL

Melanie Jauernick[1], Dr. Christine Schütz[2], Joachim Sterz[3], Birte Horn[4]

[1] Volkswagen AG, melanie.jauernick@volkswagen.de
[2] Volkswagen AG, christine.schuetz@volkswagen.de
[3] Technische Universität Braunschweig, j.sterz@tu-braunschweig.de
[4] Technische Universität Braunschweig, b.horn@tu-braunschweig.de

Keywords:

Cylinder crankcase, Thermoset plastic, Injection molding, Surface treatment, Leak-tightness

ABSTRACT

A hybrid engine block including a metal supporting structure and a composite plastic housing shall be designed and manufactured. The focus lies on the material selection, adapted to the engine block requirements, and the analysis of the joint area by pressure and leak tight tests. Several surface treatments are applied and compared to each other.

1 INTRODUCTION

For the foreseeable future, the combustion engine will remain a dominating component in the drivetrain, especially for plug-in hybrid strategies. Particularly for hybrid electric vehicles lightweight design is becoming a great challenge due to the additional mass, generated by the two powertrain systems.

However, the potential for further weight reduction and function integration in combustion engines by using conventional manufacturing techniques is nearly exhausted. Thus, the research project aims at new design methods and manufacturing technologies, such as the combination of conventional metals with lightweight materials (e.g. fiber reinforced polymers) to optimize the most challenging drivetrain housing – the cylinder crankcase (Figure 1).

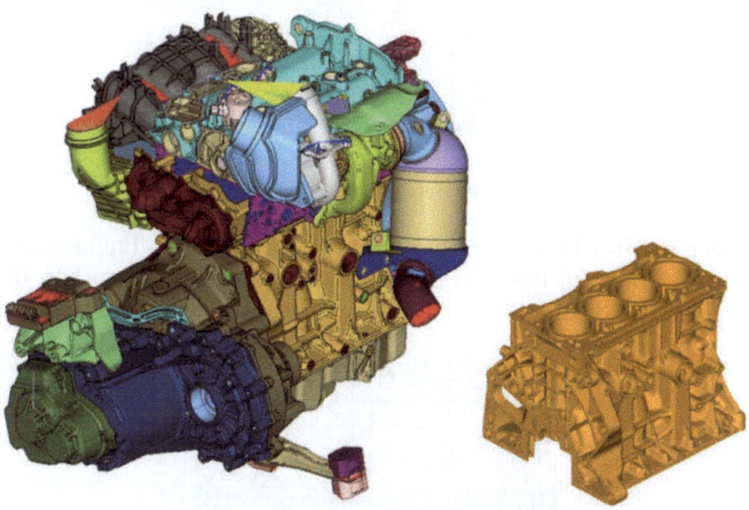

Figure 1: SI engine (left) and cylinder crankcase (right) [IAV]

K. Dröder et al (eds), *Technologies for economical and functional lightweight design*, Zukunftstechnologien für den multifunktionalen Leichtbau, https://doi.org/10.1007/978-3-662-58206-0_5

In addition to the weight reduction, these new materials and methods offer further advantages for future combustion engines, for instance a stronger damping of the vibration, resulting in better acoustics, or a thermal optimization such as faster heating up and slower cool down. These characteristics make a hybrid engine for the next-generation drivetrain an ideal application for new materials.

2 STATE OF THE ART

Hybrid materials in drivetrain components are nothing special anymore: As an example, the Fraunhofer ICT and Sumitomo Bakelite published a motorbike engine including a cylinder case with a metal sleeve surrounded by a plastic housing (Figure 2). Lightweight potentials, functionality, and endurance have been proved and acoustic measurements have been carried out [1].

Figure 2: Plastic cylinder case [Fraunhofer ICT]

Another example is an integrated plastic valvetrain module for R3 IC engines by IAV, Sumitomo Bakelite and Volkswagen AG Chemnitz. The valvetrain module has completed acoustic tests and a >100.000 km test run in a vehicle without noticeable problems.

Figure 3: Integrated valvetrain module [IAV]

The last example is a gearbox housing for electric vehicles by ARRK Engineering (Fig. 4), which is completely made of overmolded thermoplastic organosheets with local carbon fiber reinforcements (UD tapes). A weight reduction of 30 % and acoustic optimisations could be achieved [2].

Figure 4: Hybrid gearbox housing [ARRK]

3 REQUIREMENTS FOR THE DESIGN

A cylinder crankcase must be able to withstand a large number of different load conditions: An overlap of tension and pressure, bending and torsion is present. Based on the gas pressures and the cranktrain design (Figure 5), a load spectrum of up to 36 kN crankshaft vertical force, 6,5 kN piston horizontal force and 110 bar cylinder pressure during ignition have to be assumed.

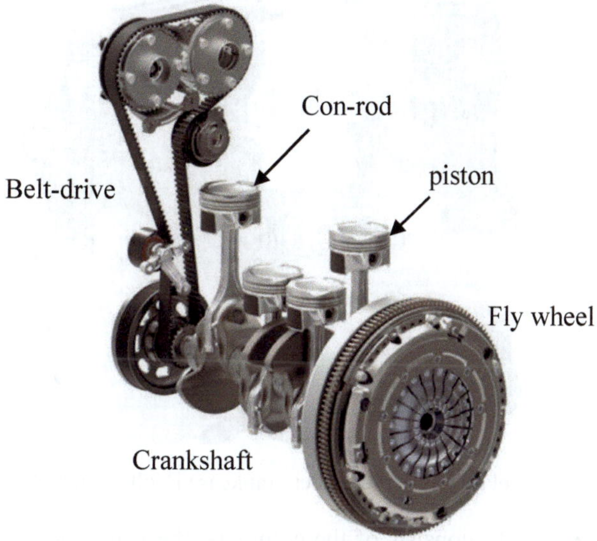

Figure 5: Schematic cranktrain [IAV]

This leads to static stresses of up to 120 MPa in the water jacket area and cyclic stresses of up to 100 MPa in the main bearing.

Moreover, high combustion temperatures and chemical influences have to be considered. The maximum surface temperatures within the cylinders are around 180 °C in the topdeck region (Figure 6).

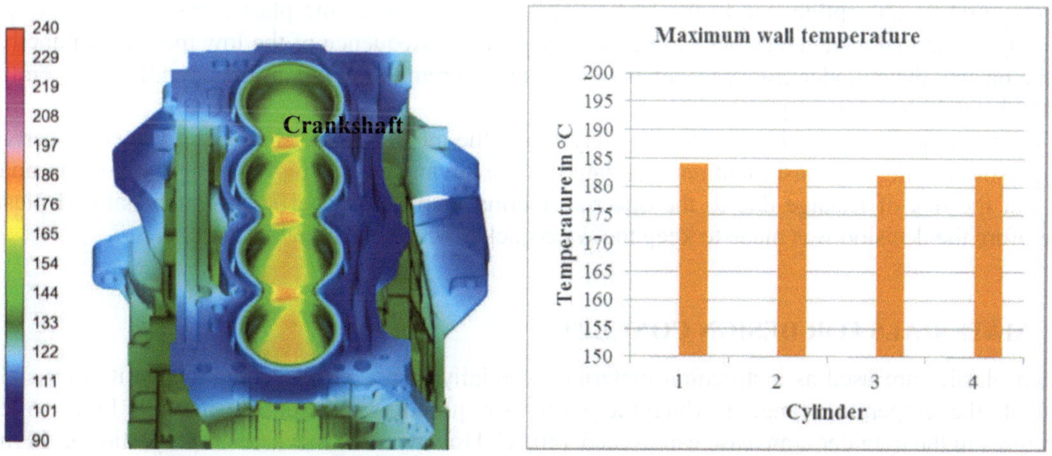

Figure 6: Surface cylinder temperatures [Volkswagen AG, Group Research]

Also temperature changes, especially due to start-stop applications have to be considered. Water-glycol mixtures and fuel-efficient engine oils with high amounts of additives lead to additional load.

Under these conditions, not only a sufficient mechanical and thermal stability has to be provided but also leak tightness in the periphery of hot fluids under high pressure is necessary.

In conclusion, when designing with hybrid lightweight materials in mind, the cylinder crankcase represents a challenging component with high requirements.

4 DESIGN CONCEPT

The design concept includes a fiber reinforced plastic housing, which comprises an inner supporting structure consisting of aluminum (Figure 7).

Figure 7: Sectional model of a hybrid cylinder crankcase [Volkswagen AG, Group Research]

The supporting structure (insert), consists of the cylinders, the water-jacket and the upper part of the main bearing. A bedplate forms the lower part of the main bearing. The thermal stresses and the operating forces are located inside the metal insert and the bedplate. A special treatment of the cylinders such as APS coating and honing is used to reduce the friction between piston rings and cylinder wall as well as improving the blowby behavior and cylinder distortion.

The surrounding plastic housing forms a portion of the cylinder crankcase which can increase the weight reduction potential as a result of the low material density. Moreover, the housing can lead to better vibration absorption due to the high absorption potential of the plastic. As a further possible benefit the thermal advantages have to be mentioned: In consequence of the low thermal conductivity of the plastic, the cylinder crankcase tends to heat up faster and to maintain the operating temperature, particularly during start-stop operation.

The water-jacket can either be placed solely inside the metal insert or between the metal insert and the plastic housing. The last mentioned variant has the advantage of a higher weight reduction potential but the risk of leakage due to the different thermal expansion coefficients of metal and plastic. Therefore the decision was made to keep the water-jacket within the metal insert.

5 MATERIALS FOR DESIGN CONCEPT

When plastics are used as a structural material, especially in powertrain components, it is crucial to evaluate the temperature range in which the plastic is required to perform. As shown in Figure 6 temperatures in the cylinder crankcase can exceed 180 °C. However, the mechanical properties of thermoplastics like polyamide 6 (PA6) tend to decrease significantly above 100 °C and the residual young's modulus only reaches around 40 % of the value it achieves at 20 °C. Thermoset materials are, therefore, the only alternative.

Phenolic resins are the material of choice because of their high temperature as well as media and aging stability. Furthermore, the relatively high tensile strength of phenolic resins increases when mineral fillers and glass fibers are added to the mixture. When intending a hybrid material application it is essential to consider the thermal expansion of the linked materials. The advantage of phenolic resin compounds is that the thermal expansion can be adjusted by the number of fillers added. Thus the differences of thermal expansions in the hybrid material can be balanced.

The process for high volume production is a thermoset injection molding process. The phenolic resin compound, which consists of a non-crosslinked polymer, fillers, a crosslinker and processing additives, is plastified in a screw at relatively low temperatures. Crosslinking of the compound to the final phenolic resin is performed in the mold with a temperature between 170 and 190 °C. A tempering process leads to the final mechanical strength as well as media, temperature and aging stability.

The application of a suitable material with an adapted manufacturing process is useless if the joining zone between plastic and metal – the contact area – isn't designed properly. Weaknesses of the contact area lead to fractures and/or displacements and finally to failures of the composite system. Therefore the analysis of the contact area is focused in this article. Therefore pressure tests and leak tightness tests of specific hybrid samples were performed. Regarding the contact area within the samples, different concepts are available: A laser-structuring of the joining zone shall increase the effective joining area and the formation of undercuts which leads to form locking. Furthermore, an adhesive bond is initiated by applying amino-groups on the aluminum surface by silanization. It is assumed that as a result hydrogen bonds could be formed between the silanized surface and the phenolic resin.

For the analyses and the proof-of-concept the sample geometry shown in Figure 8 is chosen:

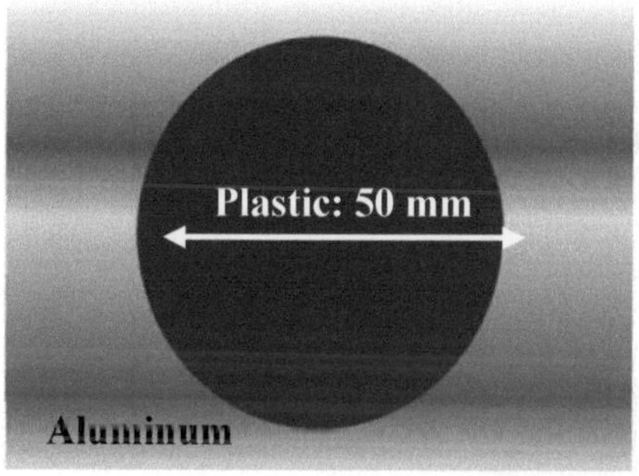

Figure 8: Sample geometry for proof-of-concept investigations [Volkswagen AG, Group Research]

The aluminum insert is placed in the tool of the injection molding machine and the phenolic resin is injected into the geometry. One part of the produced samples are climate aged according to the test specification PV 1200 "Testing of Resistance to Environmental Cycle Test (+80/ –40) °C" [3]. Another part of the samples are stored in oil as well as glycol respectively according to the Group standard TL 52682 "PA66, Glass Fiber-reinforced, for Finished Parts Carrying Coolant" [4]. The compressive strength, as well as the media density (leak tightness) of the aged and non-aged samples is evaluated. A significant innovation of the analyses is a specific testing setup which has been developed together with the TU Braunschweig and is used to determine the leak tightness.

6 EVALUATION

6.1 Surface analysis

Before primer-application and injection molding process, the surfaces are prepared by laser-roughening. The scanning electron microscope (SEM) analysis of the laser roughened surface shows several rows of similar melting-structures side by side in the shape of small ovals placed one below the other (Figure 9, left). The melted material provides the clamping structures for the plastic.

After the primer-application and the injection molding process, a cross-section through the contact area was cut. The microscope-analysis shows that the clamping structures have been undercut by the plastic (Figure 9, right), successfully creating a form fit between the two materials. The maximum achievable roughness is up to 200 µm.

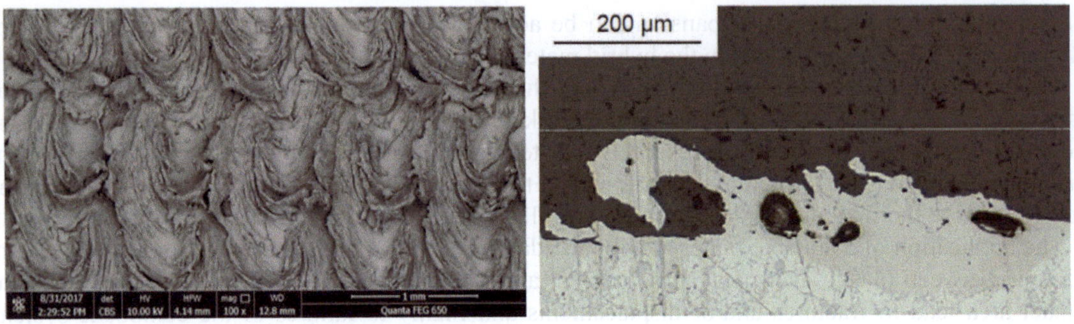

Figure 9: SEM analysis of the roughened metal surface (left) and cross-section through the metal-plastic contact area (right) [Volkswagen AG, Group Research]

6.2 Effect proof of the silanization

The effect of the silanization can be shown by the contact angle of a water drop, applied on the metal surface without and with silane primer (Figure 10, left). The higher number of hydrogen bonds, formed by the silanized surface lead to an increased wetting effect: The water drop extends more and forms a smaller contact angle to the metal surface.

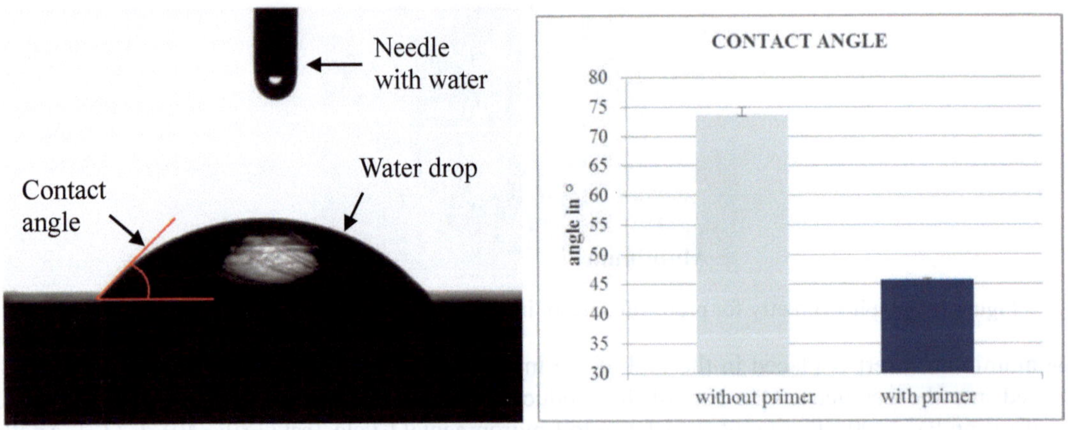

Figure 10: contact angle (left) and results for a surface with and without primer (right) [Volkswagen AG, Group Research]

Figure 10 (right) shows an explicit reduction of the contact angle due to the increased wetting effect of the primer. A significant improvement of the mechanical properties of the plastic-metal-compound by use of the primer, additional to the laser-roughening, can be assumed.

The following injection molding of the phenolic resin and the mechanical tests are carried out with laser-roughened samples with as well as without a primer to evaluate the silanization effect in combination with the plastic.

6.3 Pressure test

For the pressure test, the plastic part in the middle of the sample has slowly been pressed out by a die and the maximum compressive strength has been registered.

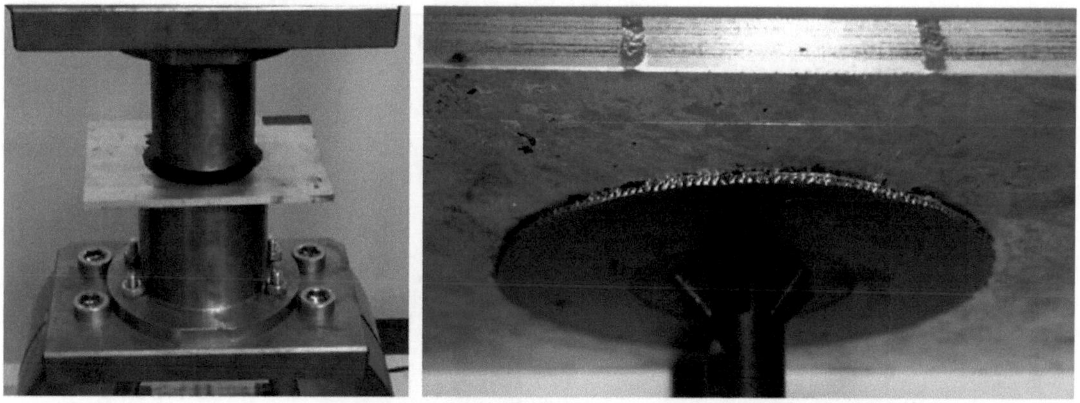

Figure 11: Pressure test (left) and pressed out plastic (right) [Volkswagen AG, Group Research]

The results of the pressure tests show a significant decrease of the compressive strength through oil aging. The coolant aging doesn't seem to have a considerable influence: The compressive strength remains nearly constant. The loss of compressive strength after the climate change test lies between the oil aging and the coolant aging (Figure 12).

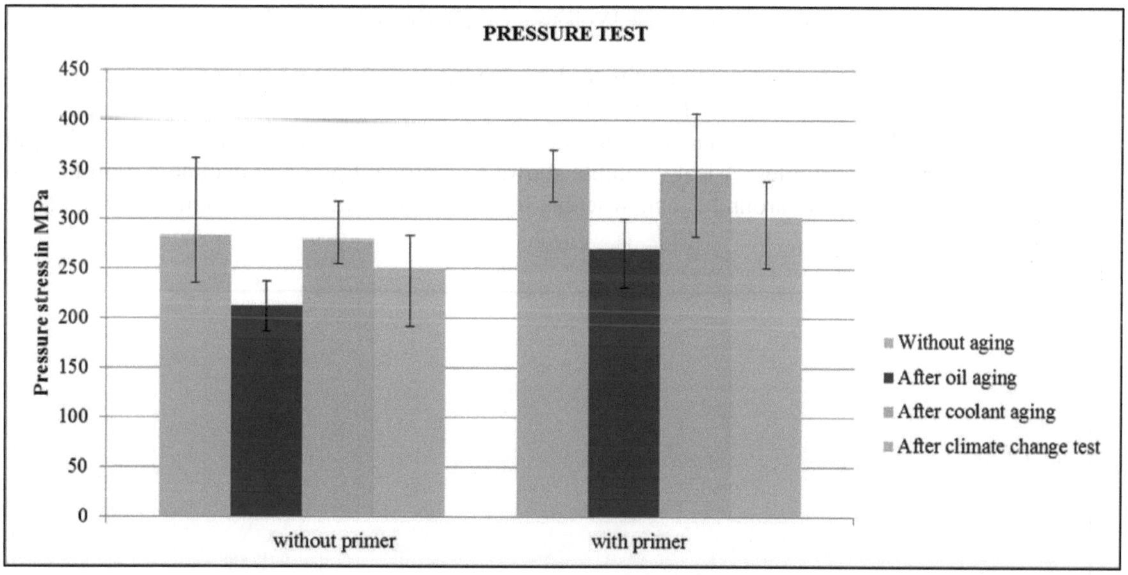

Figure 12: Results of pressure test of samples with and without primer including the min- and max-value [Volkswagen AG, Group Research]

In each case, the samples with silane primer in the contact area show an explicit increase of compressive strength in general.

6.4 Leak tightness test

The leak tightness test also demonstrates the positive effect of the primer: On average, the samples with primer show the lowest leakage (Figure 13).

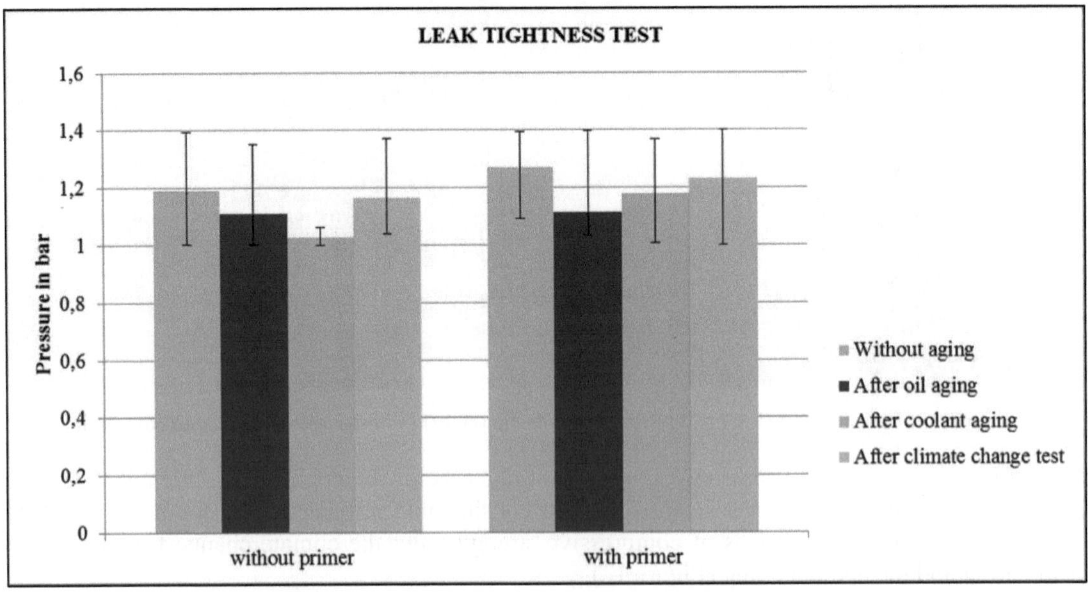

Figure 13: Results of leak tightness test on samples with and without primer including the min- and max-value [Volkswagen AG, Group Research]

Within these samples, the coolant aging seems to decrease the leak tightness the most. Figure 13 also shows a large variation within the results. In many cases, the measurement uncertainty extends from nearly 1 bar up to 1,4 bar. The explanation is that each column of the diagram includes leak-tight samples as well as leaky samples: A leaky sample shows a significant pressure decline during the test or even a constant low value around 1 bar, whereas the value of a tight sample remains on a stable high level (Figure 14).

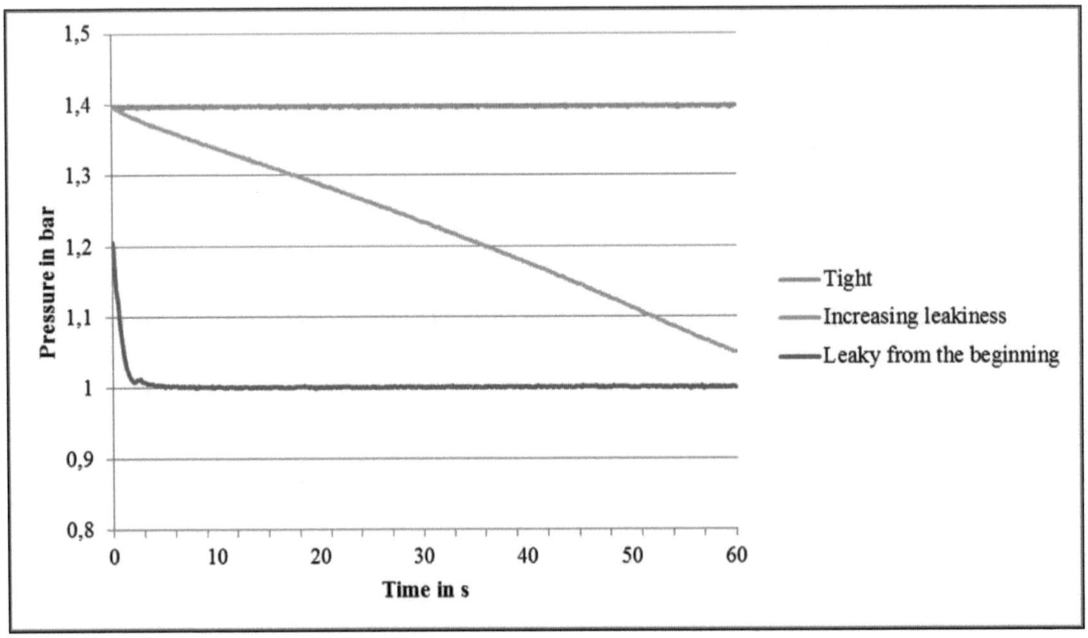

Figure 14: Pressure curve of a leaky and tight sample in comparison [Volkswagen AG, Group Research]

In addition to the average pressure values in Figure 13 it is also necessary to evaluate the number of leak tight samples and leaky samples (Figure 15):

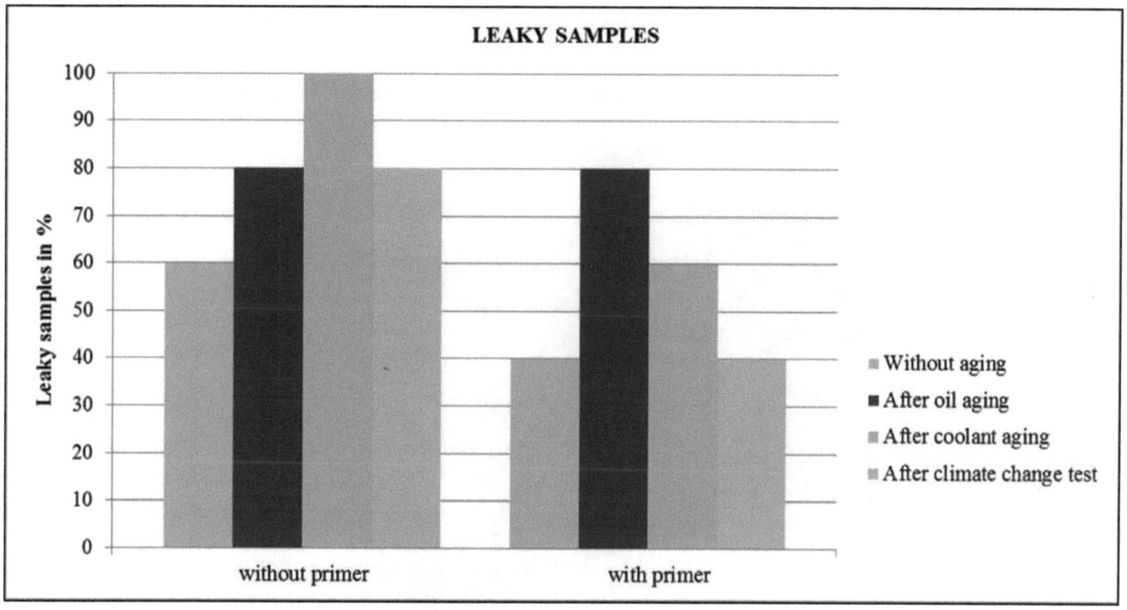

Figure 15: Amount of leaky samples [Volkswagen AG, Group Research]

Figure 15 shows an increase of the number of leaky samples without primer in each case: The climate change test and the oil aging lead to 20 % more leaky samples, the coolant aging even leads to 100 % leaky samples. For the samples with primer, the climate change test has no effect on the amount of leaky samples. The coolant aging results in 20 % more leaky samples and the oil aging in 40 % more leaky samples.

7 CONCLUSION AND OUTLOOK

In conclusion, the application of a silane primer is absolutely recommendable for an improvement of the contact area between aluminum and plastic. Regarding the surrounding media, the oil-aging has significant effects on the contact area. This has to be considered during the detailed design phase of the use case “hybrid cylinder crankcase”. As an example, the position of contact areas within oil-bearing structures of the crankcase should be avoided.

Until now, a first test block has been manufactured to determine the most challenging problems during the manufacturing process. The steps carried out so far include:

- the manufacturing of the metal support structure
- a surface roughening process (not yet by laser and primer application)
- the encapsulating with a thermoset resin
- the milling of the outer surfaces
- APS-coating of the cylinder liners (Figure 16).

Due to the suboptimal treatment of the metal surface before the plastic casting, failures of the contact area are obvious. Especially the heat developed during the APS-coating led to further displacements.

For the next prototype manufacturing a re-design considering the conclusions of the first test block and the contact area analyses is already in progress. The laser-roughening of at least portions of the metal surface and a primer application on the complete surface will be included.

Further steps for the cylinder crankcase prototype are:

- The manufacturing with optimized boundary conditions.
- An acoustic test to evaluate the effect of the surrounding plastic.
- Assembly to an engine and the motor powered operation on a test bench to evaluate friction and wear.
- If the motor powered operation succeeds, a fired operation on a test bench follows.

Figure 16: Hybrid cylinder crankcase, pre-manufactured and APS-coated [Volkswagen AG Salzgitter]

Additional to the already discussed pressure and leak tightness tests more evaluations of the contact area are necessary. These further investigations are part of a following OHLF-project and include the work packages showed below:

- A tensile test of roughened samples with and without primer
- A deeper effect-analysis of the primer. Within this test series, samples with primer but without roughening shall be produced and tested.
- Aging tests of samples with manufactured surfaces: In some cases, it cannot be avoided to manufacture hybrid material surfaces by milling or grinding. As an example, the hybrid cylinder crankcase has to be milled/grinded on each side to provide proper leak tightness to the ancillary components. In comparison to a "closed" plastic surface, a milled/grinded plastic surface can show different properties regarding aging stability, which has to be evaluated.

REFERENCES

[1] Dr. Lars-Fredrik Berg, Dr. Hans-Peter Kollmeier, Gewichtsreduktion und verbesserte Akustik – Mögliche Vorteile eines Zylinderkurbelgehäuses aus Faserverbundkunststoff", 9. Symposium Motor- und Aggregateakustik, Magdeburg 2016.

[2] Monika Kreutzmann, Leichtbau: Erstes Composite-Getriebegehäuse mit lagenoptimiertem Organoblech spart 30 Prozent Gewicht gegenüber vergleichbarem Aluminium-Bauteil, OEM& Lieferant, Ausgabe 01/2017.

[3] Volkswagen PV 1200: "Vehicle Parts – Testing of Resistance to Environmental Cycle Test (+80/ –40) °C", Wolfsburg 2004.

[4] Volkswagen TL 52682: "PA66, Glass Fiber-Reinforced, for Finished Parts Carrying Coolant", Wolfsburg 2015.

DEVELOPMENT OF A LIGHT-WEIGHT SEAT STRUCTURE USING A HYBRID MATERIAL APPROACH

Claudia Creighton, Mandy de Souza, Russell J. Varley, John Antony

Deakin University, Geelong, Australia, Carbon Nexus, Institute for Frontier Materials.
claudia.creighton@deakin.edu.au
http://www.carbonnexus.com.au/

Keywords:

Light-weighting, Seat structure, Carbon fibre composite, Thermoplast, Adhesive bonding

ABSTRACT

Carbon fibre composites are lightweight, and provide a high strength solution for automotive applications, however, the associated material and processing costs are relatively high. This project focused on the development of a composite front seat back (FSB) by primarily reducing cost through materials and component design without compromising safety. The use of cheaper, lower strength materials and low-cost production methods was explored. A hybrid light-weight material system consisting of carbon fibre non-woven and a low-cost long fibre reinforced thermoplastic was developed. Consequently, bonding solutions for dissimilar materials including the exploration of surface preparation techniques and their impact on the composite bond strength were investigated. The employment of flame and blown ion plasma treatment resulted in a 300 % increase in shear strength between inner and outer seat material. As a result, a unique materials and production solution that met the performance criteria with continued reduction in weight and cost was achieved.

1 INTRODUCTION

Increasing demand for light-weighting in the automotive industry is driving the replacement of metal components within vehicle structures. In addition to meeting new emission mandates, light-weighting also enables the implementation of new technologies that will improve the function of future vehicles. Significant effort is being taken to reduce the mass of the vehicle body itself and now attention has turned to the interior of the vehicle. The front seats of a vehicle can weigh up to 20kg each and, as a result, represent a significant portion of the interior trim mass.

A composite seat back structure was developed at Carbon Nexus, Deakin University (Australia) with a composite car seat manufacturer, successfully producing a two-piece carbon fibre composite front seat back (FSB) that achieved a 69 % mass reduction compared to the current steel seat frame due to its integrated part design. Along with improved performance, the composite FSB enables improved styling and design freedom that enhances aesthetic appeal and comfort. However, the cost of the composite seat back is relatively high due to material costs and its current manufacturing approach. In order to decrease the cost of the composite FSB, a multi-material approach was taken. Cheaper material and low-cost production methods for the inner structure were explored and a hybrid light-weight material system, consisting of carbon fibre non-woven and a low-cost fibre reinforced thermoplastic, was developed.

Consequently, bonding solutions for dissimilar materials including the exploration of adhesion materials, adhesive promoters and the optimum fabrication technologies were investigated. A strong thermoplastic–thermoset interface is critical to ensure structural performance. The interactions between thermoset and polyamide (PA) are driven by several functional groups. Covalent bonds are

K. Dröder et al (eds), *Technologies for economical and functional lightweight design*, Zukunftstechnologien für den multifunktionalen Leichtbau, https://doi.org/10.1007/978-3-662-58206-0_6

formed between amide hydrogens and epoxy groups, and hydrogen bonds form between amide hydrogen and oxygen in epoxy resins [1].

Different strategies have been presented in the past to increase the surface properties of thermoplastics in order to enhance the adhesion strength to a dissimilar substrate. The thermoplastic substrate surface might be roughened through physical treatment such as sandblasting to increase the bond by mechanical interlocking. Chemical surface modifications include acid etching, oxidising flame treatment, corona discharge and graft polymerisation. Flame treatment, corona discharge and plasma treatment have shown to be the most effective and widely used methods to functionally activate thermoplastic surfaces by increasing the surface energy and functional groups [1, 2, 3].

The aim of this project was to investigate suitable bonding solutions to join PA66 and PP thermoplastics to epoxy thermosets, including the exploration of surface preparation techniques and their impact on the composite bond strength. To provide final confirmation of a successful outcome on coupon level, fully sized two-piece front seat backs were fabricated using the most promising bonding method identified during the experimental part of this work. Using an iterative approach, including the selection of surface treatment, fibre architecture and fabrication technology, a level of performance that met industry standards to produce a low-cost composite seat back that could be realistically considered for commercial application was achieved.

2 SURFACE TREATMENT FOR ADHESIVE STRENGTH ENHANCEMENT

In order to achieve the maximum adhesive strength between the dissimilar materials two surface treatment techniques, propane flame and ion blown plasma, were explored.

2.1 Propane flame treatment

PA66 samples were flame treated with a propane torch (PRI2000D, Primus) in the bonding area. A distance holder was attached to the torch to ensure a reproducible treatment. The flame was first adjusted to maximise the main reaction zone (luminous zone), the hottest and strongly oxidising region of the flame. While the overall flame length was kept constant at 50 mm with a luminous zone of 15 mm, the distance between luminous zone and samples was changed from 5 to 45 mm in 10 mm steps in order to maximise the treatment efficiency. The results suggested that a luminous zone tip-to-substrate distance of 25 mm modifies and prepares the PA66 surface most effectively showing an average contact angle of 22.4°. Closer flame distances lead to thermal degradation of the substrate. By increasing the distance to 45 mm, the oxygen rich zone moves out of contact with the PA66 surface resulting in a sudden increase in contact angle (54.2°). Table 1 lists the process parameter used during flame treatment.

Table 1: Propane flame treatment parameters

Treatment parameters:	Propane flame	
Luminous zone-substrate distance	[mm]	25
Nozzle velocity	[mm/s]	10
Path width	[mm]	15
Repetition	[-]	2

2.2 Plasma treatment

The surface of fibre reinforced PA66 was activated using an ENERCON (USA) plasma technique. The technology bombards surfaces with a high speed discharge of ions to clean, etch and functionalise the surfaces. Without changing the substrate morphology, plasma treatment successfully changes the mechanical and chemical properties of a surface and initiates cross-linking effects. Figure 1 displays the plasma treater used in this project and the plasma jet as it was applied at a constant jet tip-to-substrate distance of about 10 mm (suggested by the supplier). The sample was quickly moved across the plasma flame with an approximate retention time of 2 seconds. Table 2 shows the process parameters used during plasma treatment.

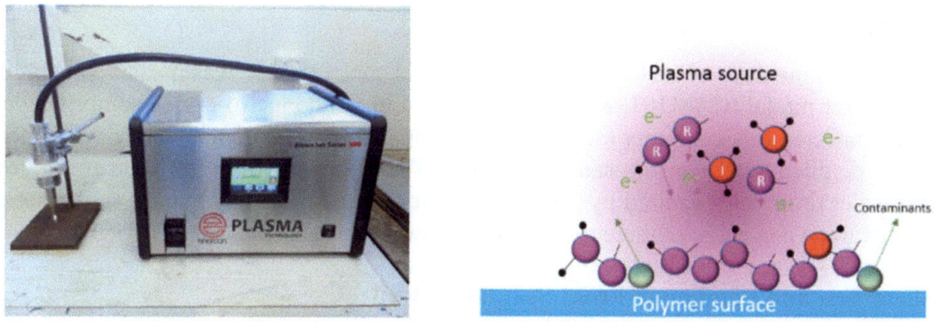

Figure 1: ENERCON plasma technique and schematic of surface treatment using blown ion plasma

Table 2: Plasma treatment parameters

Treatment parameters:	Plasma	
Frequency	[kHz]	40
Jet tip-substrate distance	[mm]	10
Nozzle velocity	[mm/s]	10
Path width	[mm]	10
Repetition	[-]	2

3 MATERIALS AND CHARACTERISATION METHODS

3.1 Materials

Three different long fibre reinforced thermoplastic materials were provided as PA66 moulded material plates reinforced with 40 % and 60 % glass fibre (PA66/GF40, PA66/GF60) and 40 % carbon fibre (PA66/CF40), and a polypropylene sample filled with 40 % glass fibre (PP/GF40). A brief overview of the material properties for the various fibre reinforced substrates is shown in Table 3. The carbon fibre reinforced epoxy samples were manufactured from carbon fibre non-crimp fabric and epoxy resin (Hexion™ RIM 935/RIMH 936) with a peel ply surface for improved bonding performance.

Table 3: Fibre reinforced thermoplastic materials to be bonded to carbon fibre/epoxy composite

Sample ID	Thermoplast	Fibre reinforcement	Fibre loading [%]	Density [g/cm³]	Tensile strength [MPa]	Tensile modulus [GPa]
PA66/CF40	PA6.6	Carbon fibre	40	1.34	294	30.9
PA66/GF40		Glass fibre	40	1.45	215	13.3
PA66/GF60			60	1.67	260	19.0
PP/GF40	PP	Glass fibre	40	1.21	150	10.0

Fibre reinforced thermoplastic substrates were adhesively bonded to the CF/epoxy composite with an epoxy-based high-strength adhesive Loctite Teroson EP 5055 (Henkel) on the day of treatment. The adhesive was cured three days at room temperature before lap shear testing.

3.2 Contact Angle Measurements

In order to provide a measure of surface polarity, the water contact angle was measured using a Theta Optical Tensiometer modal (CAM101, KSV Instruments, Ltd.) and electro-optics comprising a digital video camera connected to a computer (software Attention Theta). The contact angle was recorded by depositing 3 µl distilled water on the sample surface. The contact profile between the investigated substrate and water drop was captured by a high-resolution camera and the contact angle measured in the system software. An average of three to five samples were measured.

3.3 Lap Shear Test and Bonding of Front Seat Back (FSB)

Lap Shear Testing was performed according to ASTM D5868-01: Standard Test Method for Lap Shear Adhesion for Fiber Reinforced Plastic (FRP) Bonding. Samples were cut into specimen with 25.4 mm width and length of 100 mm using a water jet cutter. Lap shear samples were prepared from fibre reinforced PA66 and PP and carbon fibre/epoxy composite using an epoxy-based adhesive Henkel Teroson EP5055. According to the ASTM standard, the bond length was 25.4 mm. Bonded samples were conditioned at 21 °C at 55 % relative humidity for at least 24 hours prior to testing.

The lap shear test was performed on a 30 kN Instron 5567/5967 machine. A minimum of five lap shear samples were tested with a 75 mm grip separation and at a cross head speed of 13 mm/min. The individual peak load values and averages were recorded along with the type of failure.

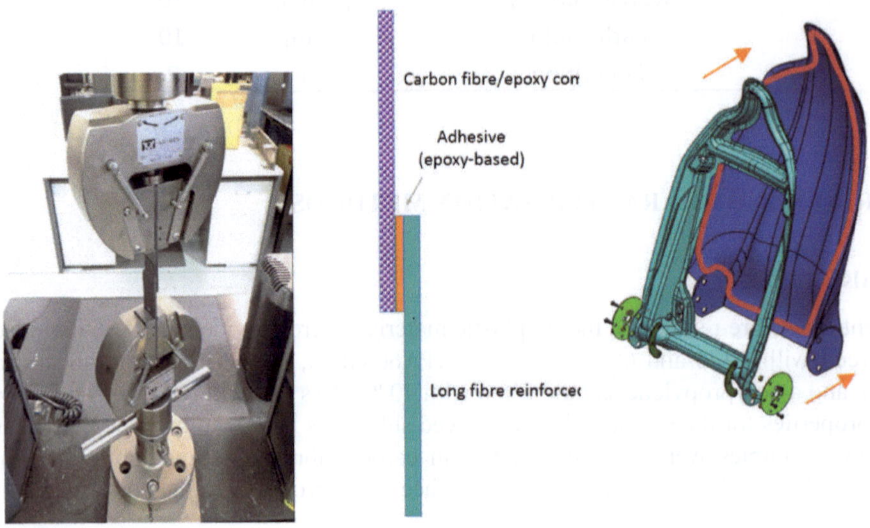

Figure 2: Lap shear test set-up and two-piece carbon fibre composites seat back with adhesive bond line (orange)

4 RESULTS AND DISCUSSION

4.1 Effect of time after surface treatment on surface functionality

Prior to bonding, the effect of time between surface treatment and adhesive application on the surface polarity was investigated. Specimen for each material were treated in two different ways (plasma vs. propane flame) and the contact angle recorded immediately after the treatment and again after 1, 2 and 7 days. A low contact angle refers to a higher surface polarity that is more likely to have polar functional groups that will interact well with the adhesive and/or resin.

The contact angle as a result of time after the treatment and associated water drop images are displayed below in Figure 3. In Figure 3a) the results of flame and plasma treatment for the PA66/CF40 % scenario are graphed over time. The results show a remarkable initial drop in contact angle after surface treatment from about 70° to approximately 20°. This was followed by a general trend of increasing contact angle as time between treatment and measurement progressed, suggesting a degradation of surface functionality. In contrast the PP/CF40 % forms an exception, where the contact angle only decreases a small amount from 90° to 72–84° after treatment (Figure 3b). This phenomenon might be caused by melting of the polymer that caused physical roughening and exposure of fibres on the substrate surface. After another small decrease in contact angle 57–66° after one day, the surface polarity remained almost unchanged. For all materials the plasma treatment generally led to smaller contact angles compared to the use of propane torch.

In order to achieve the most effective bond between adhesive and substrate surface, immediate application of adhesive after surface treatment is required.

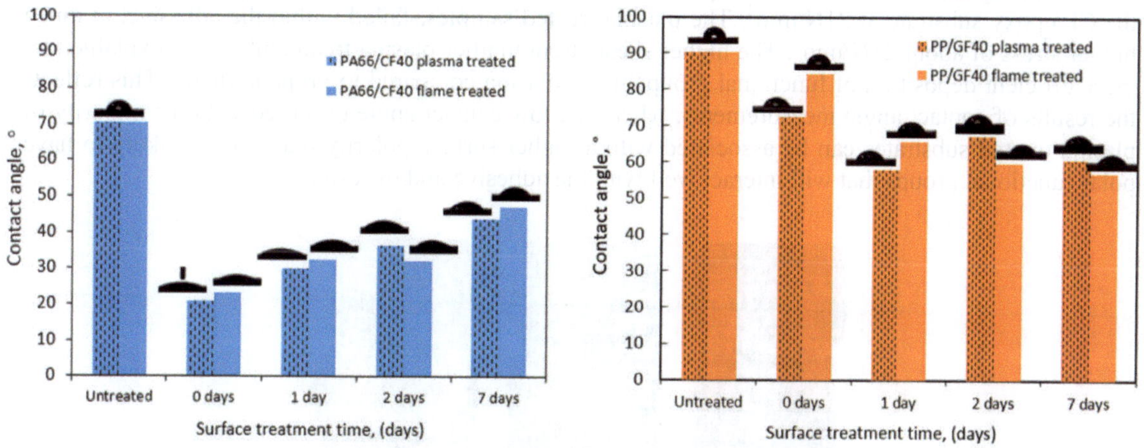

Figure 3: Contact angle (and representative droplet images) vs. time after treatment for a) PA66/CF40 % and b) PP/GF40 %

4.2 Mechanical performance of fibre reinforced PA66 and PP substrates

A series of fibre reinforced PA66 and PP substrates with plasma and flame treated surface modification were adhesively bonded to CF/epoxy substrates. Figure 4 depicts the lap shear strength of samples that underwent propane flame and plasma surface modification in comparison with non-treated samples. While both treatment methods led to 300–400 % increase in bond strength, plasma surface modification proved to be more efficient (+12–21 %) for PA66 matrix samples. Much less significant improvement in shear strength was observed through surface modification of fibre reinforced PP. Another exception with PP matrix composites was that the plasma treatment led to 25 % lower shear strength when compared to the flame treated substrates.

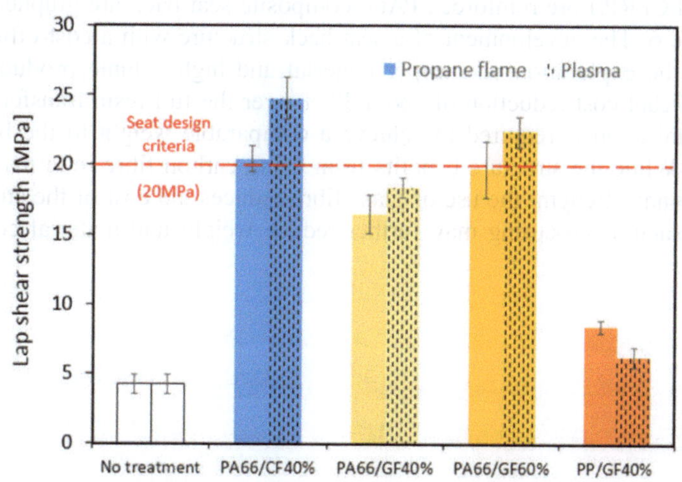

Figure 4: Lap shear strength of samples with propane flame and plasma treatment

The carbon fibre filled PA66/CF40 % material following plasma surface modification performed best in the lap shear test and provided a shear strength of 25N/mm^2. When comparing the bond strength of glass fibre filled substrates PA66/GF40 % and PA66/GF60 % to CF/epoxy, the sample with 60 % fibre load gave the higher values. This indicates that the surface treatment may have a greater effect when applied to substrates with higher fibre content in the PA66 matrix. During the treatment process, more fibres may be exposed per area with 60 % glass fibre loaded substrates and consequently more fibres act as mechanical interlockers to the adhesive.

As an example, the failure and fracture surfaces of propane flame and plasma treated PA66/CF40 % lap shear samples are compared in Figure 5. Untreated lap shear samples failed without any remaining adhesive on the thermoplastic surface. Propane flame treated samples disbonded from the CF/epoxy substrate at 21N/mm^2. The plasma treated samples, failed within the adhesive, at much higher stress of about 25N/mm^2. The higher shear strength after plasma treatment can be explained by more efficient deposition of functional groups using plasma compared to propane flame. This reflects the results of contact angle measurements, where the low contact angle observed with the majority of plasma treated substrates can be associated with a higher surface polarity that is more likely to have polar functional groups that will interact well with the adhesive and/or resin.

Figure 5: Fracture surfaces of lap shear substrates showing adhesive failure for untreated and cohesive failure for flame and plasma treated samples

Out of the tested materials for the FSB inner, plasma treated PA66 filled with 40 % carbon fibre and 60 % glass fibre met the seat design criteria of >20 MPa shear strength. Considering the propane torch is the less cost-intensive treatment solution, it was the treatment method of choice to prove the hybrid-material seat bond and design. The thermoplastic long-fibre reinforced inner was bonded to the outer seat shell and tested in a rear-pull test and passed the test and required seat design criteria.

Structural weight and comparative cost of the first generation CFRP (2-piece) and second generation 2-piece hybrid CFRP/fibre reinforced PA66 composite seat back are graphed against the original steel FSB in Figure 6. The development of a seat back structure with a cost-efficient hybrid material approach allowed the exploration of cheaper material and high-volume production methods, which resulted in a significant cost reduction of about 30 % over the full resin transfer moulded CFRP seat back. Further optimisation is required to achieve a comparable weight to the benchmark FSB fully made from CFRP. While the structure benefits from using carbon fibre in terms of weight due to the lower fibre density and strength, the use of glass fibre reduces the cost of the structure. In the future, bonding methods such as co-curing may further reduce weight and material cost by elimination of adhesive.

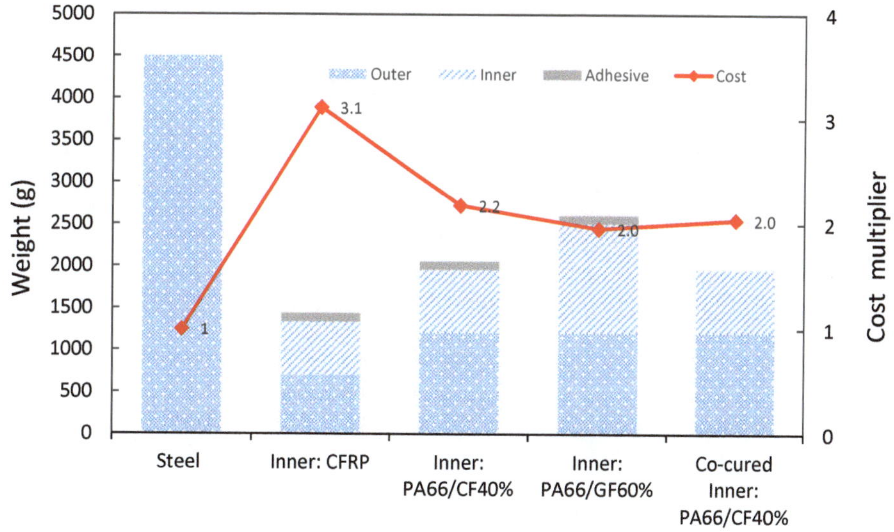

Figure 6: Weight and cost comparison of a front seat back made from steel, CFRP and hybrid CFRP/FR-PA66 (with different fibre loading)

5 CONCLUSIONS

Cheaper material and low-cost production methods for a composite seat back structure were explored and a hybrid light-weight material system, consisting of carbon fibre non-woven and a low-cost fibre reinforced thermoplastic, was developed.

Surface preparation techniques, such as propane flame and blown ion plasma, to join the thermoplastic inner to epoxy reinforced composite outer structure were investigated and compared using lap-shear testing. It was found that the surface functionality gradually degrades by each day. Hence, immediate bonding after surface modification is key to success.

While both treatment methods lead to 300–400 % increase in bond strength between inner and outer seat material, plasma surface modification proved to be more efficient. Out of the tested materials, plasma treated PA66 filled with 40 %CF and 60 %GF met the seat design criteria of >20 MPa shear strength. A higher glass fibre content (60 % compared to 40 %) provides improved adhesion by mechanical interlocking with the adhesive. Trials on fully sized two-piece front seat backs provided evidence of a strong multi-material seat design.

ACKNOWLEDGEMENTS

The authors would like to thank Excellerate Australia Ltd for their financial support for this research.

REFERENCES

[1] S. Deng, L. Djukic, R. Paton, L. Ye, Thermoplastic-epoxy interactions and their potential applications in joining composite structures – A review. *Composites Part A: Applied Science and Manufacturing* **68** (Supplement C), 2015, pp. 121–132.

[2] P. Bruckbauer, "Morphology Development and Mechanical Properties of Interfaces between Epoxy Resin and Thermoplastic Films". *A Comprehensive Approach to Carbon Composites Technology, Symposium on the Occasion of the 5th Anniversary of the Institute for Carbon Composites,* 2014.

[3] S. Farris, et al., The fundamentals of flame treatment for the surface activation of polyolefin polymers – A review, *Polymer* **51**(16), 2010, pp. 3591–3605.

Smart Production/Smart Components

SMART SYSTEM INTEGRATION – POTENTIALS AND CHALLENGES IN THE INTEGRATED CONDITION MONITORING OF LIGHT-WEIGHT STRUCTURES

Michael Heinrich[1,2], Ricardo Decker[2], Marco Walther[2], Lothar Kroll[1,2]

[1] Fraunhofer-Institute for Machine Tools and Forming Technology, Research Centre STEX, Chemnitz
[2] Chemnitz University of Technology, Institute of Lightweight Structures, Chemnitz

Keywords:

SHM-systems, Textile sensors, Embroidery technology, Injection moulding, Piezo-electric sensors

ABSTRACT

In the field of mobile applications, the use of lightweight structures based on carbon fibres is increasing more and more. Due to its high fracture toughness, the material reacts to mechanical overloads usually with invisible initial damage, such as delamination, intermediate fibre breaks or fibre breaks, which can lead to failure of the component. Such damage substantially affects the structural integrity of the structure, which requires its detection, monitoring and repair. At present, such fracture behaviour of fibre composite structures can only be measured with great effort and usually requires long inspection times (e.g. computer tomography, infrared spectroscopy or destructive testing methods). The monitoring of the structural integrity in real time by means of material-integrated sensors and signal processing electronics is therefore imperative and requires novel material and technology concepts for the large-scale functionalization of fibre-reinforced composite structures.

Due to the laminate shell construction, high-performance composite structures based on fibre-reinforcement allow the functionalization by sensors, actuators and microelectronics and thus an improvement in the performance and functional density of such lightweight structures. Continuously innovative manufacturing technologies for active systems based on micro- and nano-effects offer special advantages. These enable the integration of functional elements into textile-based, fibre-reinforced semi-finished products and preforms. In order to achieve a reliable integration of additional functional diversity, methods are being developed for the design and integration of active transducer elements in lightweight structures. For this purpose, a combination of in-situ and in-line processes is used, which includes the injection moulding process with functionalized polymer layers to apply electrical contact and mass printing processes.

In this contribution, concepts and strategies for the integration of sensors and their processing electronics into fibre-reinforced lightweight structures will be presented and their challenges in technological implementation as well as their potentials in terms of resource saving will be demonstrated and discussed.

1 INTRODUCTION

The integration of microelectronic components into hybrid structures allows for the functionalisation by sensors, actuators and electronics, and thus, the further improvement of the performance and functional density of hybrid components. Innovative continuous manufacturing technologies for active systems based on micro- and also nano-effects offer special advantages that enable the integration of functional elements into semi-finished products and preforms. In order to achieve the reliable integration of additional functionality, methods for the design and integration of active transducer elements in lightweight structures will be developed. The final goal is to enable the components to exhibit their

K. Dröder et al (eds), *Technologies for economical and functional lightweight design*,
Zukunftstechnologien für den multifunktionalen Leichtbau, https://doi.org/10.1007/978-3-662-58206-0_7

intrinsic actuator and sensory effect. A combination of in-situ and in-line processes, including injection moulding with functionalised textile layers for electrical contact and mass printing technology will be used.

Due to an increased use of FRP components for the reduction of energy consumption in mobile applications, the condition monitoring of these lightweight structures is of increasing importance. One highly innovative approach is that of in-situ functionalisation during production by using textile techniques. The integration of transducers and electronics into load-adapted FRP components requires novel interconnection, attachment and housing technologies. The major research objectives are the performance and reliability of signal transfer from hybrid structures to sensors and actuators, as well as the energy supply and response data linkage regarding cost-efficient production processes.

2 CHALLENGES AND OBJECTIVES

One focus in the further development of intelligent lightweight structures is the integration and increase of the functional density, for instance with integrated sensors and actuators in complex function-oriented systems. For this, a wide variety of materials such as piezo elements, fiber-optical fibers (so-called fiber Bragg gratings), shape memory alloys or strain gauges are available for this purpose. This can result in active composites or "smart composites" with specific properties that are particularly suitable for the use in loaded components made of fibre-reinforced composites. Although these solutions could prove their functionality on a laboratory scale, they could not be supported by suitable manufacturing strategies for competitive series production of components in vehicle and machine technology. Most of these components have been developed for aerospace applications and use integrated piezo sensors and actuators in conjunction with equipment and control electronics to adapt their geometry or structural dynamics to the acting loads. As a rule, prepreg-based fiber-reinforced composites are used as the carrier material, whereby the associated technologies were not considered from the point of view of suitability for mass production.

A decisive obstacle preventing the breakthrough of such active lightweight components is the lack of production technologies suitable for mass production. These complex systems are installed with prefabricated layer materials or as individual elements by means of a manual insertion process into the fibre composite or by application to the surface of the fibre composite component, thus counteracting automation. A decisive prerequisite for the suitability of a fiber composite solution as a mass product is the ability to achieve the highest possible degree of prefabrication and automation.

3 EMBROIDERED SENSORS FOR THE INTEGRATION INTO LIGHTWEIGHT STRUCTURES

The urgent need to implement innovative approaches in automotive engineering with series technologies is often countered by insurmountable obstacles due to a lack of mass manufacturing strategies. As a result, many innovative ideas and applications – despite impressive laboratory results – cannot be realized, so that their growth potential remains untapped. Effective production of intelligent composite components takes place if function-determining components are integrated directly into the textile structure or if they are applied before they are introduced into the matrix. The preform technology itself, together with further developments in textile technology, has made significant progress in recent years and is now a series product.

The production of function- and structure-integrated sensors promises a remedy here. The textile-technological embroidery process was developed for the production of complex systems during preform production ([1] and Figure 1). Therefore, concepts have been tested on simple functional samples, which can be produced close to mass production.

Embroidering, micro system technology and polymers combined in a closed process chain is quite unusual so far. However the research results showed that there are novel and future orientated fields of applications by embroidery technology. This technology is actually well known in industrial applications like Tailored Fiber Placement (TFP) for selective reinforcement of high duty plastic systems.

In both ways, the TFP and embroidery method, fibers or wires can be places in individual patterns on flexible base materials. Furthermore, this technique is known for producing micro system structures in the field of flexible materials. Thin metallic wires (e.g. constantan® or isaohm®) or conductive coated or rather conductive yarn are used as sensor materials. Embroidery technology are used to attach the sensor material on a non-woven. In principle the shape and dimension can be designed individually; see also Figure 1.

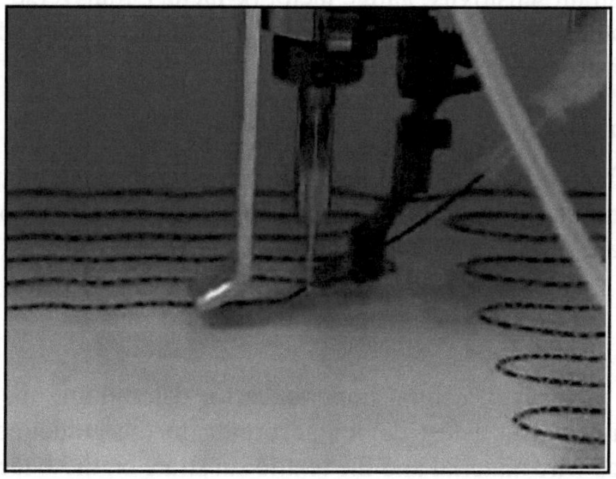

Figure 1: Tailored Fiber Placement technology for the application of functional wires

The first pilot tests show a high reproducibility of the measurement results and the possibility of differentiating between the load directions in the measuring plane. Depending on the desired property profile, conductive materials are integrated precisely into the semi-finished textile according to geometric specifications. The resulting sensor module acts directly in the semi-finished product, which is integrated three-dimensionally by textile technology. Therefore, a commercial strain gauge and an embroidered sensor, which were integrated in a laminated bending beam, were compared. Both systems were installed in a demonstrator and measured simultaneously. The signal characteristics are well matched (see Figure 2).

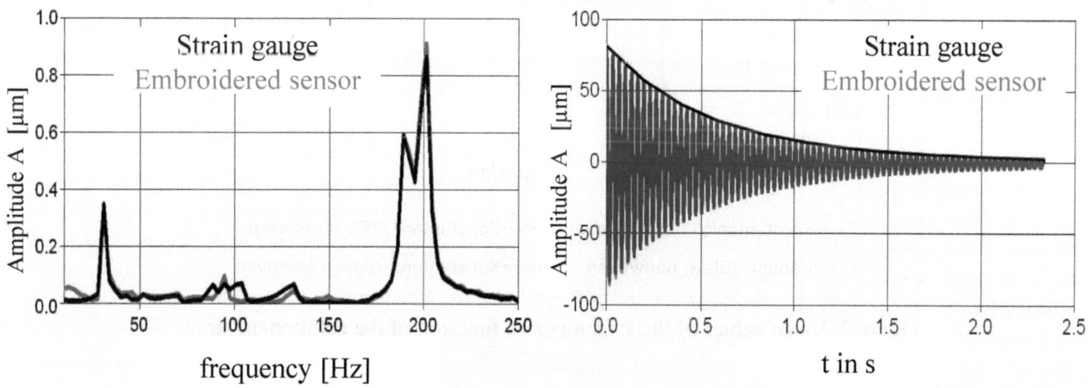

Figure 2: This is an example of a figure

4 STRAIN GAUGE MEASUREMENT

Strain measurement within fibre-reinforced plastics is limited by the bearable strain of the sensor systems, which range from 0.1 to 0.2 % (e.g. resistive methods). Thermoset based parts of lightweight structures often bear strains up to 2 % and more, which would lead to early failure of strain gauges e.g. in the rotor blades of wind turbines after a few load cycles.

In components subjected to bending loads, the integration of strain gauges close to the neutral fiber leads to an increasing alternating load resistance, due to a lower elongation. Further reasons for the integration of DMS are a measuring point protection, no aerodynamic influence on structural parts and free selection of the measuring point (also in the component). In order to use the embroidered sensors for strain measurement, it is necessary to know the following sensor characteristics established by the strain gauge technology: strain sensitivity, gauge factor (*k*-factor), transverse sensitivity and the fatigue behavior. To assess the integration suitability of the sensors in the fibre-reinforced plastics, investigations in the interlaminar energy release rate were made.

4.1 Gauge factor (*k*-factor)

Each strain gauge has a fixed proportionality factor which describes the relationship between the mechanical strain ε of the strain gauge and the electrical resistance change ΔR of the strain gauge in relation to its electrical resistance R_0:

$$\frac{\Delta R}{R_0} = k * \varepsilon \tag{1}$$

Knowledge of the gauge factor is of central importance for determining the component strain ε. The gauge factor was determined with a test device according to the guideline VDI/VDE/GESA 2635. Influences of the resistance wire material and the textile substrate on the *k*-factor were observed.

The most stable results were achieved with the wire material ISOTAN® on a glass nonwoven (Figure 3). For this sensor configuration, a gauge factor of 2.0 is resulting. It can be shown, that a stable *k*-factor is achieved with a suitable material combination [2].

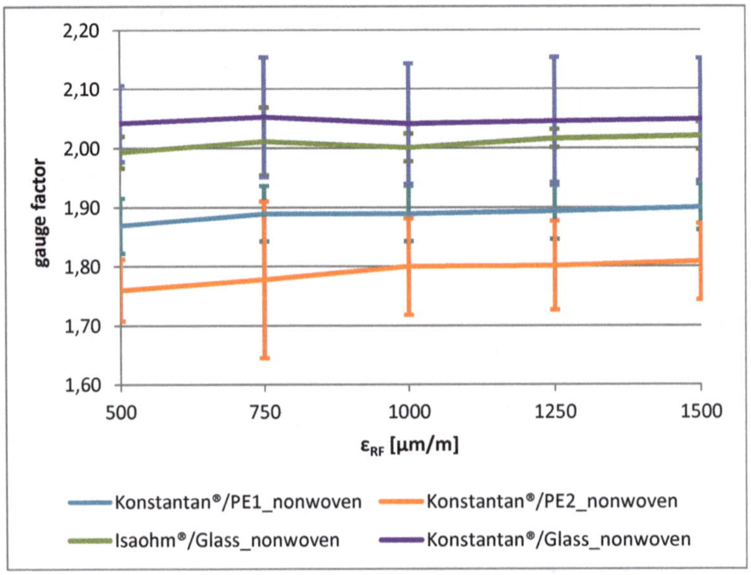

Figure 3: Mean values of the k factors as a function of the component strain

4.2 Transverse sensitivity

The strain conditions on the surfaces of structures are generally biaxial. This means that, in addition to the strains acting in the measuring direction, lateral strains also occur with bonded strain gauges. These transverse strains influence the resistance of the strain gauge and distort the measuring result of the longitudinal strain. The inverse loops of the embroidered sensors lead to a relatively large change in the electrical resistance of the sensor in the event of transverse strain. This change in resistance is therefore a not to be neglected source of error for the exact determination of the actual longitudinal strain [3].

The transverse sensitivity *q* of a strain gauge is defined as the ratio of the strain sensitivity transverse to the measuring direction and longitudinal to the measuring direction [4]:

$$q = \frac{k_{transverse}}{k_{longitudinal}} \tag{2}$$

The result of q = 3.7 % is feasible [5], as transverse sensitivities of up to 4 % are specified for wire strain gauges [3].

4.3 Fatigue behaviour

The fatigue tests of the embroidered sensors were carried out with a vibration testing system (Figure 4). For this investigations, the embroidered sensors were integrated into fibre-reinforced plastic bending beams.

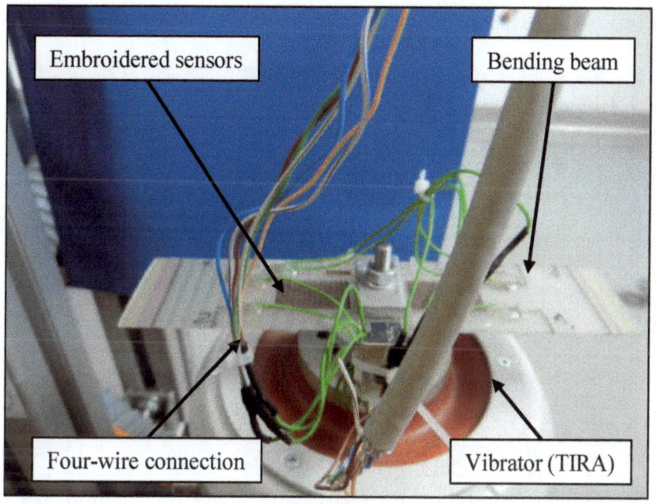

Figure 4: Embroidered sensors on a vibration testing system for investigations in the fatigue behaviour

The samples were loaded to cyclic bending stresses. As reference sensors strain gauges from HBM (1-LY48-6/120) were used. The tests showed no fundamental difference in fatigue behaviour between commercially available strain gauges and the embroidered sensors. The fatigue behaviour is a material property of the strain gauges [6]. Since the embroidered sensors are made of the same material, the same fatigue failure (Figure 5) compared to the strain gauges has been expected and also occurred.

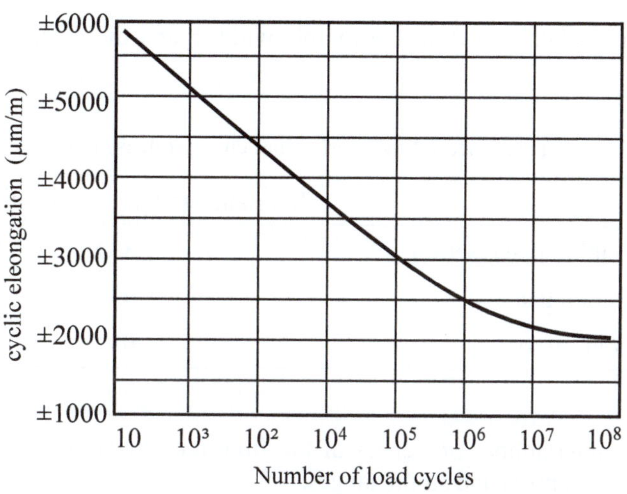

Figure 5: Fatigue behaviour of strain gauges as reference

4.4 Interlaminar energy release rate

In order to quantitatively evaluate the integration capability of the embroidered sensors in the laminate, the characteristic value of the interlaminar energy release rate G_{1C} for laminates with and without integrated embroidered sensors was determined and compared. The interlaminar energy release rate G_{1C} indicates the resistance of fibre-reinforced composites to interlaminar crack propagation [7], whereby the quality of the bonding and the influence of the embroidered sensors on the fibre-reinforced composites are characterized.

The samples consist of 24 layers of glass fiber prepreg [PR-UD EST 250/635]. The embroidered sensors and strain gauges were placed in the middle layer of the laminate. The sample width is 20 mm and the sample thickness 5 mm. The tests were carried out in accordance with DIN EN 6033 and the application of force in the tensile testing machine was realized by means of hinges. The test speed of the tensile test was 10 mm/min (Figure 6).

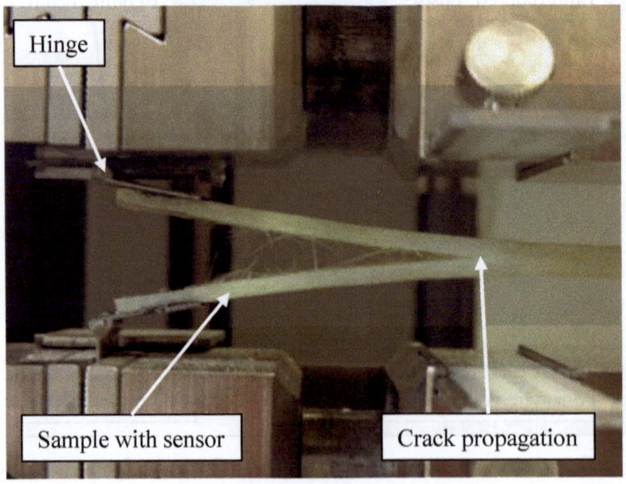

Figure 6: Experimental set-up for investigations in interlaminar energy release rate

The interlaminar energy release rate G_{1C} is calculated by dividing the applied energy w by the delamination area:

$$G_{1C} = \frac{w}{a * b} \tag{3}$$

The delamination area is calculated from the sample width b and the crack length a. The following results were achieved:

Table 1: Resulting interlaminar energy release rate

	Interlaminar energy release rate [J/m²]
Laminate with embroidered sensor	893
Laminate with strain gauge	935
Reference without sensor	902

It can be seen that the interlaminar properties of the fibre-reinforced composites were hardly influenced by the embroidered sensors and strain gauges.

5 CONCLUSION

Embroidered sensors showed the potential to measure strain in fibre-reinforced plastics. Compared to commercial strain gauges, the transverse sensitivity of the embroidered sensors is higher, but with q = 3.7 % within an acceptable range. For certain combinations of functional wire and textile substrate material, stable *k*-factors can be achieved. The fatigue behaviour is comparable for embroidered sensors and commercial strain gauges.

Considering strain measurement using embroidered sensors, for each configuration (functional wire, textile substrate and sensor geometry) existing sensor characteristics of the strain gauge, such as *k*-factor, transverse sensitivity and temperature response must be determined in accordance with VDI/VDE GESA 2635.

Compared to commercial strain gauges, the embroidered sensors have the advantage of shape freedom. Furthermore, the functional material can be positioned where desired and the semi-finished products can be handled robustly when integrated with plastic processing technologies.

6 PROSPECTS

Composite materials will continue to prevail over conventional materials if they succeed in integrating greater functionality into the components by means of mass production and thus become competitive. At the same time, however, the lack of material ductility of fibre composites (glass, carbon, aramid fibres, etc.) must be compensated for by means of intelligent strategies. So far, technical textiles have made a very significant contribution to the successful implementation of such strategies.

The most important lightweight components are made available using well-known textile technology processes. In contrast, technical embroidery plays a rather subordinate role in fiber composite component technology due to the lack of manufacturing strategies in the series of textile manufacturing processes. The best-known application to date has been achieved with the Tailored Fibre Placement process. With this process, extremely lightweight construction can be achieved in composite components with minimum use of material through maximum utilization of the reinforcing fiber.

The change in shape of the fiber composite structure (e.g. curvature/bending) generates a signal by means of the structure-integrated sensors, which can be used for any control functions, for example in vehicle and machine technology. Embroidery technology thus provides the basis for the development of new ways of mass production of competitive products in fiber-reinforced plastic composite construction with integrated sensor properties. This is the only way for new product families to enter series production. The decisive contribution here lies in creating the conditions for component developments that for the first time meet the conditions of a mass market. Thus, the current production technology deficits can be solved for a variety of problems using embroidery technology, e.g. manual controls, structure monitoring devices or automatic transmission controls for the automotive sector.

ACKNOWLEDGEMENTS

We gratefully acknowledge the cooperation of our project partners and the financial support of the German Research Foundation (DFG) within the Federal Cluster of Excellence EXC 1075 "MERGE".

REFERENCES

[1] Walther, M., Kroll, L., Stockmann, M., Elsner, H., Heinrich, M., Wagner, S.: Investigation in development of embroidered strain measurement sensors. (2011) 10th IMEKO TC15 Youth Symposium on Experimental Solid Mechanics 2011, pp. 117–118.

[2] K. Vogel, Ermittlung des strukturmechanischen Einflusses von gestickter Sensorik in Federelementen aus faserverstärkten Kunststoffen für die Anwendung in Fahrzeugsteuereinrichtungen, Diplomarbeit, Chemnitz, 2011.

[3] S. Keil, Beanspruchungsermittlung mit Dehnungsmessstreifen, CUNEUS Verlag Zwingenberg a. d. Bergstr., 1995.

[4] Autorenkollektiv: Experimentelle Strukturanalyse Dehnungsmessstreifen mit metallischen Messgitter Kenngrößen und Prüfbedingungen, VDI/VDE/GESA 2635, Beuth Verlag Berlin, 2007.

[5] R. Tirschmann, Charakterisierung von neuartigen gestickten Sensoren in Mehrschichtverbunden aus Faser-Kunststoff-Verbund (FKV), Projektarbeit, Chemnitz, 2011.

[6] C. Rohrbach, N. Czaika, Über das Dauerschwingverhalten von Dehnungsmeßstreifen, Deutscher Verband für Materialprüfung, Materialprüfung. Bd. 3, Nr. 4, Seite 125–168, Düsseldorf, 1961.

POTENTIALS OF LOAD CARRYING CONDUCTOR TRACKS IN NEW VEHICLE STRUCTURES

Alexander Pototzky[1], Daniel Stefaniak[1], Christian Hühne[1]

[1] Deutsches Zentrum für Luft- und Raumfahrt e.V. (DLR)
Institut für Faserverbundleichtbau und Adaptronik – Funktionsleichtbau
Lilienthalplatz 7 | 38108 Braunschweig
Alexander.Pototzky@dlr.de
Tel.: 0531 295-3096

Keywords:

Function integration, Structure integrated conductor tracks, Fiber metal laminates, Multifunctional structures

ABSTRACT

Digitization, autonomous driving and lightweight construction are the major future challenges in automotive engineering. This means that more and more complex driver assistance systems, engine control units, infotainment systems, actuators, sensors, etc. must be installed and wired. However, from a lightweight point of view, these cables are additional weight without any structural benefit and only affect the weight balance.

Within this paper a new approach to integrate conductor tracks directly into composite structures is presented. In contrast to conventionally integrated conductor tracks, the conductor tracks which are presented here are designed for load carrying purposes. As a result, the wiring costs, the assembly costs and the weight can be reduced significantly. Carbon Fiber Reinforced Polymers (CFRP) are used for this purpose, which show great potential for lightweight construction and, due to their layered structure of individual layers, enable the integration of load-bearing conductor tracks.

Instead of conventional copper wires, different metal foils are inserted into the CFRP vehicle structure stack and used as a conductor track. The single layers can be stacked and arranged individually. In this way, the efficiency of the overall structure can be controlled and optimized.

In order to be able to analyze and evaluate the potential of CFRP with structurally integrated conductor tracks, analytical calculations, mechanical tests and investigations of the electrical properties are carried out. Finally, a demonstrator is manufactured to prove the power supply and the bus communication within the CFRP-structure.

1 INTRODUCTION

The increasing number of driver assistance systems, engine control units and complex infotainment systems in the automotive industry also increases the cabling effort for the various sensors and actuators. The cable harness connects all these systems with each other and ensures both the power supply and the data connection between the individual components and control units.

In a modern automobile structure, there are up to 1,500 cables which have to be fastened and contacted, compare to Figure 1 [1]. These cables are summarized in the wire harness, which is one of the most complex elements. The harness can hardly be processed automatically and thus 95 % manual work [2].

The electrical system of a car not only comprises the main cable harness, but also various self-sufficient sets of wires, which are installed in doors, seats, roof or engine. Overall, the cable lengths adds up to about 3 km [2]. Additional there are also various connector housings, contacts and splicepoints.

K. Dröder et al (eds), *Technologies for economical and functional lightweight design*, Zukunftstechnologien für den multifunktionalen Leichtbau, https://doi.org/10.1007/978-3-662-58206-0_8

In order to attach the wiring harnesses safely in the vehicle and to prevent scrubbing on the body, the conductor tracks must be fixed with hoses, guides, cable ducts and clips. As a result, an additional weight of 60 kg for the entire vehicle electrical system is not unrealistic [2].

Figure 1: Highly complex wiring in an automotive vehicle [1]

Not only the production of the wiring harnesses but also the installation of the wiring harnesses into the vehicle is a time-consuming work step, which cannot be automated. This increases both the personnel costs as well as the downtime of the vehicle during the installation. The just-in-time delivery of the harnesses is also a logistical challenge, which is eliminated by a structural integration.

With the help of conductor tracks which are integrated into the fiber reinforced plastic structures, cabling and assembly costs in an automotive vehicle are reduced and the additional weight is decreased. The tracks are integrated directly in the supporting structure and thus, able to realize both the load transfer, as well as the electrical signals without any loss of functionality.

The aim of this paper is to summarize the investigations and research results for the structural integrated, load carrying conductor tracks and to implement them at a vehicle underbody demonstrator on a suitable model scale

2 CONDUCTOR GEOMETRY

In order to be able to transmit any electrical signal or power, a cable is used as reference. (Cordless transmission systems are not considered at this point). A direct integration of these cables into the structure would show a disadvantageous behaviour since the laminate`s smoothness is impaired by the cable. The conductor tracks must therefore be adapted to the structure. From the requirements of an electrical system, the minimum cross-section of a conductor is specified in general. Under this boundary condition, the shape of the conductor cross-section remains an arbitrary variable. In Figure 2 (left) the geometry change of different conductors with the same cross-section is shown. Different conductor heights are applied, which, with a constant cross-section, only depend on the conductor width. The ratio of height to width of an elliptical cross-section is depicted in blue. A special case of this cross-section is the circular cross-section in which the ratio of width to height is exactly $w/h = 1$. This point is shown separately in the diagram on the curve. All electrical wires and cables conventionally used to transport information, have a circular cross-section. The height`s influence of a rectangular conductor on the laminate as a function of its width, is shown in red. The special case with the aspect ratio $w/h = 1$, a square, is also marked. In addition to these basic geometrical bodies, it is theoretically possible that an electrical conductor can assume any desired cross-section. This cross-section can be represented as a simple polygon. Each polygon (n-corner) with a real number of corners is located in the gray-marked area between the two red and blue colored curves. With an increasing number of corners, these polygons approach a circular or elliptical cross-section.

The graph shows that the influence of the conductor height on the laminate decreases continuously.

According to Schmidt et al. the integration of round electrical conductors leads to a massive ondulation of the fibers. This may lead to massive delaminations in the laminate under bending load conditions [3].

In Figure 2, the circumference of a circle and square conductor cross-section is shown at the right. The circumference or the surface area of the conductor increases with an increasing ratio of width to height. It becomes clear that, although the negative influence of the conductor on the fibers decreases with the change in the geometry (fiber constriction), the bonding surface increases to the same extent. A good connection between the conductor tracks and the CFRP structure is absolutely necessary.

With a conductor height of 0.1 mm, the conductor tracks are approximately equal to the fiber layer.

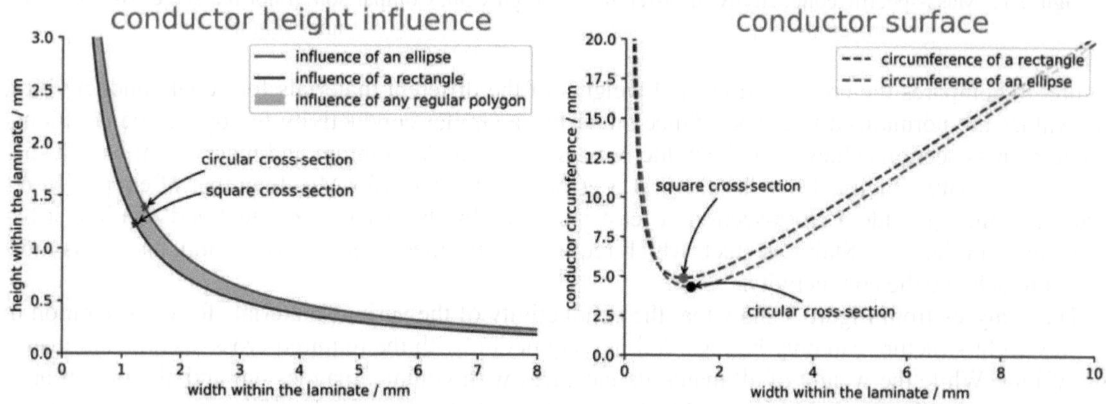

Figure 2: Conductors with a constant cross-section: As the width increases, the influence on the laminate (blue) decreases as the surface increases (red)

These investigations show that it is useful to integrate conductor tracks that are as high as a single CFRP layer. Assuming that the conductor tracks are designed to be load-carrying, the selection of sensible materials is severely restricted. In order to be able to ensure electrical conductivity, metallic materials are chosen in this paper. In combination with a fiber composite construction, one approach is to use fiber metal laminates (FMLs). FMLs such as GLARE are well known and have been introduced to increase the material's fatigue resistance [4].

Kolesnikov has proposed a local FML structure, consisting of several metal foils and a carbon fiber structure, in order to increase the joining strength in the joining area [5]. This has been extensively investigated, as shown in [6]. Furthermore, the mechanical performance of unidirectional CFRP has been improved by using metal layers as presented in [7]. An increased crash safety, especially under bending crash loading conditions, is also expected by several authors [8,9].

In addition to the already studied mechanical advantages, a further function is now to be added to the material: the electrically conductive layers are used to guide the electrical signals through the component. As already described, the mechanical properties of fiber composite structures and FMLs are adequately studied, so that the electrical properties are discussed in the following.

3 ELECTRIAL CHARACTERISTICS

The electronics industry uses silver and copper alloys as standard to ensure electrical conductivity. For cost reasons, especially with high material usage as it is needed for cables and leads, the choice often falls on copper alloys.

Figure 3a shows the conductivities of selected materials in descending order. In order to evaluate the lightweight potential of the different materials, the conductivity was set in relation to the density of the corresponding material in order to determine the mass-specific conductivity of the material.

In this picture, aluminium and magnesium stand out as lightweight construction materials, as they have significantly higher mass-specific conductivity than the classical conductor materials such as copper and silver. In general constructions and high-performance materials such as steels, titanium and stainless steel (X10CrNi) are by a factor of 12–40 worse than copper, because the electrical conductivity is significantly lower.

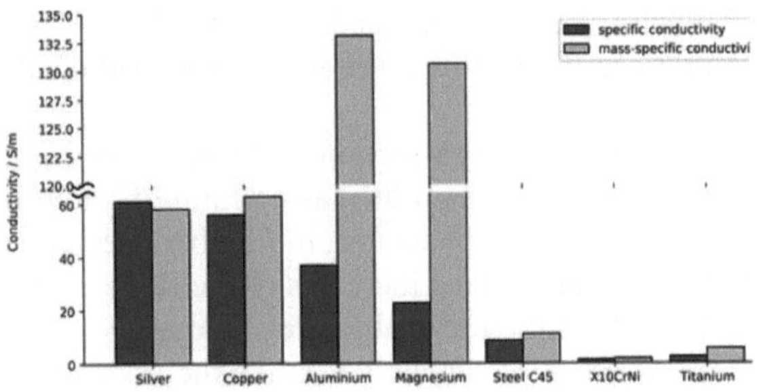

Figure 3a: Mass-specific conductivity of different materials

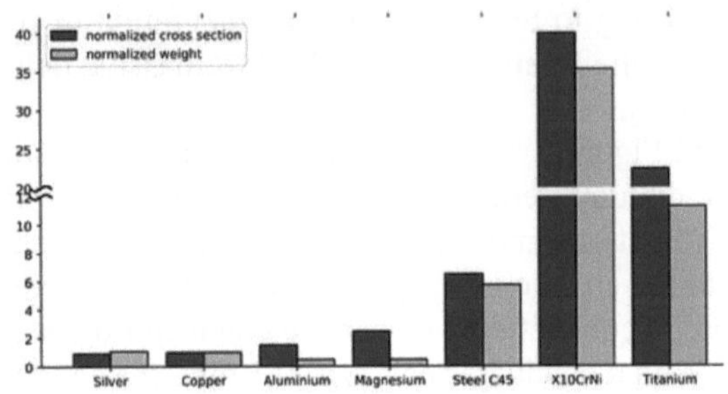

Figure 3b: Comparison of normalized cross-sections and weights

Figure 3b compares the cross-sections and weights of the different materials for equal conductivities. The values are normalized to copper. Since silver has a similar conductivity to copper, the additional required cross section achieve equal conductivity is marginal. Aluminium and magnesium have a specific conductivity which is lower by factor 1.51 (for Al) and 2.47 (for Mg) than that of copper. In return, they must provide a cross-section at least increased by this factor. For steel and titanium it behaves in a similar way. Stainless steel, which requires a 40 times larger cross-section, also adds most weight to achieve the conductivity.

The analyses from Figure 3 show that the conductivity of the various materials for consideration in a lightweight structure can only be regarded in conjunction with the required cross-section and resulting weight. While the weight of all materials increases with volume, magnesium and aluminium have a larger cross-section but a 52 % and 53 % lower weight at the same conductivity compared to copper. Therefore, from a strength perspective, aluminium and magnesium are more suitable than electrical conductors like, copper or silver.

4 MECHANICAL CHARACTERISTICS

Although the material selection based on electrical properties, as performed in the previous chapter, provides a first glimpse on suitable materials, it does not consider its compatibility to the fiber reinforced plastic. For this reason, the materials are examined for their suitability in a CFRP-design in this chapter. For this purpose, three scenarios are examined to be able to estimate the suitability of the different metallic materials. The investigations were made by using the classical laminate theory.

In a first study, a scenario is considered in which the fiber-metal laminate cures at 180 °C during fabrication and is then cooled down to room temperature. Subsequently, a laminate structure is considered, which on the one hand is loaded on tension and on the other hand loaded on bending. The laminate which is necessary to hold the strains and bend in a given framework is analyzed in the following.

The mechanical and electrical properties of all materials used are summarized in Table 1 below.

Table 1: Used material parameters

	density	young's modulus	yield strength	shear modulus	thermal expansion	specific electrical conductivity
	[g/cm³]	[GPa]	[Mpa]	[GPa]	[10^{-6}/K]	[S/m]
Silver	10.49 [10]	80 [11]	130 [11]	29 [11]	18.9 [12]	61 [13]
Copper	8.94 [10]	125 [11]	160 [11]	47 [11]	16.5 [12]	56 [13]
Aluminium 2024 T3	2.78 [14]	70 [15]	345 [15]	27 [11]	23 [12]	37 [13]
Magnesium	1.73 [10]	45 [11]	250 [11]	18 [11]	24.8 [12]	22.7 [13]
Steel C45	7.70 [16]	210 [16]	490 [17]	81 [11]	11.1 [12]	8.6 [13]
X10CrNi	7.90 [18]	190 [18]	1700 [17]	81 [17]	16.4 [18]	1.4 [13]
Titanium 15-3-3-3	4.50 [16]	130 [16]	1290 [18]	43 [11]	8.6 [18]	2.5 [13]
GFRP	2.50[19]	50[19]		4 [19]	7 [18]	0

4.1 Residual stresses in composite materials

As already mentioned, a material system is considered, which cures at 180 °C. Prussak et al. [20] have shown in their investigations that the crosslinking of the resin system and the associated interaction between the individual layers already begins during the warm-up. Hence, they determine a shear stress-free condition for a specific FML at a temperature of 163 °C. [20]. Due to different coefficients of thermal expansion, stresses occur within the laminate, when cooling the laminate to room temperature (T = 20 °C).

Three different laminates are considered for this study. The first laminate consists only of two metal layers which are each enclosed by a fiberglass layer. This corresponds to a metal volume fraction (MVF) of 40 %. The second laminate is a unidirectional laminate with a total of 15 fiberglass layers and two embedded metal layers. The metal volume content in this stack is about 11.8 %. Stefaniak shows in [21] that FMLs with a MVF up to 12.5 % have higher specific properties compared to a conventional laminate. For comparison purposes, the two metal layers are embedded in a quasi-isotropic structure and isolated only with a glass fiber layer. The laminate has an MVF of 18.2 %.

The safety factors of the metal layers, after curing process are evaluated. Regarding to Prussak the range from curing point to room temperature is set to a temperature of 143 °C.

It can be seen clearly that the safety factor only insignificantly depends on the laminate structure, but is much more influenced by the material combination. Silver and copper, which are the materials with the highest specific conductivity, are already at the limit of their elastic load capacity after cooling down due to their low yield strength. Depending on the laminate structure, the elastic limit is partial even significantly exceeded. The reserve factors of X10CrNi and Titan stand out clearly in the figure with about 6 and 10, respectively. These materials still show significant load carrying reserves.

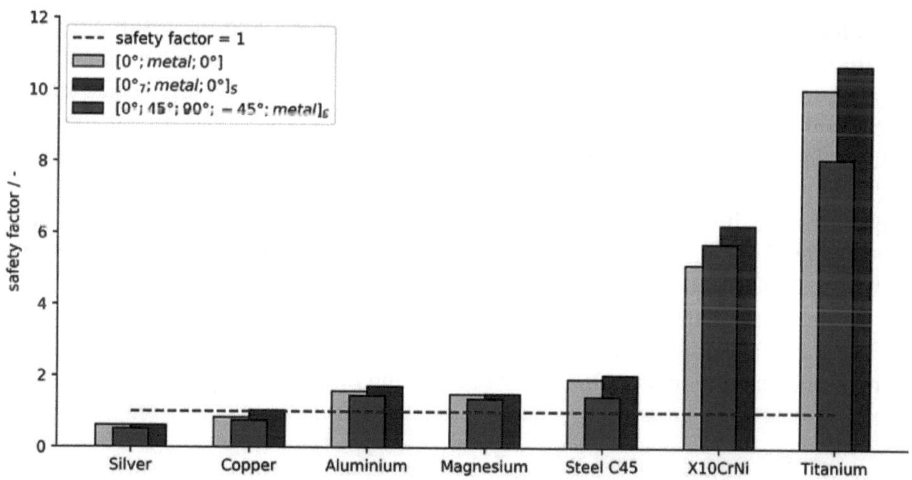

Figure 4: Safety factors of different stacking with integrated conductor tracks

4.2 Laminate under tensile load

A laminate is considered, which is firmly clamped on one side. This laminate is loaded in the x-direction with a force n_x. Subsequently, the resulting weight and the reserve factor of the laminate are analyzed and evaluated. The starting point for the laminate construction is the laminate 2 from Chapter 4.1. For the investigation of the structure integrated conductor tracks, two fiber composite layers are replaced by two metallic conductor tracks. The position of the metal layers has no influence on the elongation or the reserve factor in a pure tensile load.

Figure 5 shows the resulting weights of the fiber metal laminates and the glass fiber reference structure.

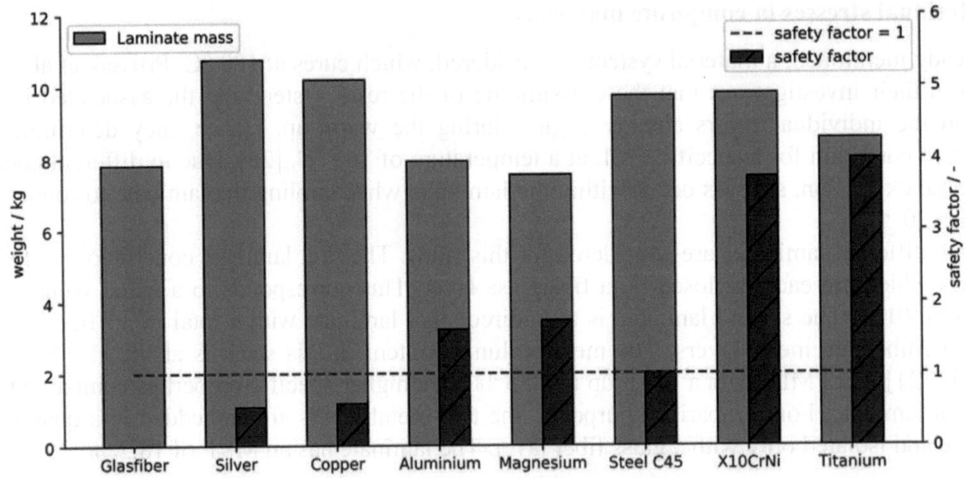

Figure 5: Weight of FMLs with tension load and corresponding safety factors

It becomes clear that the materials silver and copper cannot withstand the stresses and fail early – or rather deform plastically – which results in irreversible damage. The metals aluminium and magnesium are able to substitute glass fiber layers and at the same time they are able to produce two electrically conductive layers. Both with approximately the same weight compared to the glass fiber reference.

4.3 Laminate under bending load

A bending moment m_x is applied on the laminate from chapter 4.2 and the resulting curvature κ_x and the weight are evaluated. Comparing to the previous Chapter two fiber layers are replaced by two metal layers regarding the reference fiber composite structure. Due to the low elastic strain of the metals compared to the fracture strain of the fibers, the metal layers are placed near the neutral fiber to prevent failure of these layers. The metal layers are separated by one layer GFRP to ensure insulation. In this way, neither the metal layers nor the fiber layers fail for the complete examination.

As a special feature, it can be noticed that X10CrNi and titanium are able to withstand the stresses on the outside of the laminate due to their high yield strength and that they can contribute to increased bending stiffness. As a result, the number of glass fiber layers inside the laminate can be reduced by 3 or 2 layers without significantly changing the laminate bending. This leads to a weight reduction of the individual fiber metal laminate. Figure 6 shows the weights of the FMLs. For the FMLs with X10CrNi and titanium component, two masses are shown depending on the metal position.

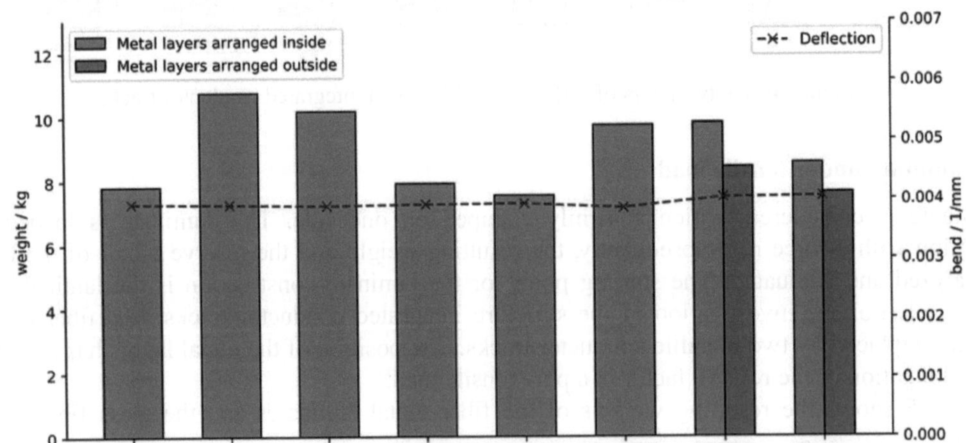

Figure 6: Weight of FMLs with bending load

By comparing the weights from Figure 5 and Figure 6 it becomes clear that with the substitution of single glassfiber layers by magnesium or titanium it is possible to produce a reliable structure which is lighter than a FRP reference structure with the same performance (magnesium & titanium). In addition, two electrical conductor tracks were inserted which are able to conduct a voltage or a current through the laminate. With the right stacking the laminates with layers of aluminium or X10CrNi show a slightly increased structural weight. This weight must be compared to a wiring harness attached to the FRP laminate, depending on electrical requirements. In addition to the cable, which consists of the copper component and the insulating sheath, the weight of an externally routed cable harness also includes corresponding fastening elements, guides and abrasion protection elements. This mass can quickly add up to a size, so that the use of X10CrNi in a composite fiber laminate pays off. The saved installation effort to mount the wiring harness is difficult to quantify at this point.

5 ELECTRICAL TESTS

When designing a component with integrated conductor tracks, it is necessary to determine the minimum spacing between the conductor tracks to prohibit any exchange of charge.

A charge exchange between two conductor tracks, which function as electrodes, leads to a voltage breakdown which occurs as a spark or as an arc. Such an arc destroys the insulation layer irreversibly because of its heat development. When designing the component, this scenario must be avoided in all circumstances. According to Grote et al. [16], the dielectric strength is equivalent to the electric field strength E. The field strength can be expressed by equation (1):

$$E = \frac{U}{d} \tag{1}$$

(E = electrical field strength; U = voltage; d = distance of the conductor tracks)

In addition to the geometric distance between the electrodes (d) and the applied voltage (U), the dielectric between the conductor tracks plays a decisive role in the value of the dielectric strength.

Typically, the average breakdown strength of air is 3 kV/mm [16]. The breakdown strength of a fiber composite is strongly dependent on the fibers and the matrix. In this paper glassfiber reinforced plastics were investigated in detail. In contrast to carbon fiber, fiber glass rovings are electrically non-conductive. Nevertheless they are able to bear mechanical forces.

For identification of the breakdown voltage, samples were produced which consist of the two electrodes for applying the voltage and of different glass fiber layers to adjust the distance between the electrodes. Among one and five layers of glass fiber fabric were inserted between the electrodes. A layer of glass fiber (MTM44-1/GF0903-40 %RW) has a nominal thickness of 0.101 mm, so that the breakdown strength can be derived by the number of layers. The electrodes were made out of carbon (MTM44-1/CF5804A-40 %RW-DC) on the one hand and on the other hand out of steel (X10CrNi18-8). Figure 8 (left) shows the results of the investigation. Two results are displayed in this diagram, while the dark blue bar is showing the breakdown strength for carbon fiber layers, the light blue bar shows the dielectric strength of the specimen with integrated steel layers.

It can clearly be seen that one layer of GFRP does not yet have an insulating effect. By adding another glass fiber layer, a voltage can be set between the two electrodes. The maximum voltage is 650 V for the steel conductor strips, which corresponds to the maximum allowed output of the power supply unit. The mean dielectric strength of the three samples with the CFRP electrodes is 400 V. The samples vary in their dielectric strength between 250 V and 650 V. Here, a phenomenon occurs that a single roving of the CFRP woven fabric sticks into the glass fiber layers and reduces the insulation distance.

According to equation (1), the dielectric strength of the glass fiber fabric with the corresponding matrix is more than 3.2 kV/mm.

Although the breakdown voltage could not be accurately specified, it has been shown that there is a risk of short which must be avoided during production. For this, at least two layers of GFRP must be used to create a working insulation.

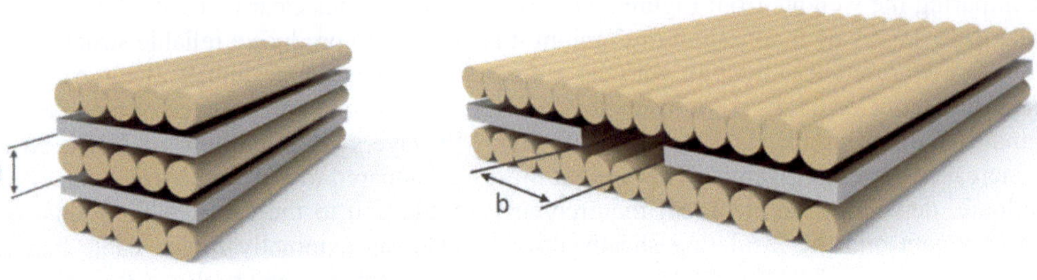

Figure 7: Examination of the dielectric strength in the vertical and horizontal direction

In contrast to the previous scenario, the conductor tracks are arranged side by side in order to identify the required distance from one another. With this arrangement of the electrodes, the spacing of the conductor tracks can be set more individually. This, in turn, also leads to an increase in the manufacturing effort, since the conductor paths are precisely cut and aligned with each other by means of a measuring instrument. Figure 7 visualizes the difference in a sketch.

The distances between the electrodes were increased in steps of five tenths of a millimeter and were tested after consolidation. The results are shown in Figure 8 (right). The values for CFRP electrodes are shown in dark blue and the values for electrodes made of steel are displayed in light blue. For carbon fiber electrodes, a voltage can be applied only at a distance of 0.2 mm. At small distances little rovings and fiber filaments again lead to a short circuit. Similar to the electrodes arranged vertically above each other (Figure 8 (left)), the voltage can only be applied to a part of the samples, while other samples exhibit a short-circuit fault. Safe insulation up to 6.5 kV takes place at a distance of 0.5 mm.

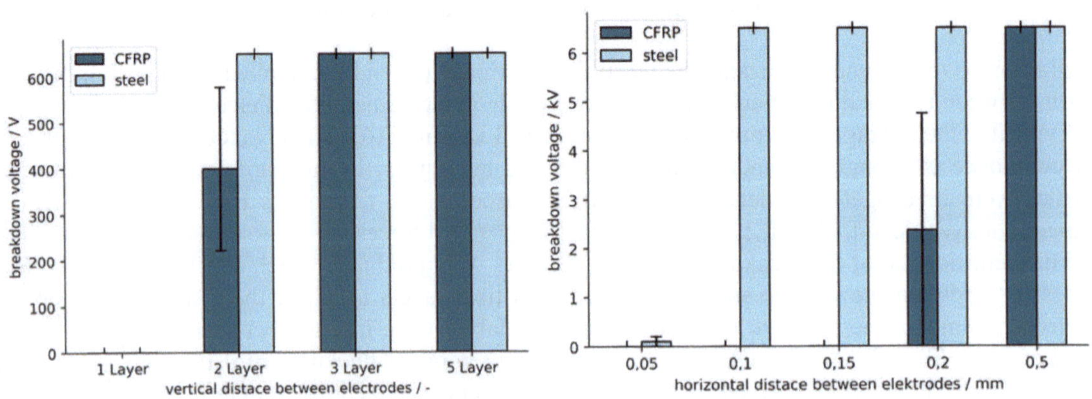

Figure 8: Breakdown voltage as a function of the glass fiber layers (left)
Breakdown voltage as a function of the horizontal distance (right)

The steel foils are safely insulated above a distance of 0.1 mm. According to (1), the breakdown strength of the pure epoxy matrix is thus more than 6.5 kV / mm.

The results show that the steel foils can be characterized very well, the influences of the CFRP electrodes have to be examined more detailed.

6 SCALED DEMONSTRATOR

In order to be able to analyze and assess the potential of fiber reinforced composite with regard to structure-integrated conductor paths, a suitable demonstrator for the integration of the conductor was developed and built.

In present project at DLR named Next Generation Car (NGC) a total of three different car concepts are developed. One of these car concepts covers an upper class model, which should compare with vehicles such as the Mercedes Benz S-Class or Audi A8. In addition to comfort, this vehicle focuses on an intensive fiber-composite-construction in order to achieve the desired structural mass of ≤ 250 kg (BIW). The drive concept will feature four separate electric motors, powered by hydrogen tanks and integrated batteries. Figure 9 shows the concept of the NGC IUV on the left side.

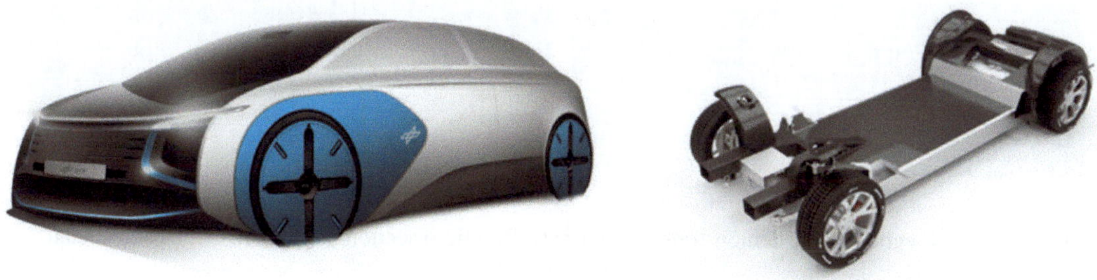

Figure 9: The concept of the NGC Interurban Vehicle (IUV) (left) and the more detailed vehicle underbody with mounted chassis (right) [22].

In order to supply the motors of the IUV with the required power, the power supply should be integrated into the CFRP structure. As a target component for this purpose, the underbody region between the axes has been identified. (Shown in black, on the right, in Figure 9.)

Since the requirements for the conductor tracks are not yet finalized, a first demonstrator was manufactured as a proof of concept, which has the standardized power supply and the bus communication of conventional automobile. For this purpose a voltage supply and a data communication integrated in the CFRP structure are necessary. This demonstrator contains the four required foil conductors: the 12 V supply voltage, the two CAN-Bus channels and the ground signal. Figure 10 shows the schematic structure with its layered structure. The individual conductor layers are labeled in the figure.

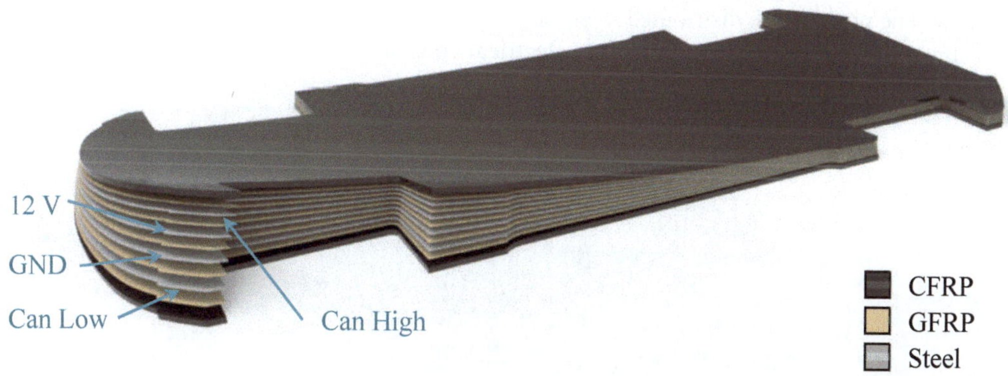

Figure 10: The sectional view of the demonstrator consisting of CFRP cover layers, GFRP insulation and conductive metallic intermediate layers

The demonstrator with the structure-integrated load carrying conductor tracks consists of the CFRP cover layers as well as the alternately layered GFRP insulation (two layers) and the conductive metallic intermediate layers. The metal layers represent the global cable harness. The inner layers of the harness are voltage supply, the outer layers the BUS architecture. (see Figure 10) Any module can be connected to this wiring harness.

The scale is chosen in such a way that the resulting dimensions of approx. 660 mm × 270 mm represent a good compromise between the handling and the demonstrability of the structure-integrated conductor tracks.

The two top layers (top / bottom) consist of a 1.6 mm quasi-isotropic CFRP structure. The middle section consists of an alternating structure of glass fiber and steel layers. The steel foils are blasted directly in front of the deposit in a special developed blasting system and treated with sol-gel to ensure good interlaminar strength (see [21]). The micro section of the structure is shown in Figure 11.

Figure 11: Micro section of the demonstrator structure (left); micro section of an FML with a significant larger MVF (right)

For the demonstrator, only four electrically conductive layers were needed. Figure 11 (b) shows that it is easily possible to integrate significantly more layers into the laminate to increase the metal volume content and to adjust mechanical properties.

During the investigation of the demonstrator, no errors could be detected in data communication or in the power supply. All connected components have been supplied with power and have received or sent commands via the CAN-Bus.

7 ELECTRICAL CONNECTION CONCEPT

In order to ensure the full functionality of the structure-integrated, load-bearing conductor tracks, the tracks must be contacted individually without making any short circuit to the other layers. For this purpose, a concept was selected in which the connection points must be determined beforehand and planned at design level. Figure 12 shows the eleven attachment points selected in the car underbody demonstrator. The points were distributed over the structure at more or less regular intervals in order to ensure the most flexible contacting of possible electrical consumers with the structure.

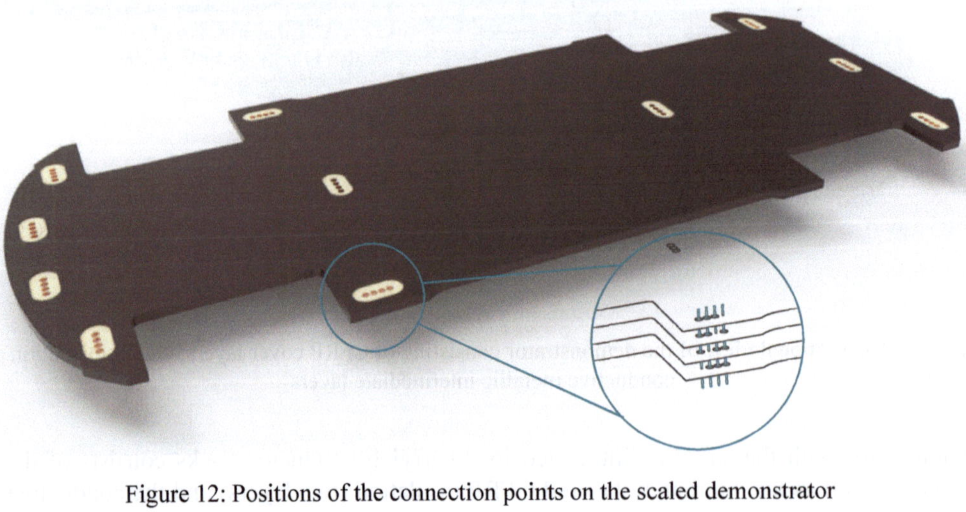

Figure 12: Positions of the connection points on the scaled demonstrator

Each plug connection consists of four connection points. The connection takes place at each area over the complete thickness of the underbody. The connection regions are geometrically designed so that only one layer is affected when contacting the foils (compare detailed view in Figure 12). In this way the connection points get to the surface and can then be processed with conventional contacting-concepts.

For the vehicle underbody, a concept was chosen in which the foils were drilled perpendicular to the surface. Subsequently, a tubular rivet was pressed into the bore and then soldered with a plug. The result is shown in Figure 13 (a). Other studies have also shown that a robust way of realising the contact is riveting a cable shoe with a blind rivet. (Figure 13 (b)).

Figure 13: connection concepts: Discretely soldered connector plug (a); Blind rivets with cable shoe (b); Rivet nuts (c)

The mounting of a rivet nut is a variation of the method (b) but has the advantage that a fixed connection area is created, to which the supply line of an electric load can be flexibly connected.

8 CONCLUSIONS

The development goals for structurally integrated conductor tracks in an automotive vehicle are the reduction of cable and assembly costs and the associated cost and weight savings by replacing the classic cable harness with an FML lightweight construction.

First investigations have shown that it makes perfect sense to use the basic structure of fiber metal laminates to integrate conductor tracks into the laminate. Subsequently, various metals were examined in relation to their electrical conductivity. Due to their highly conductivity and their low density, aluminum and magnesium have proved to be electrical conductors with a high potential for lightweight construction.

The analytical investigations have shown that the classical materials for transmitting current very quickly reach their limits of mechanical strength inside an FRP structure. Four different materials were identified, by which the integration of the conductor tracks into the laminate makes sense. Aluminium and magnesium are useful for highly conductivity requirements and titan and X10CrNi for mechanical and electrical performance.

Finally, the idea of integrating conductor tracks directly into the load-bearing structure was successfully demonstrated and tested on a scaled demonstrator. The structure is able to realize the power supply and the CAN bus communication without any loss.

When integrating conductor tracks in fiber metal laminates, the electrical conductivity of the load-bearing metallic layers is used as an additional function. In this paper it could be shown that it is possible to reduce the weight of the wiring harness significantly or in special cases completely.

The technology which is presented here is used most meaningfully for high voltage applications, as the savings potential of heavy copper cables is largest for this application.

REFERENCES

[1] T. Form. Vorlesung Fahrzeugelektronik – TU Braunschweig. [October 12, 2017]; Available from: https://www.ifr.ing.tu-bs.de/static/files/lehre/vorlesungen/efs1/ Folien_FE1_Teil6.pdf.

[2] BMW. Technology Guide: Kabelbaum; Available from: http://www.bmw.com/com/de/insights/technology/technology_guide/articles/wiring_harness.html?source=categories&article=wiring_harness. [September 07, 2015].

[3] J. Schmidt. Entwicklung eines Interfacekonzepts für integrierte Funktionselemente in UAV-Strukturen. Diplomarbeit. Technische Universität Braunschweig; 2011.

[4] Chen Q, Guan Z, Li Z, Ji Z, Zhuo Y. Experimental investigation on impact performances of GLARE laminates. Chinese Journal of Aeronautics 2015;28(6):1784–92.

[5] B. Kolesnikov. Composite Material with a Reinforced Connecting Area(Patent PCT/DE99/00790).

[6] E. Petersen E, D. Stefaniak, C. Hühne. Efficient joint design using metal hybridization in fiber reinforced plastics. Stade; 2014.

[7] D. Stefaniak. Improving The Mechanical Performance Of Unidirectional CFRP by Metal-Hybridization: ECCM15 – 15th European Conference on Composite Materials. Venice; 24–28th 2012.

[8] Eksi S, Genel K. Bending response of hybrid composite tubular beams. Thin-Walled Structures 2013;73:329–36.

[9] Haedir J, Zhao X, Bambach MR, Grzebieta RH. Analysis of CFRP externally-reinforced steel CHS tubular beams. Composite Structures 2010;92(12):2992–3001.

[10] N. N. Greenwood, A. Earnshaw. Chemie der Elemente. 1st ed. Weinheim: VCH Verlagsgesellschaft; 1988.

[11] Läpple V. Einführung in die Festigkeitslehre: Springer.

[12] Haynes. CRC Handbook of Chemistry and Physics. 92nd ed.: CRC Press.

[13] M. Winter. webelements – electrical_resistivity. [March 10, 2018]; Available from: https://www.webelements.com/periodicity/electrical_resistivity/.

[14] C. Kammer, H.W. Wenglorz. Aluminium-Taschenbuch – Band 1: Grundlagen und Werkstoffe. 16th ed; 2009.

[15] Schmolz+ Bickenbach. technical datasheet aluminium 2024. [January 05, 2018].

[16] K.-H. Grote, J. Feldhusen (eds.). Dubbel: Taschenbuch für den Maschinenbau. 23rd ed. Berlin; 2011.

[17] F. Richter. Die physikalischen Eigenschaften der Stähle: Teil I: Tafeln und Bilder.

[18] A. Fink. Lokale Metall-Hybridisierung zur Effizienzsteigerung von Hochlastfügestellen in Faserverbundstrukturen: Technische Universitaet Braunschweig; 2010.

[19] Hexcel. technical datasheet DLS1611-1.

[20] Prussak, R., Stefaniak, D., Hühne, C. Evaluation of residual stress development in FRP-metal hybrids using fiber Bragg grating sensors: Springer; 2018.

[21] D. Stefaniak. Improving residual strength of unidirectionally reinforced plastic laminates by metal layering. Dissertation. Technische Universität Braunschweig; 2017.

[22] C. David, S. Vohrer (eds.). Development of novel vehicle structures for automotive series production.

NUMERICAL AND EXPERIMENTAL INVESTIGATION OF FIBER REINFORCED BIOCOMPOSITES AS STRUCTURAL PARTS IN AUTOMOTIVE APPLICATIONS

Benedikt Lahl[1], Nikita Pyatov[2], Christian Busch[3], Stefan Hartmann[4], Hans-Josef Endres[5], Tim Andreas Osswald[6]

[1]Volkswagen AG, 38440 Wolfsburg, Berliner Ring 2, benedikt.lahl@volkswagen.de
[2]Volkswagen AG, 38440 Wolfsburg, Berliner Ring 2, nikita.pyatov@volkswagen.de
[3]Volkswagen Nutzfahrzeuge, 38442 Wolfsburg, Stellfelderstraße 46, christian.busch@volkswagen.de
[4]Volkswagen Nutzfahrzeuge, 38442 Wolfsburg, Stellfelderstraße 46, stefan.hartmann@volkswagen.de
[5]Institut für Biokunststoffe und Bioverbundwerkstoffe, Hochschule Hannover, 30453 Hannover, Heisterbergallee 10A, hans-josef.endres@hs-hannover.de
[6]Polymer Engineering Center, University of Wisconsin-Madison, 1513 University Avenue, Madison, WI 53706, USA, tosswald@wisc.edu

Keywords:

Biocomposites, Natural fibers, Draping simulation, Crash simulation, Viscoelastic material behavior

ABSTRACT

In the automobile industry, bio-based component structures such as natural fiber reinforced thermoplastics mostly use fiber fleeces as reinforcement. As example the interior door trim parts of the Golf VII are made of natural fiber fleeces. However, the component properties of natural fiber fleeces are not sufficient for load-bearing structures. Hence, their application has been limited to interior trim parts.

In order to expand the use of bio-based fibers for load-bearing components, new technical and ecological synergies are necessary. The combination of a thermoplastic PP matrix with a mix of glass and flax fibers does create synergistic effects on many properties. As natural fibers bleached flax fibers are used. In addition to natural fibers, also the use of recycled short fiber reinforced PP as a matrix polymer is investigated. The present study shows that these bio-based and recycled materials can be used for load-bearing component structures. Moreover the use of process simulation as a tool for developing and optimizing the manufacturing chain of a seat shell is presented.

The purpose of simulating the forming process of a structural part is to determine how the mechanical properties in the final part will be like as these properties highly depend on the fiber orientation. Therefore a draping simulation which follows the continuum mechanics approach and takes the material behavior as well as the process conditions into account was performed. Since the main objective of process simulation is to achieve the convergence between the virtual and the physical product, great care must be taken to characterize the material behavior at process relevant conditions which is the main input for an FE simulation. Therefore the material behavior was characterized in terms of thermal, mechanical and strain rate depended properties. Subsequently a crash simulation was conducted in order to analyze the crash behavior of the seat shell.

1 NATURAL FIBER REINFORCED ORGANOSHEETS

When combining reinforcing fibers and polymer matrix, the focus is on material selection and manufacturing technology. The semi-finished product consists of four monolayers in the form of a bi-directional hybrid weave of natural and glass fibers and of a surrounding low-viscosity thermoplastic PP matrix.

K. Dröder et al (eds), *Technologies for economical and functional lightweight design*, Zukunftstechnologien für den multifunktionalen Leichtbau, https://doi.org/10.1007/978-3-662-58206-0_9

1.1 Manufacturing process

In semi-finished product (organosheet) manufacturing, the polymer compound is applied directly to the fabric layers in a molten state. By applying pressure the molten matrix is then impregnated into the dry fiber reinforcement. Hereafter the continuous organosheet is cut in blanks of desired dimensions.

The process for manufacturing a hybrid seat shell involves heating the thermoplastic organosheet above the matrix melting temperature, transferring the heated organosheet into a shaped tool and forming it by closing the cavities. Subsequently the short fiber reinforced PP is injected. This overmolding material can integrate functions such as rips or visible grained areas and also enhance the quality of component edges. The schematic procedure in Figure 1 describes the manufacturing process of the composite material as well as a hybrid part.

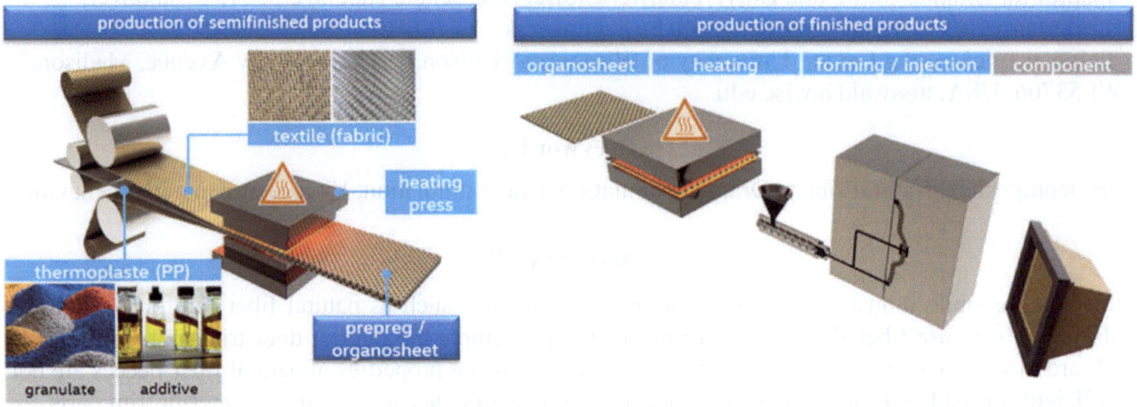

Figure 1: Manufacturing process [1]

2 THERMAL CHARACTERIZATION

In order to achieve a stable process and a homogeneous impregnation, the control of the process temperature is crucial. Therefore the temperature of polypropylene, PP-recyclate and for the additives must be above their melting temperature. On the other hand, the processing temperature is limited by the low thermal stability of the additives and the natural fibers. The natural fiber structure consists mainly of cellulose and numerous accompanying substances like hemicellulose, lignin, pectin, etc. The decomposition temperature can be determined by means of thermogravimetric analysis (TGA). In Figure 2, the filled bars on the left side describe the temperature range in which the materials have not yet melted. On the other side the decomposition temperature ranges are shown.

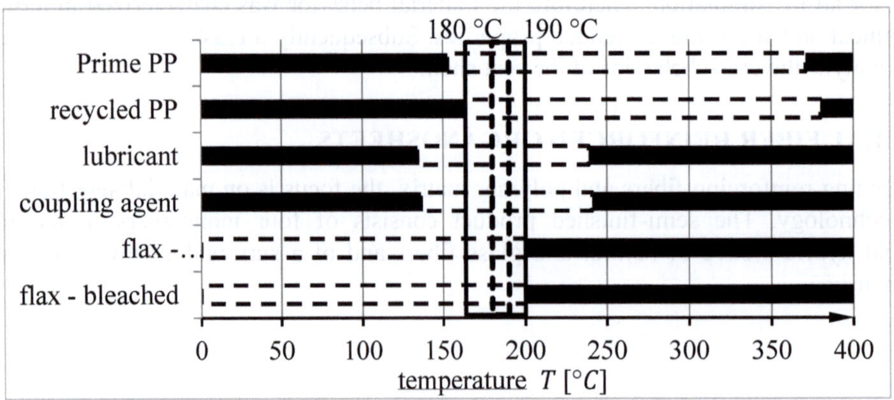

Figure 2: Processing temperatures for different materials [1, 2]

The results in Figure 2 are indicating that the process temperature is limited by the melting point of the recycled PP and the decomposition temperature of the bleached flax fibers. Considering a safety factor of 5 % the chosen process window is between 180 °C and 190 °C.

3 MECHANICAL CHARACTERIZATION

The textile fabric of the organosheets is based on glass and natural fibers. The natural fiber yarns are flax fibers. The polymer matrix is a low melting polypropylene with a low-viscosity and it is modified for improved impregnation with additives like internal release agents, lubricants and coupling agents.

3.1 Single fiber testing

The composite's mechanical properties are essentially determined by the fiber characteristics. Therefore the aim is to ensure the highest possible fiber volume fraction. Hereby the fiber fineness and the binding method play an important role. When using natural fibers as reinforcement material, the chemical fiber structure is the most important for the composite properties. While glass fibers absorb no moisture, the cellulose-based fibers are very affinitive to water due to the numerous hydrophilic groups. The moisture is absorbed from the environment [3] [4].

The water absorption of natural fibers also influences the fiber characteristics. The results of the single fiber behavior are shown in Figure 3. The single-fiber tensile tests are based on DIN EN ISO 5079 with a sample number of 50 individual fibers per variant. Therefore bleached and bleached and dried flax fibers with the same fiber fineness were used.

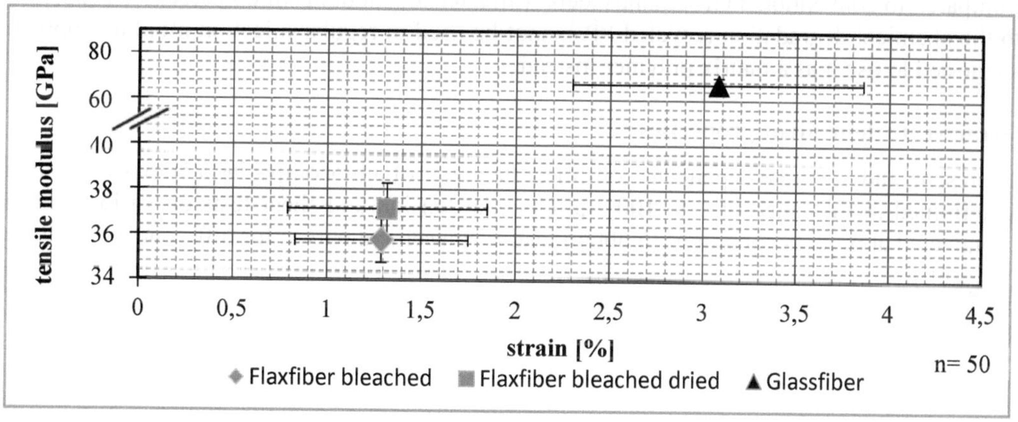

Figure 3: Single fiber tensile test

By comparing the stiffness characteristics, it can be seen that the bleached flax fibers have a lower tensile modulus compared to the glass fibers. Both bleached and bleached and dried fibers show similar ultimate strain whereas the dried fibers have a higher tensile modulus which can be a result from the amount of moisture within the fibers. Hence, the moisture in the fibers decreases the stiffness of the NF. The GF show a significantly higher ultimate strain which is about 3 %. Hence, bleached and dried flax fibers were used for the following production of the organosheets.

3.2 Organosheet testing

The organosheets consist of four layers of fabric and a surrounding PP matrix. The fiber volume content is 52 vol.-%. The matrix in this series of tests consists of an optimized polypropylene without recyclate, an internal lubricant and a coupling agent based on maleic anhydride grafted PP. In terms of the low thermal stability of natural fibers, the production temperature of the organosheets is set to 180 °C.

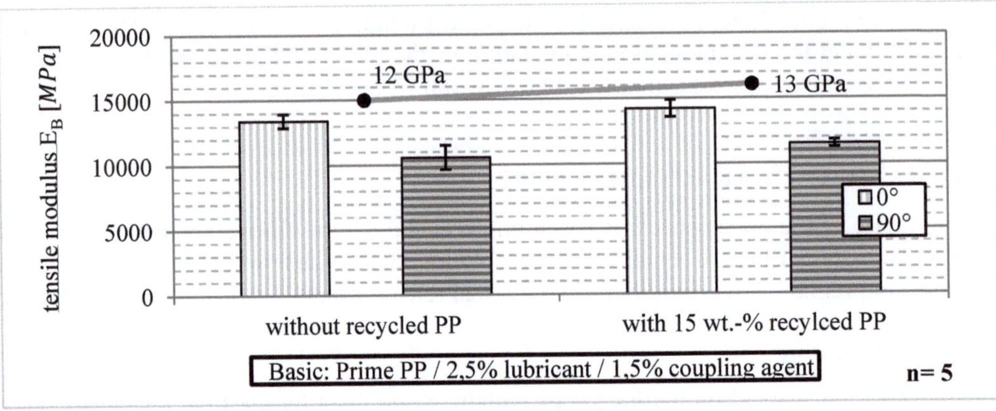

Figure 4: Tensile test of biocomposite materials

In order to evaluate the tensile behavior of the biocomposites, tensile tests have been performed with different material combinations and in warp (0 °) and weft (90 °) direction of the fabric. The results with or without recycled PP are shown in Figure 4. Both materials show a similar behavior. The standard deviation is acceptable for all four variants; hence, the material properties are not subjected to any great fluctuations. The organosheet without recycled PP has an average tensile modulus of 12 GPa, which is approximately 60 percent of the tensile modulus of a conventional organosheets with GF reinforcement. Although recycled plastics are subject to molecular degradation which causes a decrease in mechanical properties, the organosheets with the recycled PP showed a higher tensile modulus. Compared to conventional GF-organosheets with a tensile modulus of 65 percent is reached. Since, due to the chemical degradation recycled PP cannot be used in regular injection molding applications, it is suggested to use it as a matrix polymer instead.

3.3 Strain-rate dependent properties

In this present study the rate dependent non-linear behavior of a biocomposite material with PP matrix was determined in the range of 1 to 100 1/s. Uniaxial tension and shear tests on various specimens were conducted at process conditions to obtain load-displacement curves for different strain rates.

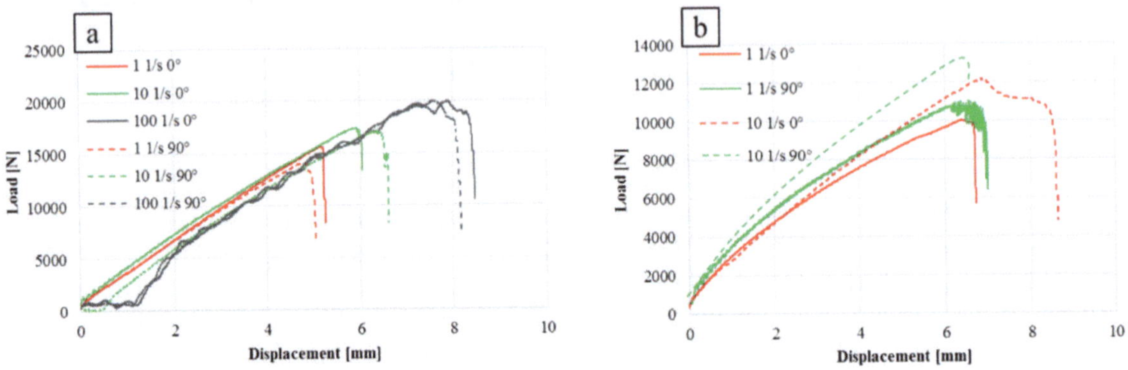

Figure 5: Strain rate dependent tensile tests a) PP-GF47; b) PP-NF/GF

Figure 5a shows the results of the tensile tests of a PP organosheet with glass fiber reinforcement. It can be seen from the figures that there is a respond time during the deformation, which becomes larger with increasing strain rate. It can also be seen that strain and force enlarge with the increase of the strain rate. The highest tensile strength was measured for a 0° specimen at a rate of 100 1/s. Between 0° and 90° specimens no difference in stress-strain- behavior could be observed. Figure 5b presents the load-displacement curves of the natural fiber reinforced organosheet at various strain rates. At the initial stage of testing, which is elastic deformation, the load enhances almost linearly. Unlike the

PP-GF47 the NF reinforced organosheet shows different mechanical behavior depending on the test direction. The specimen tested in 90° direction achieved a higher tensile strength than specimen tested in 0° direction. This behavior is caused by the different weave and ondulation of the hybrid fibers. The maximum overall capability of the natural fiber ogranosheet is lower than the PP-GF47 organosheet. Figure 6 represents the failure behavior of the NF organosheet. A typical failure behavior with a combination of normal and shearing breakage can be seen.

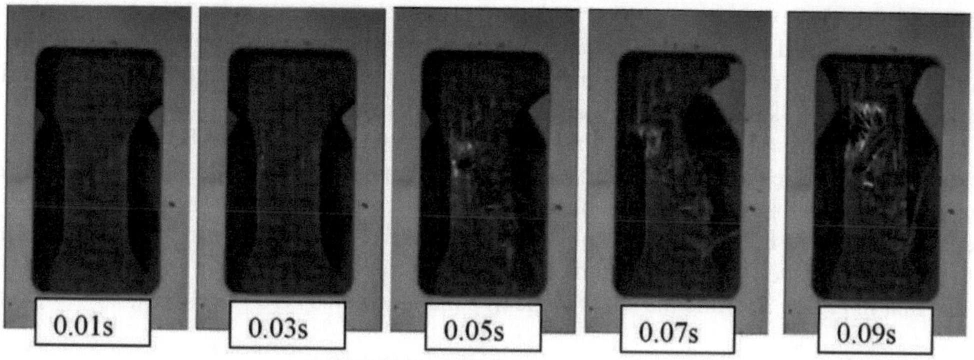

Figure 6: Failure behavior of tensile specimen of natural fiber reinforced organosheet conducted at Fraunhofer IWU

4 IN PLANE SHEAR CHARACTERIZATION

Due to the thermo-viscoelastic material properties of the organosheets, the forming behavior is highly dependent on temperature as well as forming velocities. Hence, there is a high importance of characterizing the material under the same thermal and mechanical load as in the real process. Moreover due to the deviation from melting and recrystallization point of the thermoplastic matrix, the temperature cycle the material undergoes during forming is also of great importance. Therefore it is necessary to perform mechanical testing according to the same thermal cycle which is depicted in Figure 7. Here the organosheet is heated until process temperature using an infrared heater and following transferred into the tool. Depending on transfer conditions the material already loses some heat due to convection and conduction to the grippers. Subsequently the tool is closed and the material experiences a non-isothermal cooling due to conduction to the steel tool. In this step the cooling rate of the organosheet plays an important role in the forming process as the recrystallization of the matrix is highly dependent on the cooling rate [5]. At the end of the process the part reaches almost the tool temperature and is ejected.

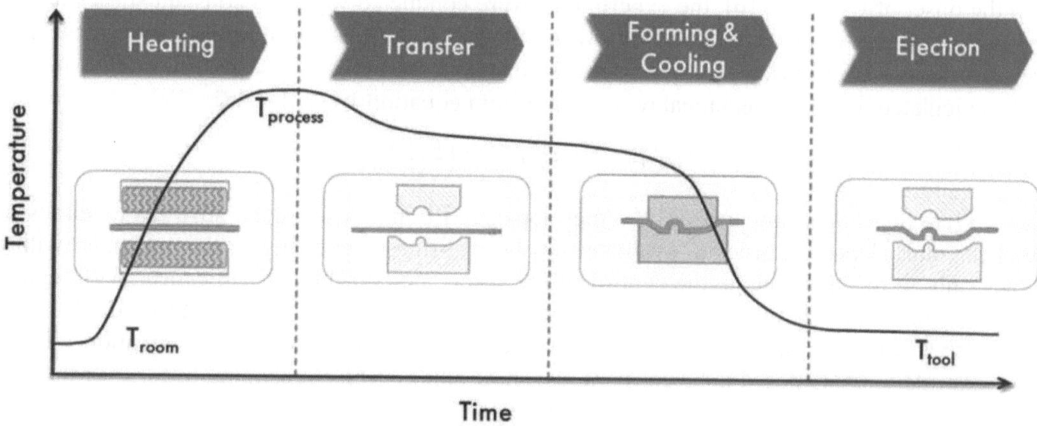

Figure 7: Schematic temperature time history of forming process

4.1 Bias extension

As the changes of the fiber orientation during thermoforming are mostly affected by the in plane shear behavior it is presumed as the most important deformation mode during forming [6]. Thus the focus of the present investigation was set to the characterization of the shear behaviour.

While there is no test standard for characterizing the shear behavior of continuous fiber reinforced thermoplastics, two methods have been widely used and investigated in the literature – the picture frame test and the bias extension test [7–11]. In the present investigation the bias extension test was chosen due to the limited capabilities of commonly used shear frames to bear higher loads which can occur as the matrix viscosity increases during the recrystallization of the thermoplastic matrix and hence, could damage the shear frame or lead to inaccurate results.

The bias extension test involves stretching a rectangular specimen with a fiber orientation of 45° to the direction of the applied axial force. Unlike the picture frame test where the specimen experience a pure shear mode of deformation, during bias extension test three different shear zones occur in the specimen, like shown in Figure 8. If the fabric is considered as inextensible, no intra-tow slip occurs within the sample and the specimen remains flat without wrinkling, the shear angle in region *A* is always twice that in region *B*, while the shear angle in zone *C* remains zero [8, 11–15].

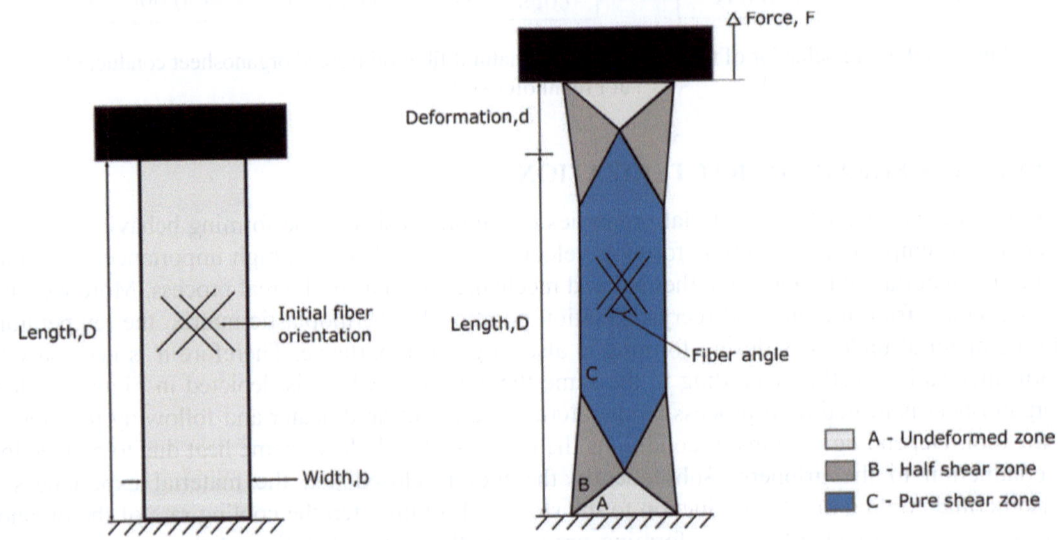

Figure 8: Schematic representation of the bias extension test

As the shear rate has a significant effect on the flow behavior of the thermoplastic Matrix and therefore on the onset of wrinkles [16], the experiments were conducted with two constant shear rates: 3°/s and 6.43°/s. Also the tensile testing software was only able to process discrete velocity values. Hence, the velocity profiles were discretized in a stepwise manner (see Figure 9). The shear angle γ in region *C* can be calculated using a kinematical relationship from equation 1 [11, 17, 18].

$$\gamma(d) = \frac{\pi}{2} - 2\cos^{-1}\left(\frac{D+d}{\sqrt{2}D}\right) \quad (1)$$

Due to the rate dependent properties of the thermoplastic PP matrix the shear stiffness increases with rising shear rate. Hence, in order to evaluate the shear stiffness occurring in the specimen with dependency only to the shear angle, the control of the shear rate is of high significance. With regard to equation 1 a nonlinear velocity profile needs to be applied to the ram displacement in order to achieve a constant shear rate. For the calculation of the velocity profile a derived form of equation 1 is used. As the shear rate ω is supposed to be constant, the shear angle $\gamma(t)$ is changing linear over the time:

$$\gamma(t) = \omega \cdot t \quad (2)$$

If equation 1 for the geometrical relationship of shear angle γ (d) is solved for d and the shear angle is replaced by the time dependent function from equation 2, a time dependent equation for the displacement is achieved. The first derivation of this equation leads to equation 3 where the testing velocity depends on time. After replacement, equation 4 where the velocity is depending on the displacements is obtained.

$$v(t) = \dot{d}(t) = \frac{\sqrt{2}}{2} D \cdot \omega \cdot \sin\left(\frac{\pi}{4} - \frac{\omega \cdot t}{2}\right) - D \quad (3)$$

$$v(d) = \frac{\sqrt{2}}{2} D \cdot \omega \cdot \sin\left(\cos^{-1}\left(\frac{D+d}{\sqrt{2}D}\right)\right) \quad (4)$$

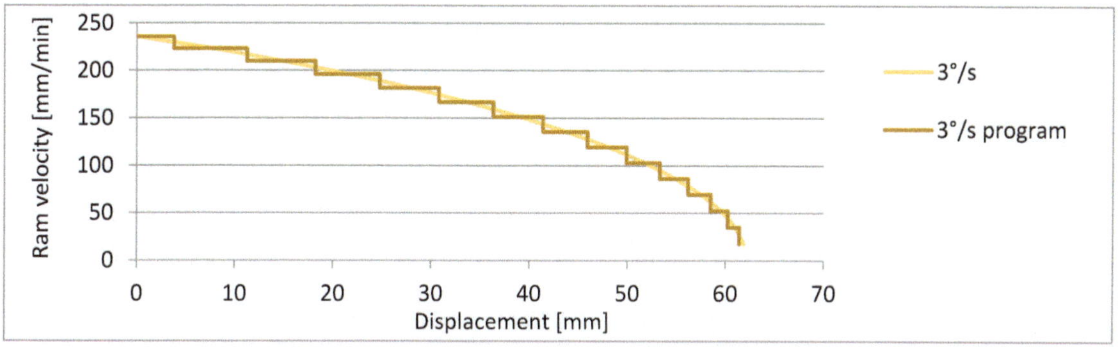

Figure 9: Velocity profiles for 3°/s and 6.43°/s

As the forming takes places in a temperature range from 190 °C until about 140 °C, 3 temperatures were chosen to perform the shear characterization test and to study temperature dependency of the material. As it is anticipated that the crystallization of the thermoplastic polymer has an unneglectable influence on the forming behavior and subsequently on the mechanical properties of the formed part, the characterizations were performed according to the part manufacturing temperature profiles. Therefore the shear behavior is analyzed according to the temperature steps during the manufacturing process which involves heating the organosheet to 190 °C and subsequently cooling it due to the tool contact until the tool temperature is reached. Therefore the specimen were fastened in the tensile testing machine and heated until 190 °C. Then the specimens were cooled down to the desired testing temperature by turning off the oven. The specimen temperatures as well as the oven temperatures were controlled using thermocouples. Figure 10 shows an example of the measured temperature history on top and in the bottom of the specimen.

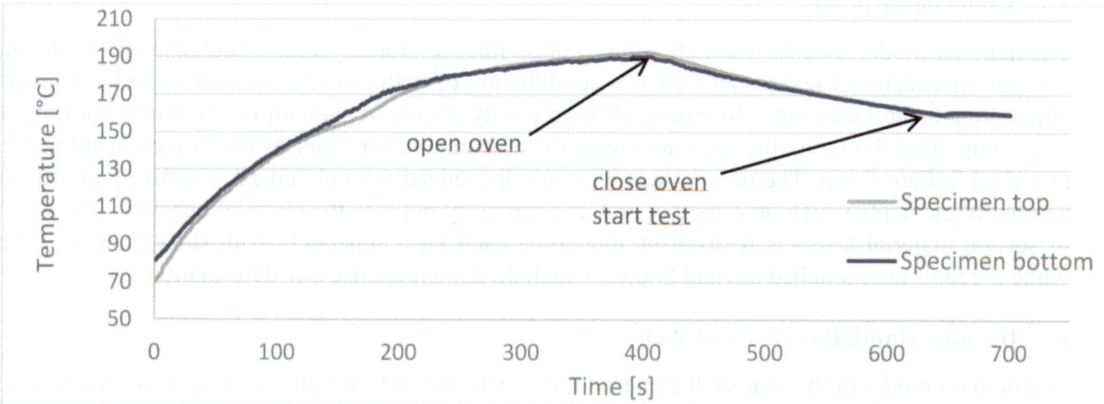

Figure 10: Temperature history of bias extension tests conducted at 160°C

The results in Figure 11 show a typical load-extension curve with progressively increasing load and temperature dependency above 55° shear angle. However, the results are not consistent regarding their temperature dependency, i.e. the average load at 160 °C is below the one at 140 °C. Also an anticipated increase in shear stiffness with decreasing temperature could not be observed. Thus measurements at 190 °C have a significantly higher load-extension curve.

As the measured forces contains regions with different shear angles and hence, do not exactly describe the shear force in relation to the shear angle, it is necessary to normalize the forces for further use in draping simulation. A detailed review of the normalization methods was presented in [8, 11, 19–22]. The conducted normalization method is described in [8].

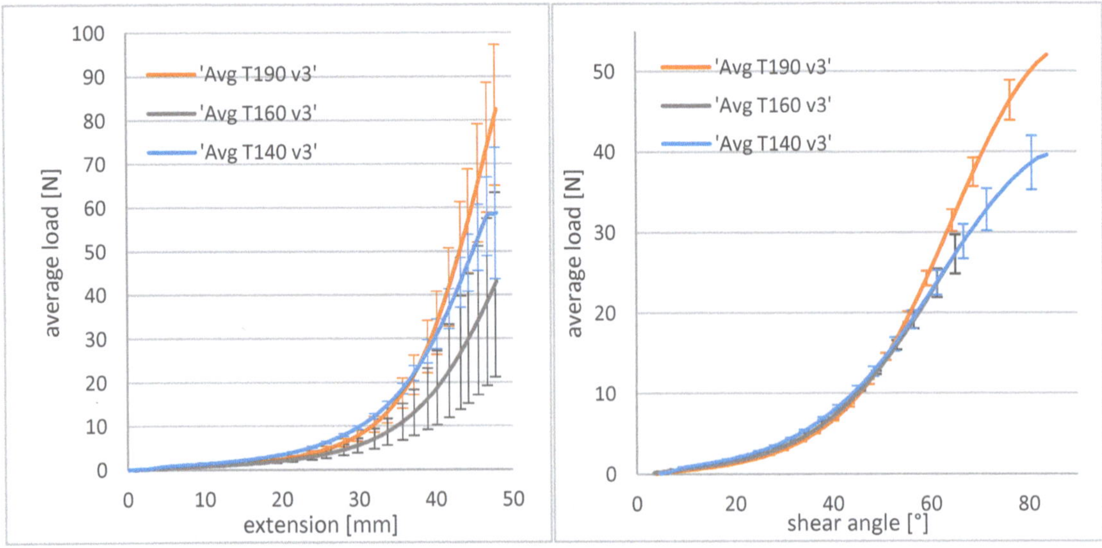

Figure 11: Bias extension results for 3°/s and different temperatures

5 FE DRAPING SIMULATION

In scope of the present work the draping behavior of a NF-PP organosheet is studied in order to optimize the blank geometry. In comparison to a kinematical draping simulation, the finite elements method does take many material and process parameters into account for modeling the complex forming behavior [23]. Therefor the used FE draping software Pam-Form 2017 as well as the model of the seat shell is introduced in following.

5.1 Model description

The material model used for simulating the endless fiber reinforced organosheet also allows the modeling of unidirectional textiles as well as a combination of both as each ply is modelled as a separate object trough shell elements. To ensure close to reality results in simulation, the whole forming process, containing the tools, the separate layers of the composite as well as the process conditions are modelled in Pam-Form. The organosheet plies are discretized through a mesh consisting of a random amount of elements which describe not only the geometry but also the material behavior. The forming of several material layers is realized by modelling each layer separately with shell elements. Meanwhile the tools are modelled as rigid bodies, which do not experience any deformation.

5.2 Draping simulation of a seat shell

The draping model of the seat shell consists of a punch, die and four plies of shaped blanks which are hold in place by two locating pins (see Figure 12). As the process delivers a finished part without the need of trimming, the blanks need to be trimmed to the desired geometry beforehand. Hence, they already contain all the necessary holes and openings for later assembly. In pursuance of achieving a stable process, after forming the holes in the blank need to meet the tolerances of the tool in order to carry the inserts. This makes the geometry very complex and difficult to form into the die.

The tools are modelled as rigid bodies whereas the blanks are discretized by mostly quadrangular shell elements. The cavity between the shaped molds has a gap of 2 mm for consolidating the 4 plies of organosheets. All objects contain several attributes like material properties, kinematical, load and contact boundary conditions which describe their behavior in the respective stages. Thus the tools are modeled with different kinematical degrees of freedom. In the first stage, the gravity stage, the organosheet material is placed on the locating pins through the predefined holes and experiences a gravity force $F_{gravity}$ (see Figure 12). During the following forming stage the translation of the die is locked in all direction, while the punch moves along the z-axis with a predefined speed v_{punch}.

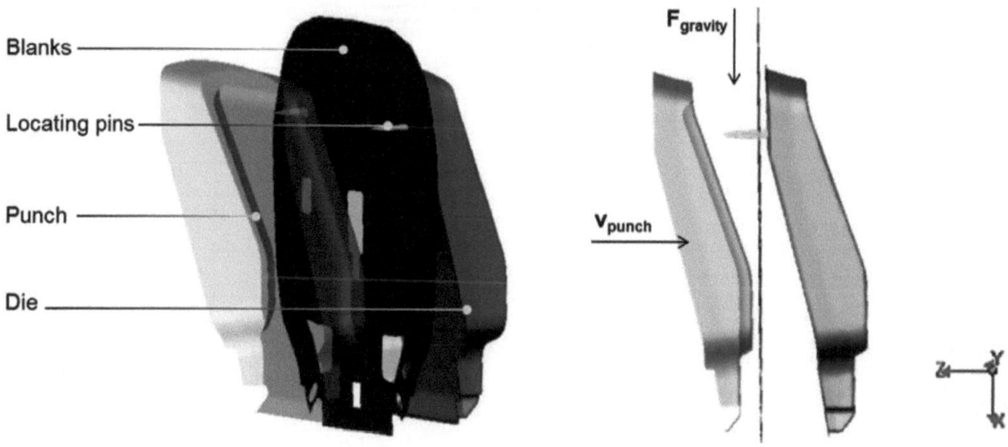

Figure 12: Model of the seat shell

5.3 Optimization

In order to achieve the desired net shape of the organosheet in the manufacturing chain, an iterative reverse engineering process is used in the forming simulation. This iteration usually starts with rectangular shaped organosheet plies which are formed according to the earlier described process. After the forming simulation reaches the end state, a 3D trim curve which represents the boundaries of the net shaped organosheet is orthogonally projected onto the plies. After the curve is projected, the plies are unfolded to their first state where there is no deformation. This process is called flattening. This step also includes the flattening of the gravity state in which the shell undergoes a certain deformation as well. After the first iterative step, the rectangular organosheet is cropped to the new projected curve and the simulation of the new shape starts all over again. Depending on the deviation from the first generic shape to the second, the organosheet does not yet fit the desired shape as the trimming influences the organosheet's in plane behavior and hence, it's forming properties. This iterative process is repeated until the desired shape is reached. Figure 13 depicts an iterative optimization process for the seat shell.

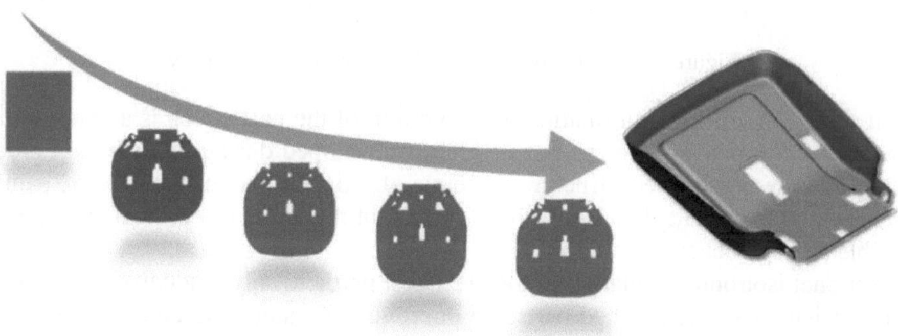

Figure 13: Iterative optimization process of blank geometry

6 CRASH SIMULATION

6.1 Modeling the NF-organosheet behavior

Within the explicit solver Pam-Crash, a rate-dependent material model was used. A three-parameter material model is applied to fit the experimental data. In Figure 14 the simulation is shown in comparison to the experimental data:

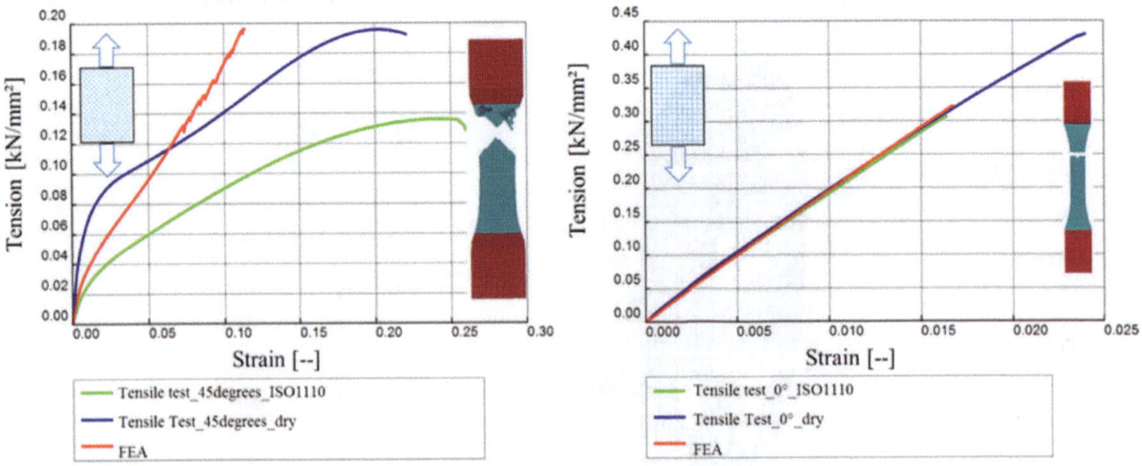

Figure 14: Comparison between simulational and experimental tensile test results

6.2 Frontal crash simulation of the seat shell

The determined material data and material cards were used to perform highly dynamic simulations of a frontal crash with a 95 percentile dummy. The crash pulse for the simulation is an ECER17 with a delta velocity of 64 km/h. In these calculations a simple failure model was used. In Figure 15 the effective equivalent plastic strain is presented as a result of the crash simulation.

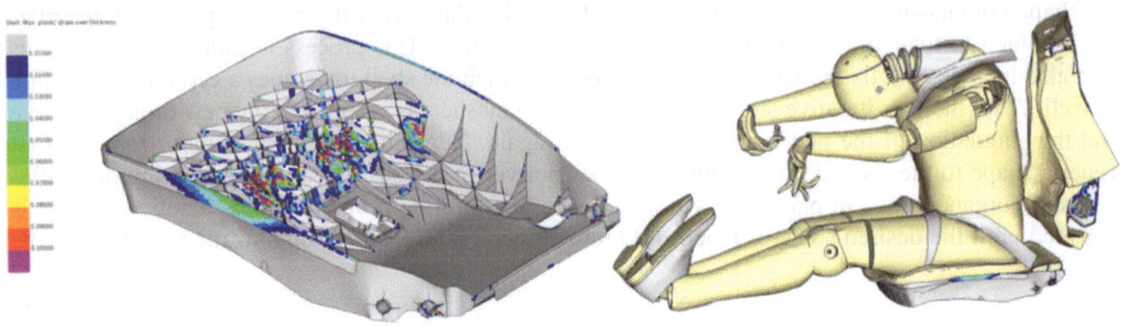

Figure 15: Crash simulation with a 95 percentile dummy

The seat shell shows significant deformation in the bottom of the part which is a result of the forward displacement of the dummy. The rip structure which is developed during the injection process, transfers the energy into the organosheet structure. In terms of failure behavior the average damage behavior was evaluated and is presented in Figure 16. It is evident that there is a critical area in the seat shell where a possible damage behavior can occur.

The conventional isotropic simulation of the crash test predicts only inadequately the plastic strains and the failure behavior in the finished parts. To improve the results, the consideration of the anisotropic behavior represents a very promising option. In many areas a more precise prediction is obtained as a result, whereas in other areas the results do not improve. One of the reasons is that the treatment of the experimental data is extensive and is partly based on non-standardized tests methods.

Furthermore, the description of organosheets has not yet been studied completely, thus making the verification of the compiled material cards difficult. There is still a great potential in the processing of the experimental data, as well as in the actual determination of the input values.

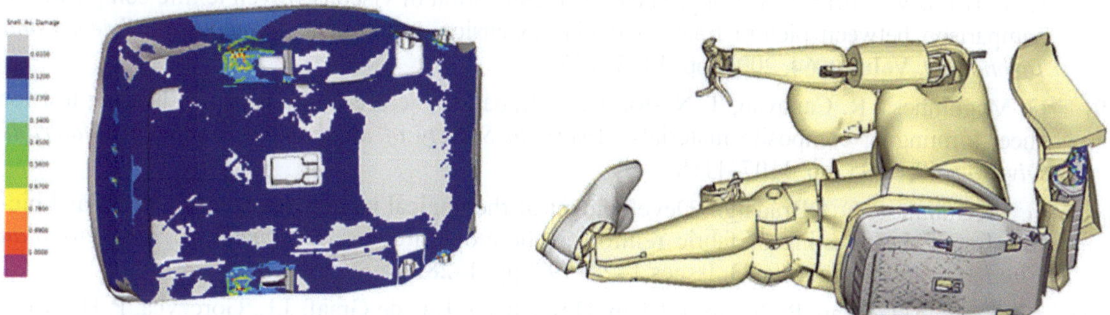

Figure 16: Damage behavior of the seat shell

7 CONCLUSIONS

The objective of the present study was to enable the use of biocomposites in structural parts in automotive applications. Hence, a set of different material characterizations has been performed in order to evaluate the mechanical properties as well as for comparison with standard composites materials. In the case of thermoplastic matrix composites the temperature is of high importance. Thus the characterizations needed to be performed in function of the temperature. Also due to the viscoelastic material behavior, the strain rate plays an important role. Therefore the composite was tested at different strain rates as well. The gathered material data was then processed for the use in FE simulations. In scope of this work the draping simulation and crash simulation was presented on a hybrid seat shell made of biocomposites. The results showed that the use of natural fibers in the combination with glass fibers results in a good mechanical behavior, which would not be possible by using natural fibers solely. It was also shown that this new biocomposite material can be used for load bearing structures. With regard to the conditioning of the natural fibers, the use of dried fibers is recommended in order to achieve higher mechanical properties.

REFERENCES

[1] B. Lahl, M. Nelles, H.J. Endres, Fiber Reinforced Composites for Structural Applications Made from Bio-based and Recycled Materials, *Biocomposites Conference Cologne (BCC), 7th Conference on Wood and Natural Fibre Composites, Cologne, 06.–07.12.2017.*

[2] B. Lahl, M. Nelles, H.J. Endres, Entwicklung von thermoplastischen endlosfaserverstärkten Strukturbauteilen aus biobasierten und recycelten Werkstoffen, *DGAW Deutsche Gesellschaft für Abfallwirtschaft e. V., 8. Wissenschaftskongress Abfall- und Ressourcenwirtschaft, Wien, 15.–16.03.2018.*

[3] H. Schürmann, Konstruieren mit Faser-Kunststoff-Verbunden, vol. 2, Springer Verlag, Berlin, 2007.

[4] C. Cherif, Textile Werkstoffe für den Leichtbau, Techniken- Verfahren- Materialien- Eigenschaften, Springer Verlag, Berlin, 2011.

[5] E. Guzman-Maldonado, N. Hamila, P. Boisse, J. Bikard, Thermomechanical analysis, modelling and simulation of the forming of pre-impregnated thermoplastics composites, *Composites Part A: Applied Science and Manufacturing*, **Volume 78**, 2015, pp. 211–222.

[6] D. Kugele, D. Dörr, F. Wittemann, B.Hangs, J. Rausch, L. Kärger, F. Henning, Modeling of the non-isothermal crystallization kinetics of polyamide 6 composites during thermoforming, *AIP Conference Proceedings 1896, 2017.*

[7] P. Harrison, M. Clifford, A. Long, C. Rudd, Constitutive modelling of impregnated continuous fibre reinforced composites: micromechanical approach, Plastics, *Rubber and Composites 31, 2002,* pp. 76–86.

[8] P. Harrison, M. Clifford, A. Long, Shear characterisation of viscous woven textile composites: a comparison between picture frame and bias extension experiments, *Composites Science and Technology*, **Volume 64**, 2004, pp. 1453–1465.

[9] G. Mcguinness, R. Canavan, T. Nestor, C. O'Bradaigh, Picture-frame intraply shearing test for sheet-forming of composite materials, *American Society of Mechanical Engineers, Materials Division*, 69, 1995, pp. 1107–1118.

[10] G. McGuinness, C. ÓBrádaigh, Development of rheological models for forming flows and picture-frame shear testing of fabric reinforced thermoplastic sheets, *Journal of Non-Newtonian Fluid Mechanics*, **Volume 73**, Issues 1–2, 1997, pp. 1–28.

[11] J. Cao, R. Akkerman, P. Boisse, J. Chen, H.S. Cheng, E.F. de Graaf, J.L. Gorczyca, P. Harrison, G. Hivet, J. Launay, W. Lee, L. Liu, S.V. Lomov, A. Long, E. de Luycker, F. Morestin, J. Padvoiskis, X.Q. Peng, J. Sherwood, Tz. Stoilova, X.M. Tao, I. Verpoest, A. Willems, J. Wiggers, T.X. Yu, B. Zhu, Characterization of mechanical behavior of woven fabrics: Experimental methods and benchmark results, *Composites Part A: Applied Science and Manufacturing*, **Volume 39**, Issue 6, 2008, pp. 1037–1053.

[12] S. Bel, N. Hamila, P. Boisse, F. Dumont, Finite element model for NCF composite reinforcement preforming: Importance of inter-ply sliding, *Composites Part A: Applied Science and Manufacturing*, **Volume 43**, Issue 12, 2012, pp. 2269–2277.

[13] P. Harrison, M.K. Tan and A.C. Long, Kinematics of Intra-Ply Slip in Textile Composites during Bias Extension Tests, *University of Nottingham – School of Mechanical, Materials & Manufacturing Engineering*, 2005.

[14] N. Pan, R. Kovar, M.K. Dolatabadi, P. Wang, D. Zhang, Y. Sun, L. Chen, Origin of tensile strength of a woven sample cut in bias directions, *Royal Society Open Science*, 2015, pp. 1–18.

[15] K. Potter, Bias extension measurements on cross-plied unidirectional prepreg, *Composites Part A: Applied Science and Manufacturing*, **Volume 33**, Issue 1, 2002, pp. 63–73.

[16] W.F. Stanleya, P.J. Mallon, Intraply shear characterisation of a fibre reinforced thermoplastic composite, *Composites: Part A,* **Volume 37,** 2006, pp. 939–948.

[17] G. Lebrun, M.N. Bureau, J. Denault, Evaluation of bias-extension and picture-frame test methods for the measurement of intraply shear properties of PP/glass commingled fabrics, *Composite Structures*, **Volume 61**, Issue 4, 2003, pp. 341–352.

[18] J. Launay, G. Hivet, A.V. Duong, P. Boisse, Experimental analysis of the influence of tensions on in plane shear behaviour of woven composite reinforcements, *Composites Science and Technology*, **Volume 68**, Issue 2, 2008, pp. 506–515.

[19] X.Q. Peng, J. Cao, J. Chen, P. Xue, D.S. Lussier, L. Liu, Experimental and numerical analysis on normalization of picture frame tests for composite materials, *Composites Science and Technology*, **Volume 64**, Issue 1, 2004, pp. 11–21.

[20] P. Harrison, J. Wiggers, A. Long, Normalisation Of Shear Test Data for Rate-Independent Compressible Fabrics, *Journal of Composite Materials*, **Volume 42**, 2008, pp. 2315–2344.

[21] F. Härtel, P. Harrison, Evaluation of normalisation methods for uniaxial bias extension tests on engineering fabrics, *Composites Part A: Applied Science and Manufacturing*, **Volume 67**, 2014, pp. 61–69.

[22] P. Harrison, F. Härtel, Corrigendum to 'Evaluation of normalisation methods for uniaxial bias extension tests on engineering fabrics', *Composites Part A: Applied Science and Manufacturing*, **Volume 80**, 2016, pp. 104–106.

[23] P. Ermanni, Composites Technologien, ETH Zürich, 2007.

Design and Simulation of Hybrid Structures

INNOVATIVE HYBRID MATERIAL CONCEPTS AND THEIR VIRTUAL SAFEGUARDING

Henning Gleich[1], Frank Meißen[2], Kim Kose[3], Christian Paul[4]

[1] inpro, Steinplatz 2, 10623 Berlin, henning.gleich@inpro.de, www.inpro.de
[2] inpro, Steinplatz 2, 10623 Berlin, frank.meissen@inpro.de, www.inpro.de
[3] inpro, Steinplatz 2, 10623 Berlin, kim.kose@inpro.de, www.inpro.de
[4] formerly thyssenkrupp AG, TechCenter Carbon Composites, Frankenring 1, 01723 Kesselsdorf, www.thyssenkrupp.com

Keywords:

Hybrid materials, Virtual safeguarding, Simulation

ABSTRACT

New innovative manufacturing technologies allow the production of lightweight structures in multi-material design. By providing a high functional integration the economic mass production becomes feasible, which is needed for their use in car production. However, numerous issues in connection with an efficient and cost-effective series production and assembly of lightweight structures in multi material design are not yet fully solved. This paper covers three publicly funded research projects with participation of inpro. All of them belong to the FOREL-Cluster (research cluster electro-mobility).

The projects LEIKA (resource-efficient mixed-material construction for lightweight car bodies), ReLei (manufacturing and recycling strategies for electric mobility for recycling of fibre reinforced lightweight structures in composite hybrid constructions) and SamPa (integral production of hybrid lightweight sandwich structures with particle foam and injection moulding for mass production) cover several aspects concerned with modern hybrid car-body designs. Common to all of them is the development of appropriate simulation strategies followed by virtual safeguarding of the manufacturing process.

1 INTRODUCTION

The mass introduction of battery electric vehicles into the consumer market poses significant challenges to the OEMs. While the customer expects the same upfront cost of ownership relative to performance as for cars with an internal combustion engine the OEMs face major changes in car design and new production methods. To drive down production costs while retaining customer experience and at the same time reducing the environmental footprint of the car new and hybrid materials as well as hybrid material combinations are developed. The resulting lightweight constructions are a way to minimize material input and usage, the developed recycling strategies secure a second life of these high-tech materials in a high-performance part as opposed to dumping it in a landfill.

2 LEIKA, RELEI AND SAMPA

The projects LEIKA, ReLei and SamPa address various aspects in simulating and safeguarding hybrid lightweight constructions. A major challenge for the simulation is the consideration of manufacturing processes as well as their influence on the performance of components. Therefore, a software is required that enables the simulation of process-chains in the virtual world. inpro has developed a software solution suitable for series production that links the different simulation systems and maps the relevant state variables, see Figure 1.

K. Dröder et al (eds), *Technologies for economical and functional lightweight design*, Zukunftstechnologien für den multifunktionalen Leichtbau, https://doi.org/10.1007/978-3-662-58206-0_10

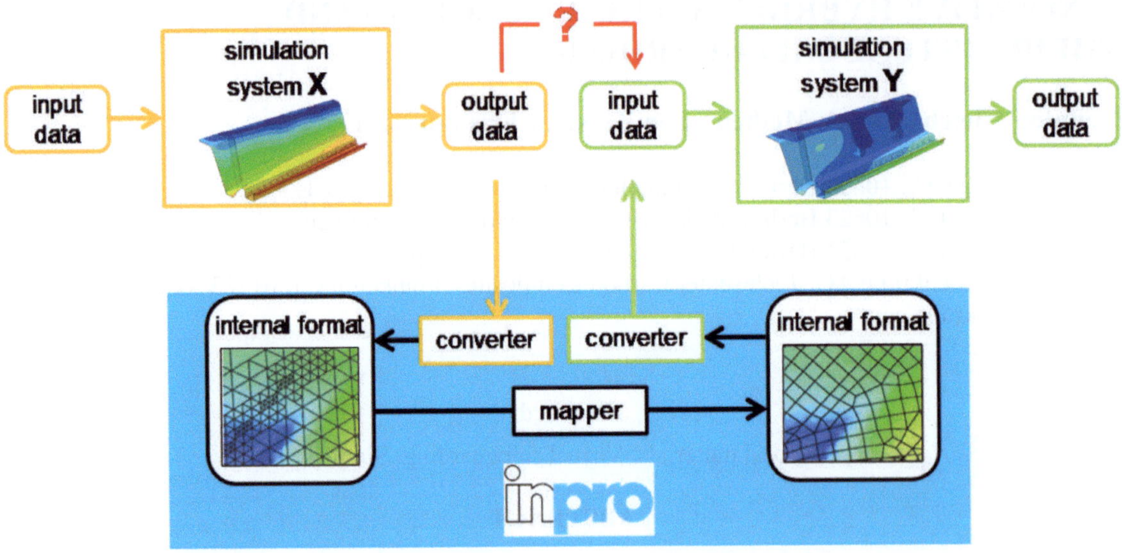

Figure 1: Inpro-Mapper enabling the simulation of complete process-chains

2.1 LEIKA

In LEIKA, the focus is on the simulation of the forming of hybrid sheet materials that consist of outer layers of metal and inner layers of carbon-fibre reinforced polymer as well as sheet materials with outer layers of carbon-fibre-composite with a metal core. After forming the sheets, stiffeners can be added through injection moulding. The forming and the injection moulding process have been combined to use only one tool. The combination with other hybrid materials was demonstrated via the construction and production of the front part of a car underbody structure, see Figure 2.

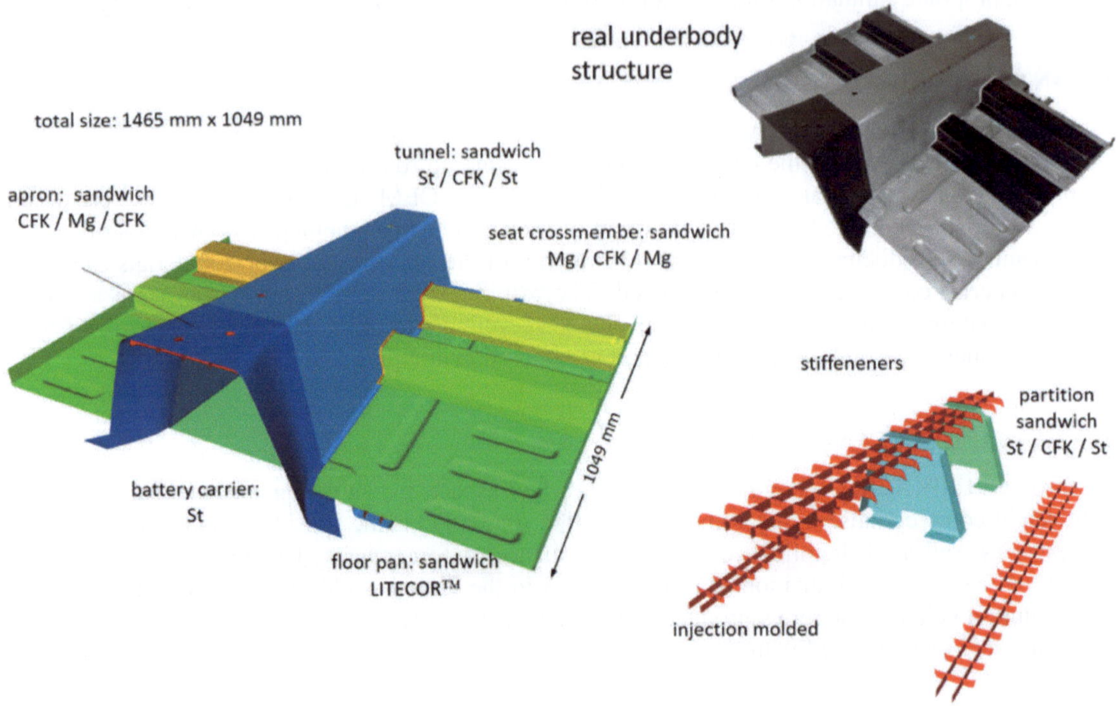

Figure 2: Parts and materials used in the LEIKA underbody structure

The simulation chain (Figure 3) safeguards the production process and crash safety of the underbody structure. Therefore, results from each intermediate simulation are provided for subsequent simulations to ensure adequate initial values. A mapping software used for this task has been developed and maintained by inpro.

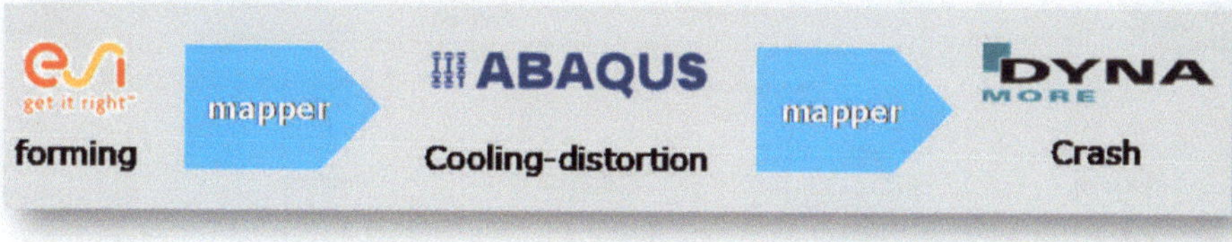

Figure 3: Simulation chain in LEIKA

Thus, an overall simulation and optimization for hybrid materials and hybrid combinations of materials becomes feasible. The LEIKA project finished with testing the crash performance of the underbody structure. This showed a very good agreement between test and simulation (Figure 4). However, since not a complete car with the supporting body-structure, but only part of an underbody was tested, the overall performance results of this test cannot be extrapolated to full car crash tests.

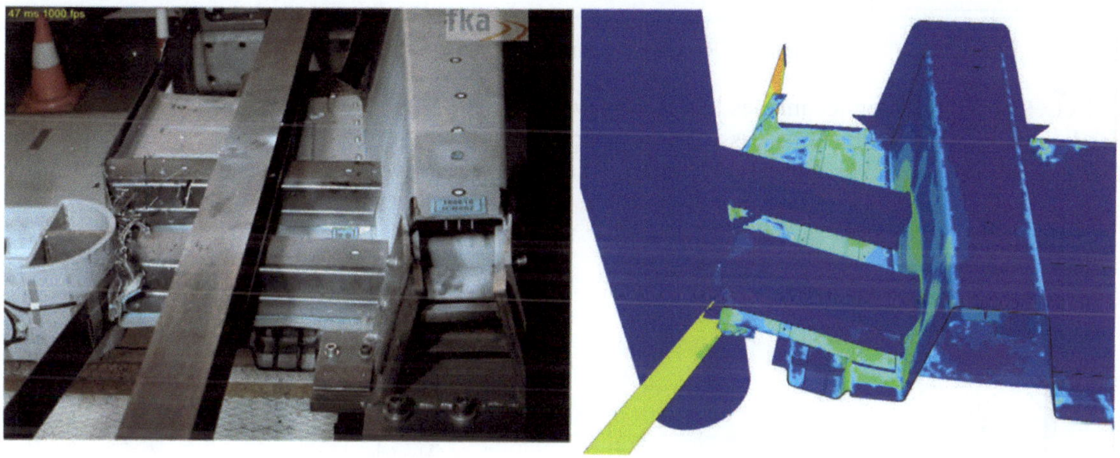

Figure 4: Crash test on real part (left) and in simulation (right)

2.2 ReLei

Carbon fibre reinforced materials usually provide the best weight-performance ratio for car parts. However, producing the carbon fibres is energy intensive and expensive. Therefore, an end-of-life strategy should encompass the reuse of the fibres. In this process the originally very long fibres get chopped and a downgrading of the material is inevitable (Figure 6). In ReLei the chopped fibres were used for making fleece mats covering a polymer core expanded by a MuCell process.

Figure 5: Simulation chain in ReLei

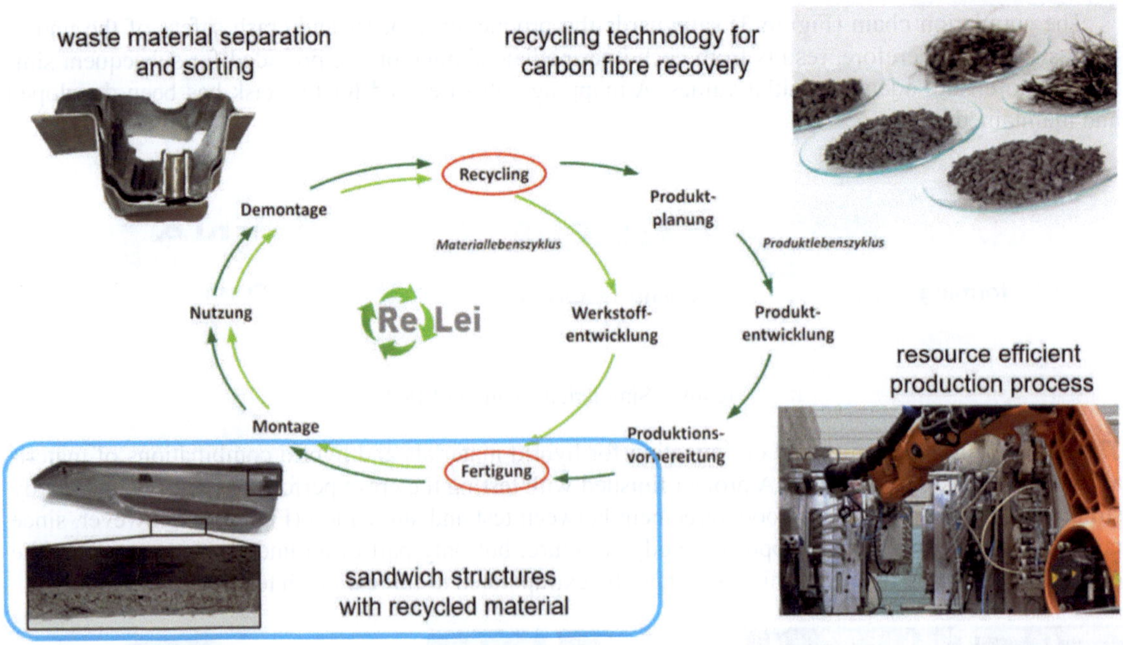

Figure 6: Recovering carbon fibers for further use

In ReLei new simulation techniques had to be developed to combine several commercial codes in order to simulate the complete process of producing and forming the fleece-mat (Figure 6) which was finally used combined with an organosheet for a rear shelf. That rear shelf fulfills all specs for crash.

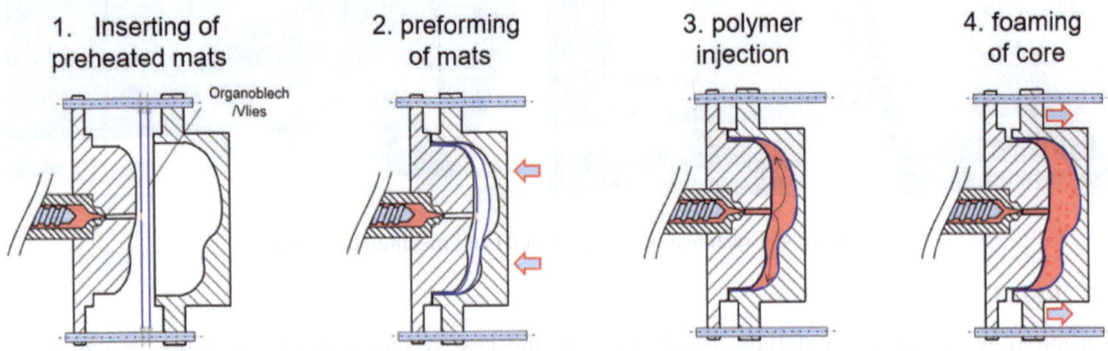

Figure 7: Combining fleece mats with the MuCell process

Central for the performance is the combination of the MuCell process (Figure 7) with the fibre mats made from recycled material. The better this works, the less organosheet (virgin material) has to be used in the part.

The part is produced in a complex single tool, first the pre-heated organosheet and fibre-mats are inserted. During tool closing the organosheet obtains its final form and simultaneously the fibre mats are preformed (Figure 8). Now the gas loaded polymer is injected between the fibre mats and at all areas marked brown in Figure 8 left. In the last step the MuCell process is triggered through pulling apart the center area of the tool.

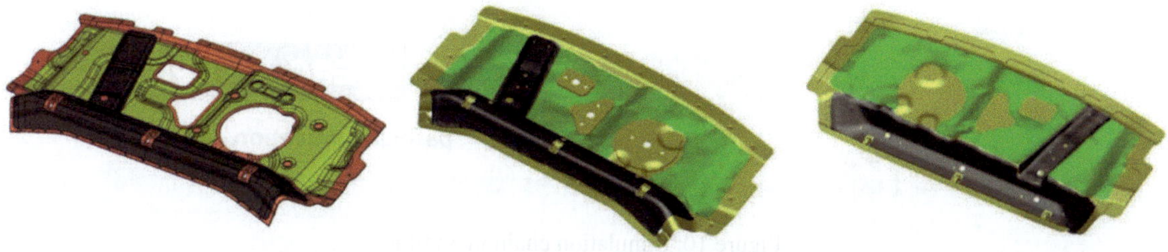

Figure 8: Rear shelf after design (left) with organosheet in black, recycled mat with MuCell process in green and injection molded polymer in brown. On the right side the upper and lower views after preforming simulation are shown

2.3 SamPa

The aim of the SamPa project is to overmold parts made of foam particles like EPP (expanded polypropylene) with the goal of a class A surface suitable for car interiors (Figure 9). The basic process of cavity filling by an airstream carrying the foam particles followed by loading of the cavity with steam in order to weld the particles together is well known. Although it is standard since about 60 years the simulation of the cavity filling has not yet been performed. However, this is needed for complex tools and parts with safeguarded properties. In these, a mutual blocking of the particles is possible, thus preventing a successful filling of the cavity. The simulation can ensure, that not only a sufficient number of injectors is used, but also that their placement is optimal with respect to fill-time and ensuing density.

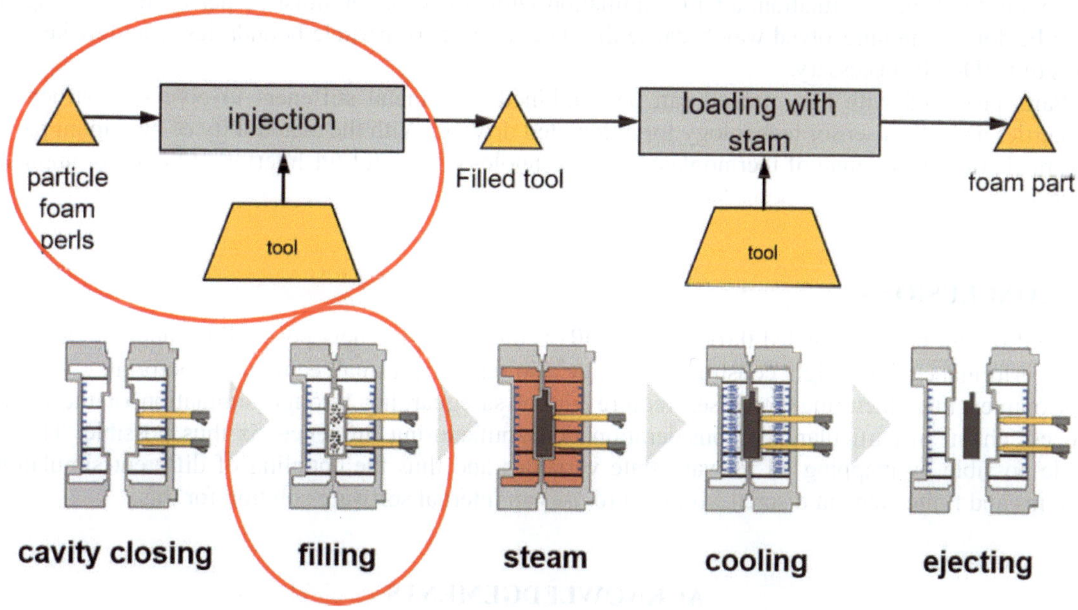

Figure 9: SamPa process, red marked are the main simulation developments

The filling simulation combines a CFD (computational fluid dynamics) with a DEM (discrete element method) simulation, (Figure 10). Particles not only collide with each other but also get influenced by the airflow. The airflow in its turn gets disturbed by the particles and the more particles are in the cavity the more the airflow gets blocked. Because the particle number is very high, the CFD simulation must be used in an unresolved way, i.e. not having exact boundaries with the particles but seeing only a medium density of them (Figure 11).

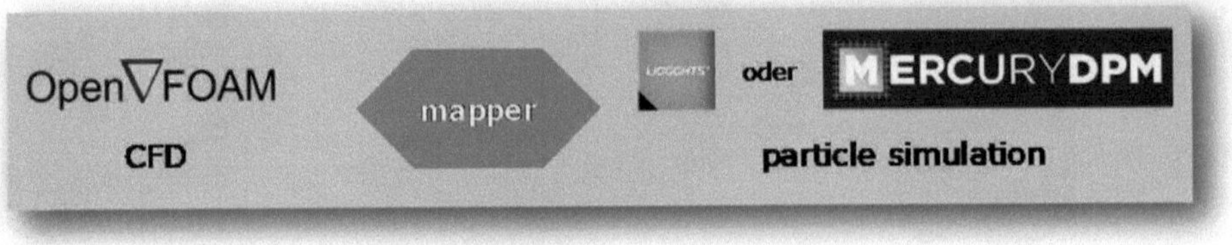

Figure 10: Simulation chain in SamPa

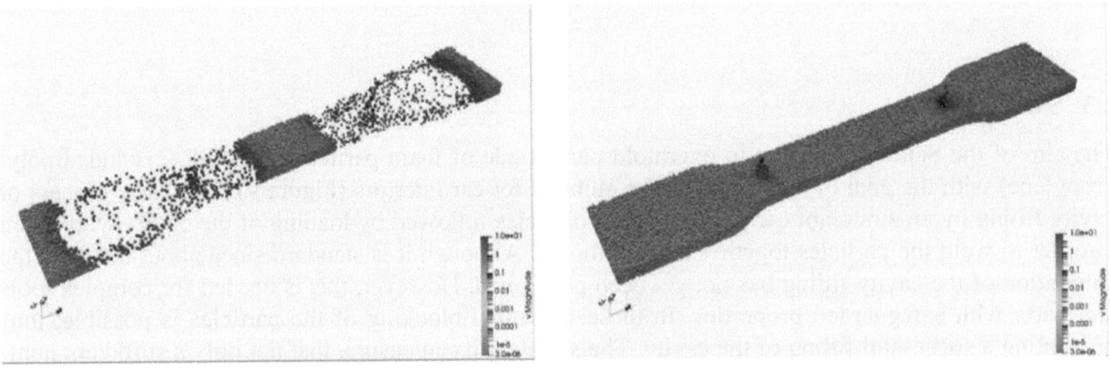

Figure 11: Filling simulation of a cavity, intermediate stage (left) and final stage (right)

Following the filling simulation, a CFD simulation with condensation must be performed. This again must be done in an unresolved way because the sheer number of particle boundaries would make millions of CFD-cells necessary.

Parts produced with this method can be combined with metal stiffeners to give semi-structural properties as well as sensor technology for automated driving. With the availability of new foam particles made of a wide range of thermoplastics the technology is useful for interior and exterior applications.

3 CONCLUSIONS

Three different projects funded through the FOREL initiative give new possibilities and solutions for high performance lightweight constructions while also retaining a small ecological footprint.

Complex manufacturing processes require virtual safeguarding through simulation of the entire process chain. In particular, the consideration of manufacturing influences is thus possible. This is made possible by mapping all relevant state variables and thus the coupling of different simulation systems and finite element discretization. inpro uses an internal software solution for this.

ACKNOWLEDGEMENTS

The research and development projects presented here are funded by the German Federal Ministry of Education and Research (BMBF) with the funding number 02PJ2770 – 02PJ2781 for LEIKA, 02PJ2800-02PJ2808 for ReLei and 02P15Z000 – 02P15Z006 for SamPa and managed by the Project Management Agency Karlsruhe (PTKA). The responsibility for the content of this publication lies with the authors.

CARBON CARRIER – INTEGRATED CONCEPT FOR INNOVATIVE INTERIOR STRUCTURES

Michael Hage[1], Benjamin Wagner[2], Frank Preller[3]

[1] Bertrandt Projektgesellschaft mbH, Ehningen, michael.hage@de.bertrandt.com, www.bertrandt.com
[2] Bertrandt Ingenieurbüro GmbH, Tappenbeck, benjamin.wagner@de.bertrandt.com
[3] Bertrandt Ingenieurbüro GmbH, Tappenbeck, frank.preller@de.bertrandt.com, www.bertrandt.com

Keywords:

CFRP, GFRP, FRP, Lightweight concepts, Instrument panel, Load path

ABSTRACT

INNOVATIVE VEHICLE STRUCTURES BASED ON LIGHTWEIGHT FIBRE COMPOSITES AND FUNCTIONAL INTEGRATION

With the common developed concept Carbon Carrier, Bertrandt and SGL show how new vehicle structures can be designed. In addition, it shows how it can be possible to make technologies visible and touchable to the end customer. The focus of the project is the use of available or soon available technologies for glass or carbon fibre reinforced plastics with a thermoset matrix or thermoplastic polymer matrix.

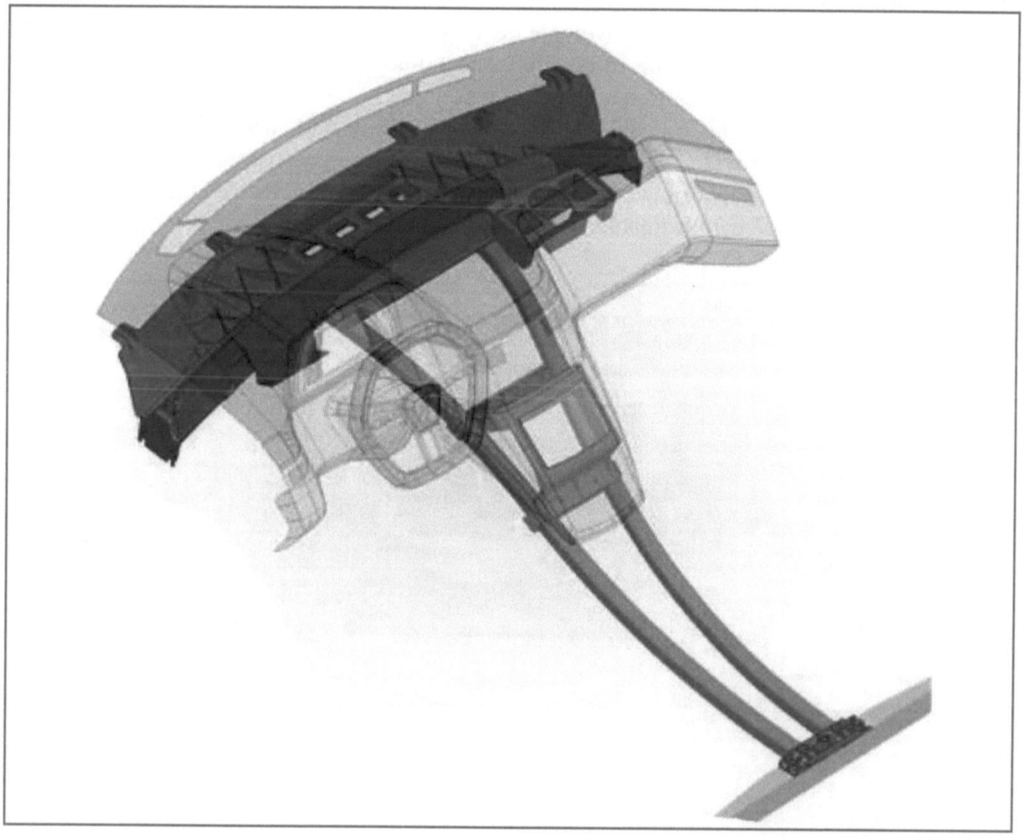

Figure 1: Carbon Carrier: the instrument panel as an important body component

K. Dröder et al (eds), *Technologies for economical and functional lightweight design*, Zukunftstechnologien für den multifunktionalen Leichtbau, https://doi.org/10.1007/978-3-662-58206-0_11

As a large, spatial interior structure, the Carbon Carrier performs additional functions throughout the interior (Figure 1) while remaining within the weight of a classic instrument panel carrier.

1 INTRODUCTION

In the field of body development, the preferred materials are high and super-high strength steels and, depending on the manufacturer, increasingly aluminum and magnesium alloys. Cast parts, sections and body panels are made of these materials. Fibre-reinforced plastics are less commonly used than steel and aluminum. The exceptions include specific well-known manufacturers who make large-scale use of CFRP and the sports car segment. Interior components are often manufactured from short-fibre reinforced materials.

Lightweight structural components in the aviation industry are made either from metals, such as aluminum, magnesium and titanium, or from plastics and, in particular, carbon-fibre reinforced materials (CFRP). The team at Bertrandt's site in Hamburg has wide-ranging experience of designing and developing CFRP components. In addition, the Bertrandt Group as a whole has increased its expertise in manufacturing and testing prototypes made from CFRP over recent years.

In the case of fibre-reinforced composite materials such as CFRP, the properties and the cost of components depend on a large extent on the process used and the composition of the material, for example, whether thermosets or thermoplastics are used as the matrix and the length, direction and proportion by volume of the fibres, together with the number of layers and the type of semi-finished product. With the aim of expanding its skills in the field of vehicle structures, Bertrandt has set up a bilateral technology product together with SGL Carbon, a global company in the field of composite materials and fibre production.

The objective is to combine Bertrandt's vehicle development expertise with the materials and technology competence of SGL Carbon and to develop a technology demonstrator. Work on the project began in October 2016 after a cooperation agreement was signed.

The project is based on a vehicle developed internally by Bertrandt, which is available as a CAD model and will be used to provide a package for the technology demonstrator. Two variants of the vehicle are available with different powertrains, conventional and fully electric, that will allow different stresses and load paths to be investigated (Figure 2).

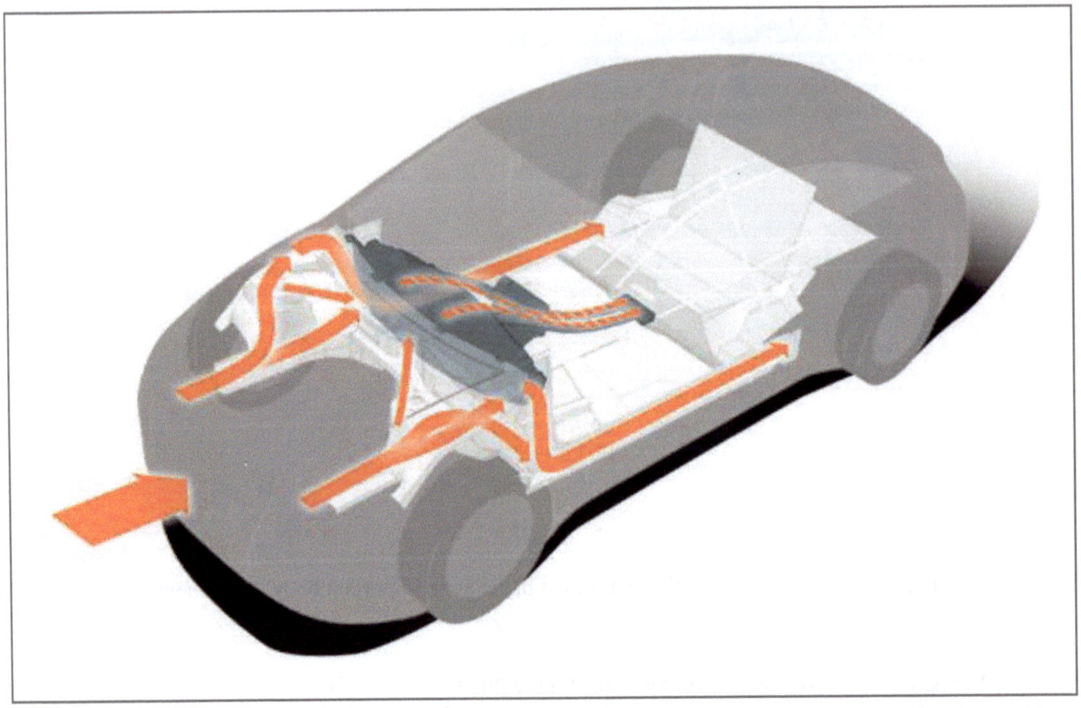

Figure 2: Integration of a stiffness-related tunnel load path into the overall Carbon Carrier concept

2 THE ASSEMBLY

An important assembly in a vehicle body is the frame of the instrument panel, which is one of three structural components that are bolted to the body. It also represents an ideal introduction to automotive manufacturing in the form of a subassembly with an intelligent mix of materials. This makes it possible to introduce new materials into the vehicle and, at the same time, to continue using large parts of existing production lines without major changes. The resulting findings can later be applied to other areas of the vehicle.

The instrument panel frame, which is made primarily from carbon fibre, represents a new concept for integrating a number of different functions. The new frame is intended to perform the functions of the current instrument panel, instrument panel frame and center console (Figure 3). A unique feature of the new component is the integration into the overall Carbon Carrier concept of a stiffness-related tunnel load path.

This new Carbon Carrier (CC) will demonstrate the potential of new structural components when used in convertibles and coupés, for example, and in particular those with electric powertrains. This is based on the principle that the convertible and coupé body variants need to compensate for the lower levels of stiffness that result from the introduction of new requirements, such as a level floor for the battery package. In order for the Carbon Carrier to provide support in this area, its structures and its connections to the vehicle body, need to be redesigned.

A significant improvement in its bending stiffness can be achieved by means of a brace on the bulkhead that can extend as far as the heelboard.

The Carbon Carrier can support all the other instrument panel components and consider their requirements. It replaces the conventional module cross beam in the body-in-white, the instrument panel frame and the tunnel, including all the relevant functions.

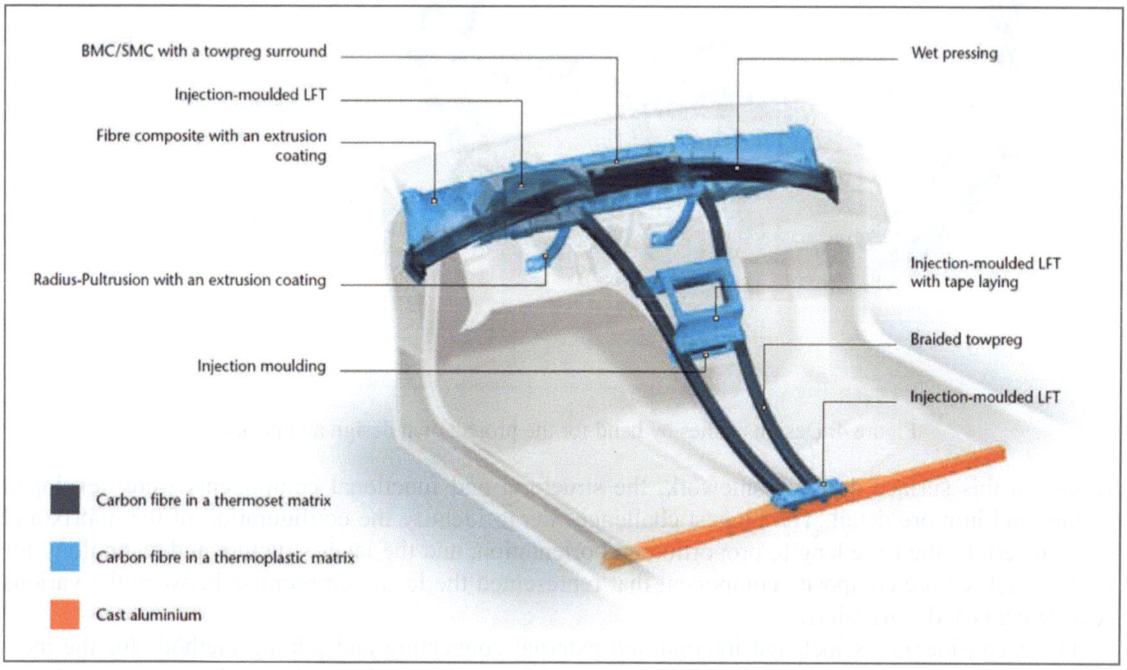

Figure 3: Configuration of matrix and fibre materials and technologies for each composite component

3 THE DEVELOPMENT PROCESS

During the initial phase of the project, a functional package model was developed for the interior of the front of a convertible or coupé based on a non-proprietary body design created in-house by Bertrandt. It includes all the important functional and trim-related components of a conventional instrument panel, but with a new and attractive design. For example, the driver will be able to see the central load paths and the CFRP components, despite the fact that no purely decorative, exposed carbon fibre elements have been used. Because of Bertrandt's many years of experience of automotive development, it was possible to develop and integrate the relevant functional components without links to a specific manufacturer. The main supporting structures were defined based on this initial package model. Because the design, simulation and technology specialists at the various Bertrandt sites are used to working closely together, they were able to transform the wide variety of early sketches into a refined 3D concept very quickly.

At the same time, design sketches were created and turned into a new surface design for this area of the interior together with the package model (Figures 4 & 5). The design of the structural components gave the interior a more open, light and airy atmosphere, with sharp curves and innovative details to create the feel of a sporty electric car.

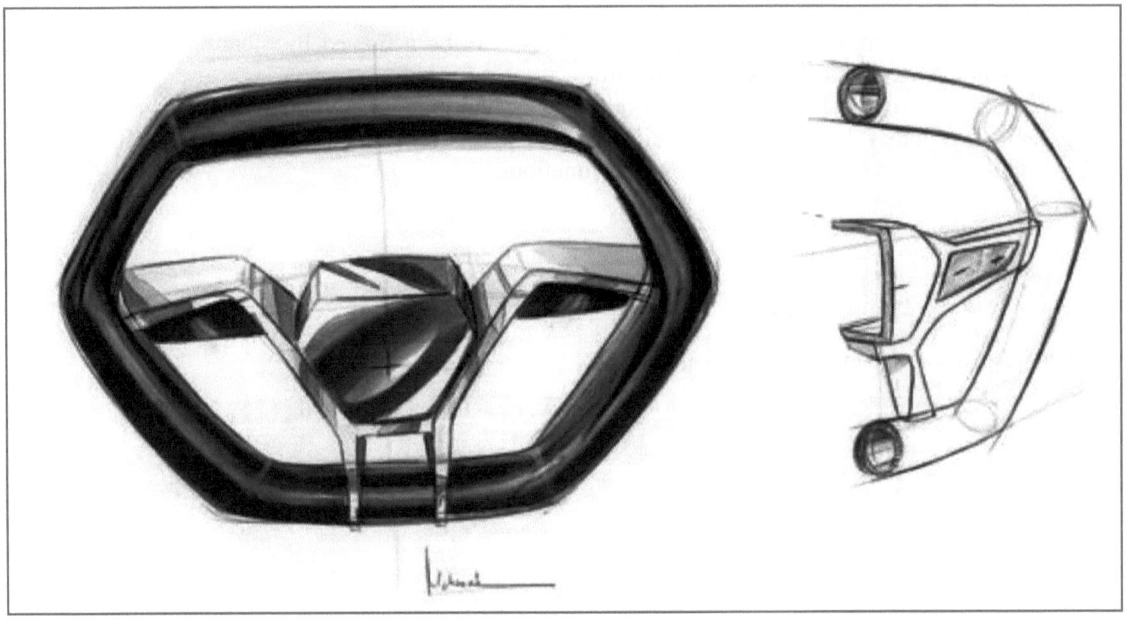

Figure 4: Design studies by hand for the project own design and package

Based on this surface design framework, the structural and functional components were developed further and in more detail. The biggest challenge was to identify the configuration of the matrix and fibre materials, the fibre length, proportion and orientation, and the layer structure and technology for each complex fibre composite component that represented the ideal compromise between the various requirements and constraints.

Other considerations included internal and external connecting and joining methods for the individual parts, together with the production and assembly conditions. These had to be determined based on visibility, stresses, the selection of materials and the sequence of joining processes.

The laminate structure was defined in a comprehensive set of specifications that incorporated legislative and consumer protection requirements. The factors that determine the size of the load-bearing structural components of the Carbon Carrier include misuse load cases, vibration behaviour and, most importantly, the forces specified in the crash load cases. The laminate structure was designed and the rib patterns were refined with the aim of creating a lightweight structure that could be easily manufactured.

The CATIA V5 Composite Part Design (CPD) module that is used at Bertrandt for designing fibre composite components made it possible to include the laminate in the CAD model. The tool enabled the transitions between the different areas to be designed in detail and allowed for clear and effective communication with the production department.

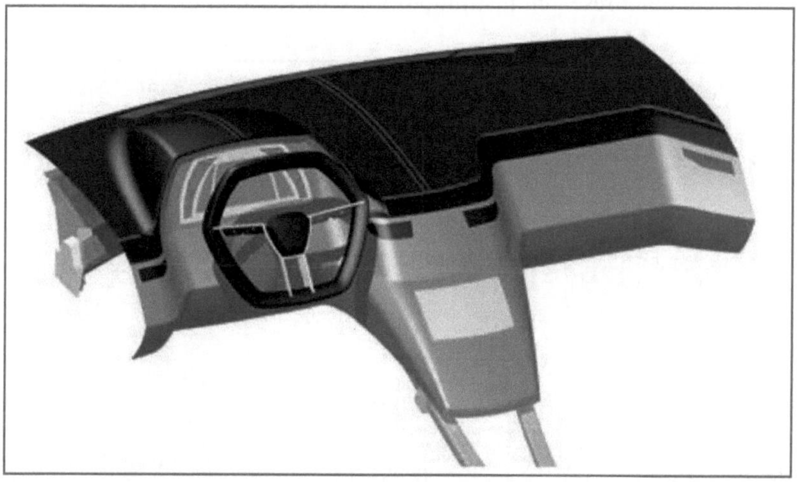

Figure 5: Surface design framework as the base for structural and functional components

The ongoing detailed development process of the new structure (Figure 6) required close cooperation between the simulation, design and production teams. An interface developed by Bertrandt that enabled the CPD data to be transferred directly into the FE model saved a large amount of valuable time.

In dynamic simulation tools, fibre composites have a much more complex, multi-layered definition than traditional materials. These material definitions depend heavily on the type of process and the materials used and therefore cannot easily be transferred from one manufacturer of materials or semi-finished products to another. For this reason, SGL's existing material cards were used in some cases in the simulation process. However, hardware tests were also carried out using predefined samples in order to identify the analysis parameters.

The development departments of Bertrandt and SGL worked closely together throughout the entire development period. Despite the long distances between the teams, the in-depth cooperation was made possible by means of regular meetings at the development sites in Meitingen, Wolfsburg and Rüsselsheim and frequent telephone and video conferences.

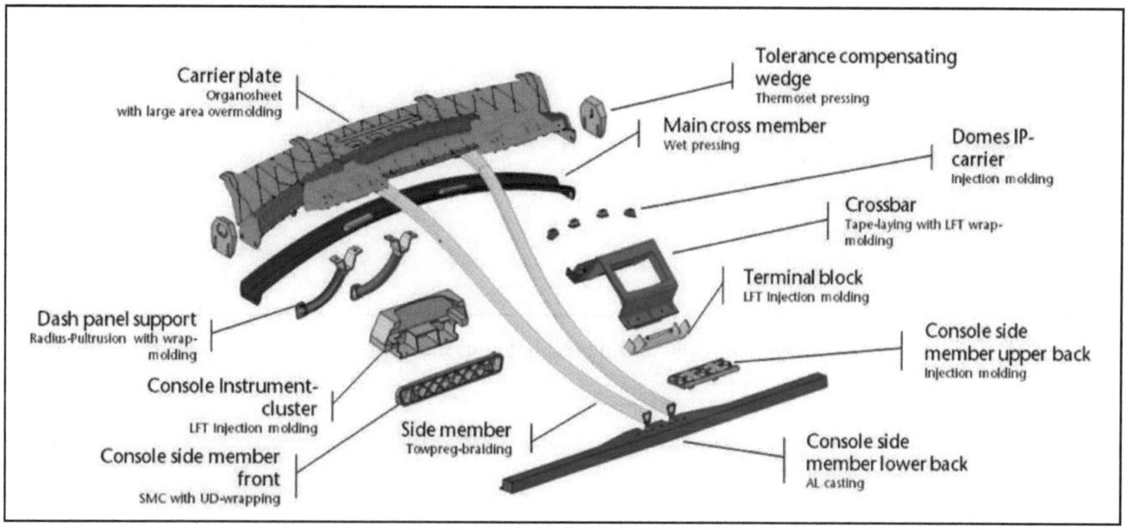

Figure 6: Designed components in detail

4 PRODUCTION

During the process of developing the Carbon Carrier, the engineers focused on ensuring that the materials, technologies and assembly concepts would be suitable for volume production either now or in the near future (Figures 7 & 8). In addition, the latest results of research and pre-development activities at Bertrandt, SGL and other research institutions were included in the process.

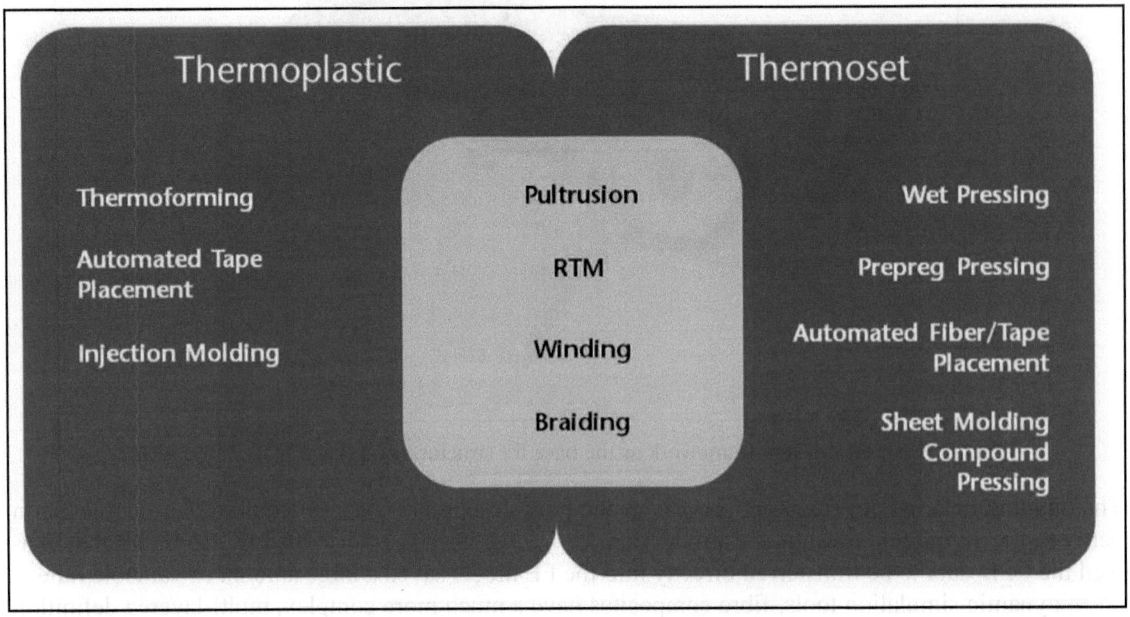

Figure 7: Coordinated technologies in the project

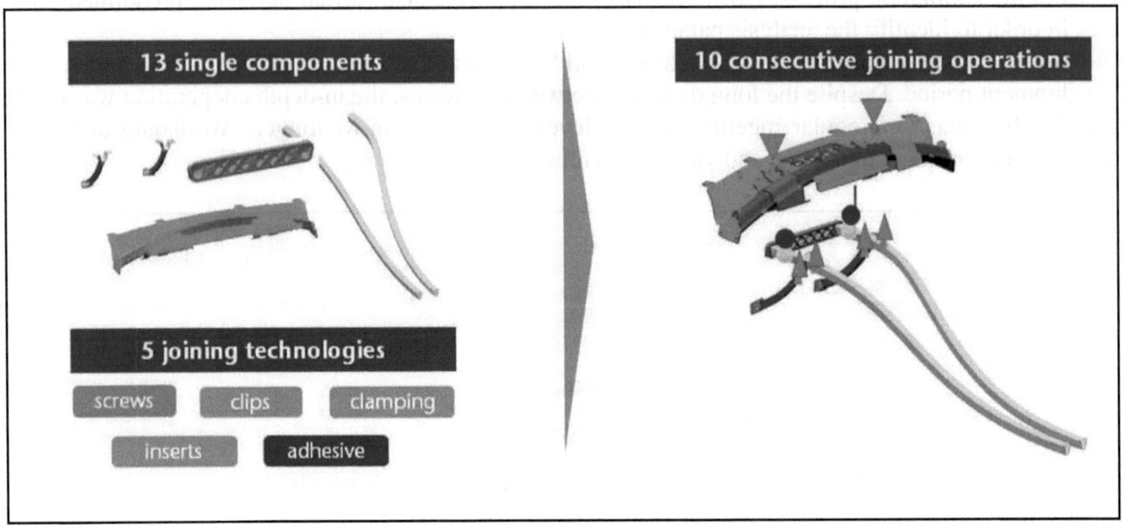

Figure 8: Cold joining technologies in the project

5 CONCLUSION

A high innovative test structure with the potential for volume production has been developed by Bertrandt and SGL working in close cooperation. It demonstrates how new structures can be incorporated into modern vehicle concepts. Both organisations have enhanced their knowledge of the volume production processes for new types of structural concepts in modern vehicle bodies, starting with the initial idea and the pre-development activities and also including the design and the close interaction between CAD, CAE and production. This knowledge has been put into practice in a demonstrator that can be used by both companies to illustrate the different challenges and approaches involved in the increased use of fibre composites in the structure of vehicles. The newly developed component has excellent structural properties and a pleasing aesthetic appearance. It also forms the basis for the use of other combinations of materials and technologies.

The Carbon Carrier is an innovative, integrated concept for use in interior structures that can be presented to large-scale vehicle manufacturers and producers of niche models and incorporated into their vehicle concepts (Fig. 9). Bertrandt and SGL regard the new structural component as an opportunity for working with users and end customers to increase their understanding of the use and behaviour of fibre reinforced plastics and, in particular, CFRP. It will also allow more sophisticated components and systems to be designed specifically for these materials. In addition, both companies gained valuable new experience for use in current and future projects.

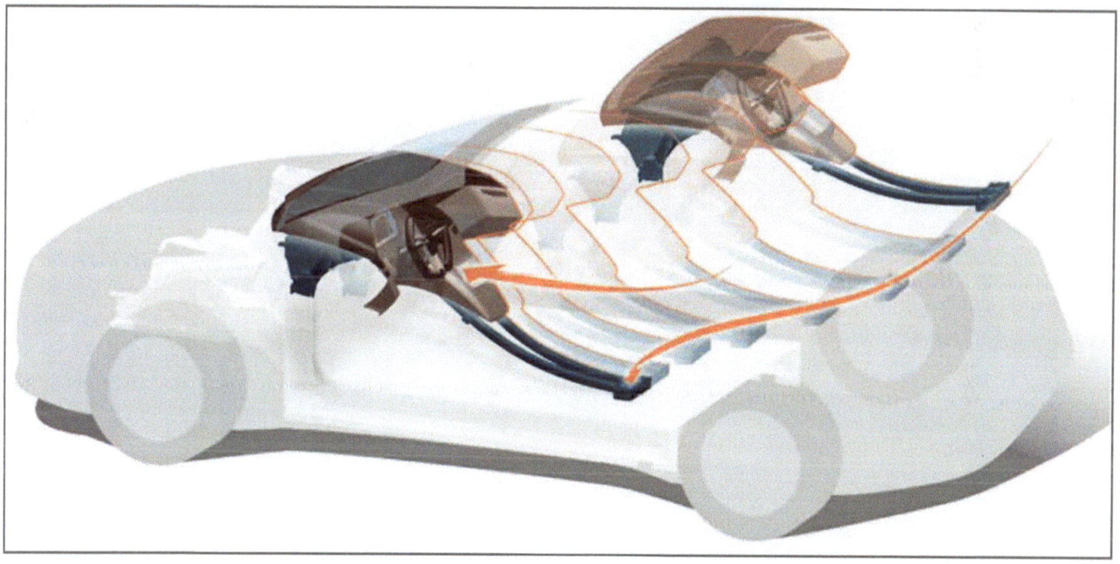

Figure 9: Installation simulation for the entire system – shown with side members

TAILORED STACKED HYBRIDS – AN OPTIMIZATION-BASED APPROACH IN MATERIAL DESIGN FOR FURTHER IMPROVEMENT IN LIGHTWEIGHT CAR BODY STRUCTURES

Alan A. Camberg[1], Ina Stratmann[1,2], Thomas Tröster[1]

[1] Pohlweg 47-49, 33098 Paderborn, Germany, Automotive Lightweight Design, Paderborn University, alan.camberg@uni-paderborn.de, www.leichtbau-im-automobil.de

[2] Seffenter Weg 8, 52074 Aachen, Germany, Chair and Institute for Rail Vehicles and Transport Systems, ina.stratmann@ifs.rwth-aachen.de, www.ifs.rwth-aachen.de

Keywords:

Hybrid Materials, Optimization, Fiber-Metal-Laminates, Material Concepts, SIMP

ABSTRACT

In latest body-in-white (BIW) concepts, engineers take into account a wider range of different materials to pursue a multi-material design approach. However, the lightweight potential of common materials like steel, aluminum or even fiber-reinforcement plastics (FRP) is limited. In keeping with the motto "the best material for the best application", a new approach for a top-down material design is introduced. With the aim to develop an application tailored material, the multi-material concept is adapted for the thickness dimension of the component. Within this contribution a new optimization-based design methodology is applied on a stiffness relevant car body part. Starting with benchmark simulations of a reference BIW structure, a critical car body component is determined by an internal energy based method and a subsequent sensitivity analysis. The identified demonstrator component is later subdivided into multiple layers and submitted to a first optimization loop in which the developed methodology varies the material parameters for each single layer. Once an optimum for the through-thickness properties of the part is found, further optimization loops with concrete material pendants and manufacturing restrictions are carried out. The result is a hybrid laminate part consisting of steel and FRP plies. To achieve a further improvement in body characteristics and lightweight, the investigated part is redesigned by the aim of topology optimization. Finally, the tailored hybrid stacks are validated in BIW simulations and compared with the reference. The optimization-based approach allows a weight reduction up to 25 % while maintaining or even improving the BIW properties.

1 INTRODUCTION

Automotive lightweight design is a considerable measure to meet the worldwide need for reducing CO_2 emissions. In the past, this led to an excessive portfolio development of conventional metals like steel or aluminum. However, the lightweight potential of common materials like steel, aluminum or even fiber-reinforcement plastics (FRP) is limited. High strength steels play a significant role in the design of safe and light car body structures. Nevertheless, the high density and buckling problems related to reduced sheet thicknesses limit the achievable mass reduction. Aluminum alloys are well known for the potential to improve the strength to weight ratio of car bodies. Nonetheless, in terms of stiffness aluminum has a clear disadvantage due to a relative low Young's modulus. Even FRP components, which have superior light-weight characteristics, show limitations for the car body design, as catastrophic failure or high production costs. Hybrid materials combine metals and FRP in a manner to

K. Dröder et al (eds), *Technologies for economical and functional lightweight design*, Zukunftstechnologien für den multifunktionalen Leichtbau, https://doi.org/10.1007/978-3-662-58206-0_12

offset the drawbacks of every single material and reach an optimum of mechanical properties and costs. Nonetheless, the achievable lightweight potential of such materials heavily depends on the loading situation, geometry or cross section of the chosen material design.

To account for these limitations, a new approach is necessary. The "LHybS" (Lightweight Design with Novel Hybrid Materials) project aims high to design a material, which different than usual is not developed in a bottom-up way, but rather in a top-down manner. Within the scope of the project a new optimization-based process is developed, which provides for the first time a methodology that allows the design of sophisticated requirement-optimal layered materials. The through thickness property profile of the material takes into account not only loading dependent requests, but also demands that are related to the direct application of the material in the vehicle. The goal of the project is to develop a lightweight hybrid material with processing characteristics similar to materials conventionally used in body-in-white production plants [1].

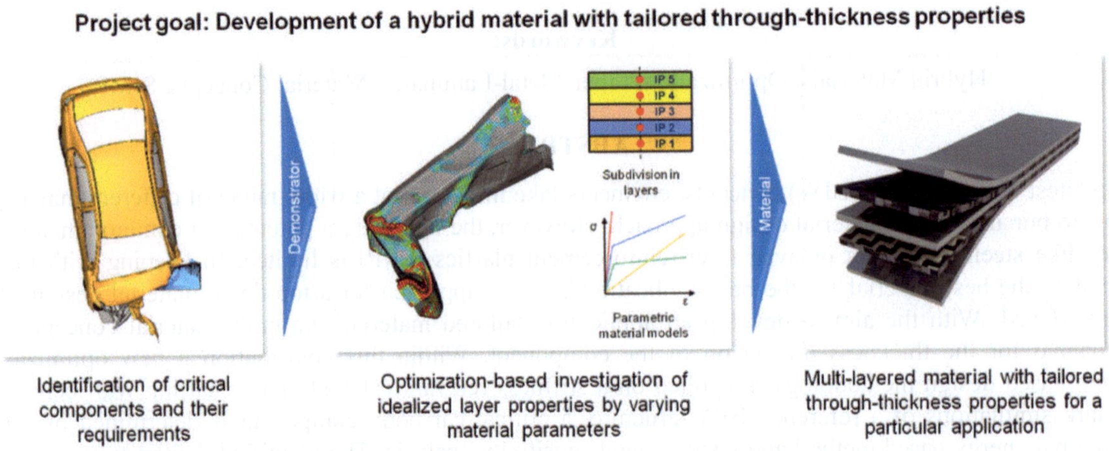

Figure 1: The approach of the LHybS project: from the application to the material profile [1]

2 PROPOSED OPTIMIZATION METHOD FOR TAILORED HYBRID STACKS

2.1 LHybS Ordered Multi-Material Interpolation

Structural optimization has been studied extensively over recent decades [2]. Among a variety of approaches for continuum topology optimization, the density-based "Solid Isotropic Material with Penalization" (SIMP) [3][4] method had gained a wide popularity and has been successfully integrated into commercial software, such as, for example OptiStruct or Tosca [5]. In SIMP, every single element within the defined problem space is coupled to an elasticity-density-scaling function with the aim to control the elemental mass and stiffness tensor. The elemental material properties are usually expressed by using a power-law function (1), where ρ is the notional density variable, E_0 the upper bound of the elastic modulus, and p the penalty factor. The possibility of an element-wise variation of density and elasticity allows finding a load-optimal material distribution. However, state-of-the-art topology optimization has been implemented for single-material design problems what limits the achievable weight savings by this method.

$$E(\rho) = \rho^p E_0, \qquad \rho \in [0; 1] \wedge p > 1 \tag{1}$$

To overcome these limitations a modified interpolation scheme for multiple materials was introduced in [6]. The proposed extension of the classical SIMP can be formulated for two materials as (2), where E_1 and E_2 are the elastic modulus for material 1 and material 2, respectively. By letting $E_2 = 0$ the classical SIMP definition from (1) can be recovered.

$$E(\rho) = \rho^p E_1 + (1 - \rho^p) E_2, \qquad \rho \in [0; 1] \land p > 1 \tag{2}$$

In general, the extension proposed in [6] requires $(m-1)$ design variables for each element to take into account m materials in the optimization. As a consequence the computational cost increases significantly.

An alternative formulation can be found in [7], where an Ordered Multi-Material SIMP Interpolation is proposed to solve multi-material optimization problems without introducing any new variables. In the proposed method, the multiple materials are sorted in an ascending order of the normalized density variable ρ_i :

$$\rho_i = \frac{\rho_i^{tech}}{\rho_{max}}, \quad i = 1, 2, 3, \ldots, m \tag{3}$$

Where ρ_{max} is the maximum density and m the number of all candidate materials. Introducing additional coefficients A_E and B_E, a scaling and a translation coefficient, respectively, and using the normalized density variable ρ_i, the single-material power function (1) can be extended as:

$$E_e(\rho_e) = \rho_e^p A_E + B_E, \qquad \rho_e \in [\rho_i, \rho_{i+1}] \land p > 1 \tag{4}$$

$$\text{where} \quad A_E = \frac{E_i - E_{i+1}}{\rho_i^p - \rho_{i+1}^p} \land B_E = E_i - \rho_i^p A_E \tag{5}$$

Where E_i and E_{i+1} are the elastic modulus of ascending ordered material i and $i+1$, respectively [7].

While the Ordered Multi-Material SIMP Interpolation in (4) is able to describe monotonically increasing trends of elasticity in respect to the normalized density (3), this may not always be the case for real lightweight applications. By a way of example, common materials taken into account for multi-material design, such as steel, aluminum and CFRP do not exhibit a monotonic increase of these properties. For this reason the approach presented in [7] is revisited in the present paper. Instead of a strictly monotonically increasing function, a coupled exponential function (6) is used to describe the relation between the normalized material properties of common lightweight materials. Analogous to [7], the LHybS Ordered Multi-Material Interpolation expresses the properties of candidate materials in respect to the independent discrete normalized density variable ρ_i (3). The elastic modulus is regarded as a continuous function with respect to density, obtained from a fit of normalized candidate material data by a two-term exponential function:

$$E(\rho_i) = a * e^{b*\rho_i} + c * e^{d*\rho_i}, \qquad \rho_i = \frac{\rho_i^{tech}}{\rho_0}, \quad i = 1, 2, 3, \ldots, m \tag{6}$$

$$E_i = \frac{E_i^{tech}}{E_{max}}, \quad i = 1, 2, 3, \ldots, m \tag{7}$$

Where ρ_0 and E_0 are the maximum density and elastic modulus of the candidate materials, respectively, and a, b, c, d are fitting parameters of the exponential functions. For simplification, all materials are assumed as isotropic and have a Possion ratio equal to 0.3.

By applying Equation (6) on a set of common automotive lightweight materials, whose properties are listed in Table 1, the interpolation curve of normalized elastic modulus with respect to normalized density shown in Figure 2 is obtained.

Table 1: Material properties of considered automotive lightweight materials and corresponding fitting parameters of Equation (6)

Material	ρ_i^{tech} [ton/mm^3]	E_i^{tech} [MPa]	ρ_i	E_i
CFRP	1.488e-09	133e+03	0.19	0.63
GFRP	1.996e-09	45e+03	0.26	0.21
Aluminum	2.693e-09	70e+03	0.34	0.33
Steel	7.829e-09	210e+03	1.00	1.00
Fit. param. Eq.(6):	a	b	c	d
	1.977e+13	–165.8	0.1571	1.852

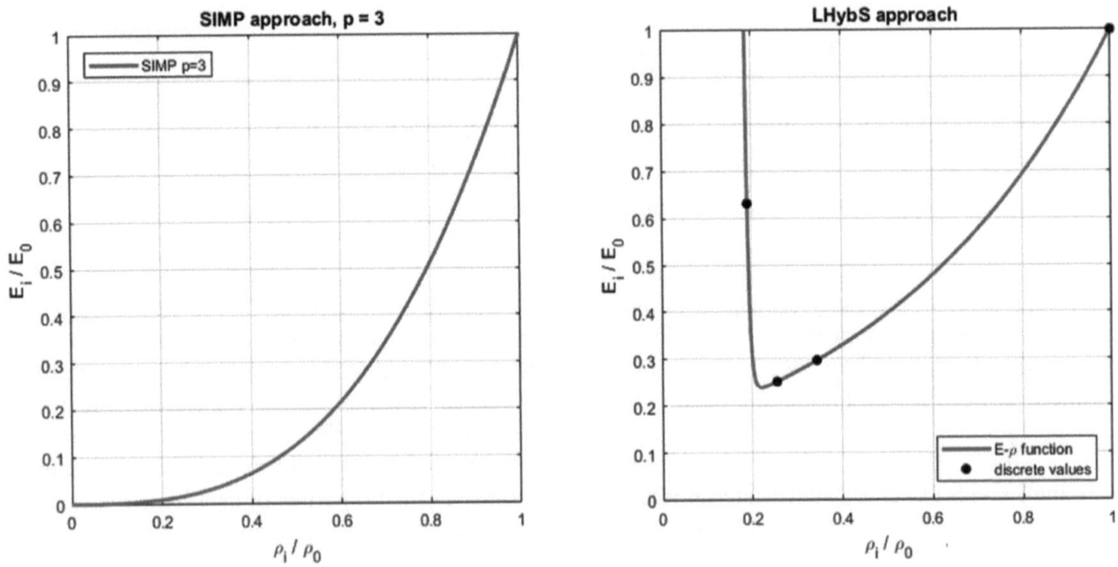

Figure 2: Comparison between the classical SIMP and the LHybS approach for scaling the stiffness and density

While the LHybS Ordered Multi-Material Interpolation is able to describe the trend of normalized properties of real lightweight materials, the zero density for void representation typically found in SIMP is lost. However, with the introduced interpolation method we rather want to choose between candidate materials than between solid or void. An alternative method for solving this kind of material optimization problems was introduced in [8] as Discrete Material Optimization (DMO), where the element constitutive matrix is expressed as a weighted sum of constitutive matrixes of each candidate material. This conception however is disadvantageous, as each candidate material demands its own design variable, while the LHybS approach uses solely the normalized density as design variable.

2.2 Optimization methodology

The LHybS Ordered Multi-Material Interpolation can be applied directly on multi-layered shell element structures. The only difference to topology optimization is that the material parameterization is invoked at the level of shell integration point layers instead on solid element level. Further, to ensure manufacturability and uniform properties at stack layers, the integration points of all elements on a given layer are clustered into plies. That in turn enables to realize layers of different properties within the investigated component and the design of an application-optimal hybrid material. However, the

LHybS interpolation approach is still limited to optimize material distribution problems and additional design variables, such as material thickness or orientation cannot be changed during one optimization run.

To account for this limitation, the developed material design process consists of two main optimization runs. After benchmarking the monolithic reference design, the chosen part is subdivided into N^l integration point layers k, while the overall wall thickness t of the part remains unchanged. The first optimization run follows in which the algorithm optimizes the material parameters for each ply following Equation (6). The optimization problem of the first run of compliance minimization can be expressed as in (8), where C is the structural compliance; K, u and P are the global stiffness matrix, displacement vector and force vector, respectively; φ_M is the mass fraction; M and M_0 is the mass of the current and the reference design, respectively; and the superscripts e and l refer to "element" and "layer".

Once a global optimum was found by choosing multiple sampling points, the still idealized material properties of each layer are compared with a material database and replaced by concrete pendants by taking into account real material properties, as for example anisotropy. Within the last optimization run the layers are optimized in terms of thickness and material orientation. Following (9), the objective is again compliance minimization, but here, since the multi-material distribution is known, only layer thicknesses t^l and orientations θ^l are optimized. The boundaries for t^l and θ^l are set in a way to meet manufacturing limitations.

Figure 3 depicts a flowchart for the task of multi-material optimization defined in the introduced optimization methodology.

$$\begin{gathered}
Objective\colon \min_{\rho} C = u^T K u \\
Subject\ to\colon \begin{cases} M \le \varphi_M M_0 \\ Ku = P \\ K = \sum_{m=1}^{N^e} \int_V K_m^e, \quad K^e = \sum_{k=1}^{N^l} \int_V \left(B_k^l\right)^T C_k^l B_k^l \, dV \\ M = \sum_{m=1}^{N^e} M_m^e, \quad M^e = \sum_{k=1}^{N^l} V_k^l \rho_k^l \end{cases}
\end{gathered} \tag{8}$$

$$\begin{gathered}
Objective\colon \min_{t,\theta} C = u^T K u \\
Subject\ to\colon \begin{cases} M \le \varphi_M M_0 \\ C \le C_0 \\ \theta_{min}^l \le \theta^l \le \theta_{max}^l \\ t_{min}^l \le t^l \le t_{max}^l \end{cases}
\end{gathered} \tag{9}$$

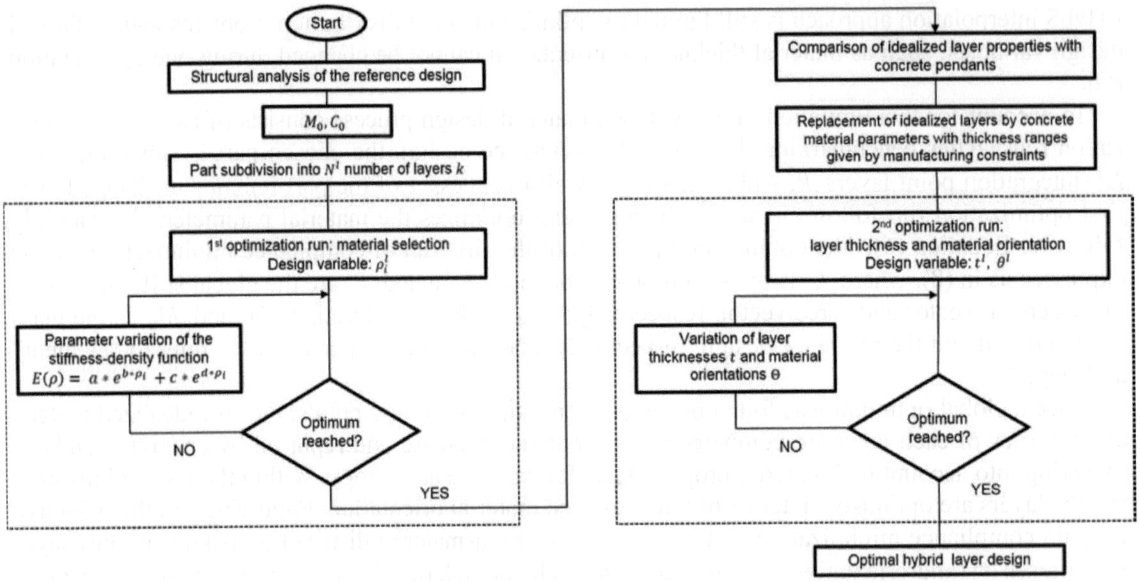

Figure 3: Flowchart of the developed method for an optimization-based hybrid design

3 REFERENCE STRUCTURE BENCHMARK

As a basis for the material development the thyssenkrupp InCar® plus model is used. The InCar® project started back in 2006 as a customer-independent car body structure project. The goal of the InCar® project was to develop a base for the potential analysis of new material grades and processing concepts. A further development and an adjustment to newest crash safety demands was realized in 2014 as InCar® plus. The steel-intensive structure of the InCar® plus represents the current state of the art by utilizing the usage of hot formed, ultra-high and high strength steels [9], [10]. The stiffness and noise-vibration-harshness (NVH) properties of the InCar® plus BIW are investigated in OEM specific methods to set a benchmark for the development of the new tailored hybrid materials. The benchmark analysis is carried out by tests described below.

3.1 Global bending stiffness

The global bending stiffness of the body-in-white is investigated by boundary conditions adapted from [11]. For this purpose, two static forces are distributed and applied to the front seat attachment points while the BIW is constrained in a static determined manner at the front and rear damper carriers. To suppress the influence of local structure stiffness on the determined global value, an evaluation approach from [12] and [13] is used. By this approach the maximum rocker deflection and a corrected reference line given by the clamping points is taken into account. The global bending stiffness follows from equation (10).

$$c_B = \frac{\sum F_z}{u_{z,max.rocker\ corr.}} \tag{10}$$

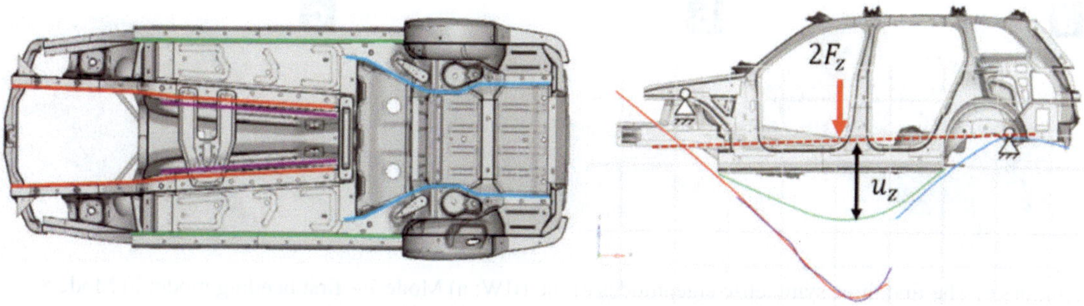

Figure 4: Evaluation of the global bending stiffness – loading conditions and the resulting deflection lines of the underbody

3.2 Global torsion stiffness

The determination of the global torsion stiffness is based on a method proposed in [14]. Here, the BIW is loaded by applying a couple of opposing forces on the strut towers while constraining the rear damper carriers. To suppress a possible bending of the BIW structure, an enhancement of the boundary conditions from [12] is introduced. Hereto, an additional constraint at the middle of the front bumper beam is defined to ensure a global torsion of the body. The torsion stiffness is then calculated from the twist angle between the strut towers and the introduced torque, see equation (11).

$$c_T = \frac{M_x}{\Phi_x} \left[\frac{Nm}{deg}\right] \tag{11}$$

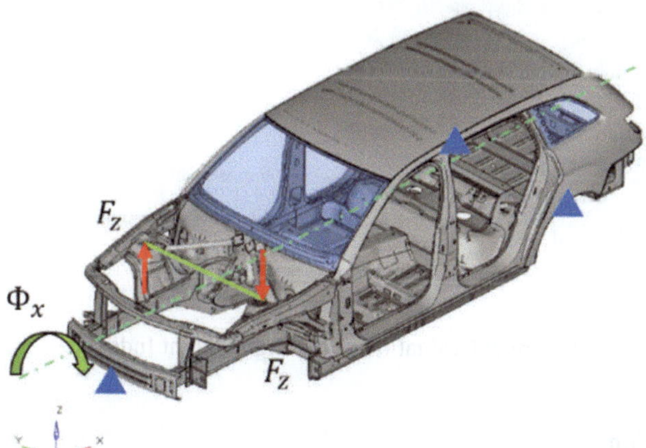

Figure 5: Loading and boundary conditions used for the determination of the global BIW torsion stiffness

3.3 Modal analysis

To characterize the NVH properties of the BIW an eigenvalue analysis is performed. As reported in [15] the eigenfrequencies should be analyzed in free-free boundary conditions. It should be note that due to the chosen boundary conditions the first six eigenmodes are related to the rigid body motion and are not relevant for the BIW design. Therefore, the first three symmetric modes are evaluated. The 7th and 9th mode correspond to the first and second bending mode respectively. The 8th mode corresponds to the first torsion mode, see Figure 6.

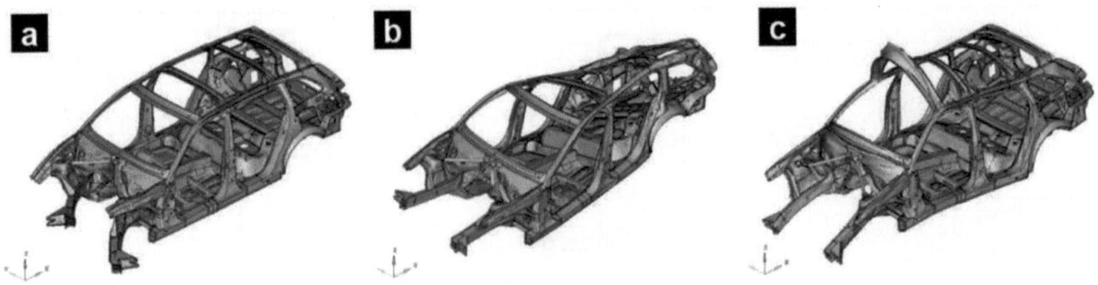

Figure 6: The first three symmetric eigenmodes of the BIW: a) Mode 7 – first bending mode; b) Mode 8 – torsion mode; Mode 9 – second bending mode

3.4 Lightweight Index

To quantify the trade-off between stiffness and lightweight the so-called "Lightweight Index" L is introduced. The Lightweight Index describes the purposeful material and package utilization within the car body structure and follows from equation (12). As shown in the equation, the global torsion stiffness is crucial for the lightweight index. However, as reported in [16], an improvement of the torsion stiffness usually leads concurrently to an improvement of other stiffness properties of the BIW.

$$L = \frac{m_{RK}}{c_T * A} \left[\frac{kg}{Nm/\deg * m^2}\right] * 10^3 \tag{12}$$

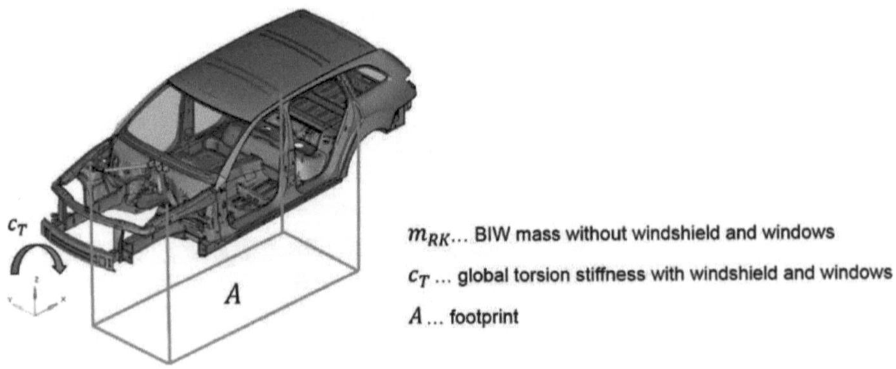

Figure 7: Evaluation of the Lightweight Index L

3.5 Component selection

The investigated body-in-white properties result from an interaction of all car body components. Every single part accomplish in terms of a body structure a particular function and contributes with its specific characteristics to the global body-in-white properties. Depending on the external load, singular parts and component groups could be more or less involved in the deformation resistance of the body structure. To classify the BIW parts into load case relevant groups and evaluate their importance, a strain energy based method introduced in [1] is used. The assessment of a potential component is further supplemented by a sensitivity analysis to identify structural parts with a significant impact on the global stiffness. With the aim to reduce the computational effort of the sensitivity analysis, the results from the prior step are used for a domain reduction. Thus, the sensitivity is analyzed only for ten components with highest prorate strain energy values. By increasing the elasticity modulus of these particular components, the influence of an improved material on the global characteristic can be evaluated. In that manner, it can be ensure that the material improvement will be realized on parts that show a high relevance for the global BIW properties and so, additional costs can be justified by better overall vehicle characteristics [1].

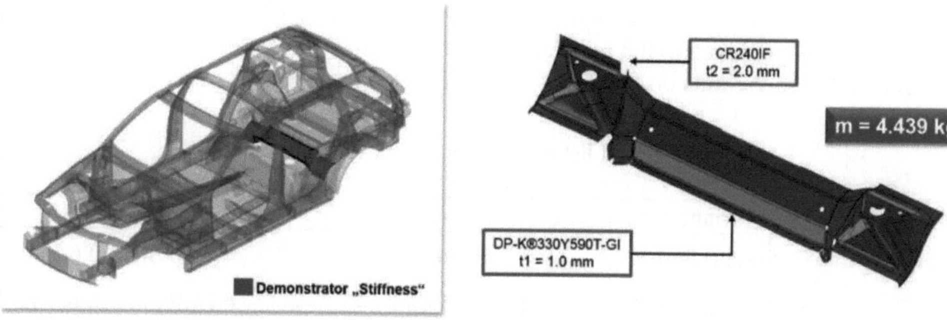

Figure 8: Selected demonstrator part: InCar® plus rear cross member

As a result of the described procedures the rear cross member is selected for a tailored material design trail, see Figure 8. In addition to mechanical requirements that have to be satisfied by the new material, there are secondary demands for this particular part, as for example damping characteristics, corrosion protection or joinability that have to be taken into account within the material design process. These needs are investigated and defined in a specific requirement catalogue.

4 BIW COMPONENT OPTIMIZATION

4.1 Hybrid material design

Since the global stiffness is directly affecting the "Lightweight Index" L of a car body, see equation (12), the load case "global torsion" is defined as design relevant. So, the investigation of an optimal layer-wise material selection within the selected component is carried out for "global torsion". To ensure a symmetric material design, one left and one right twisting "global torsion" load case is defined, what leads to a multi-objective optimization.

To access a layer-wise variation of the part material parameters, the reference part is divided into seven layers. To find the optimal material for each layer, the approach from section 2.1 is used. Based on equation (6), the algorithm varies the density and calculates the elastic modulus. It changes automatically the material parameters in each layer which allows finding the optimal material distribution for the hybrid laminate. Possible materials are steel, aluminum, GFRP and CFRP. As mentioned in section 2.1, both fiber-reinforced plastics are implemented as an isotropic material to reduce the computational effort within the first optimization loop. The optimization results in a stacking sequence that consists of steel top layers and a multiple layer CFRP core.

In the second optimization step, the variation of the thickness of each layer and the orientation of every FRP-layer is automatized, while complying with the following constraints. The minimal deliverable thickness of steel ranges from 0.5 mm to 0.6 mm, the thickness for one layer of unidirectional CFRP is set to 0.05 mm and the thickness for one layer of CFRP-fabric is set to 0.35 mm. This second step includes a loop, in which the routine automatically varies the thickness and the orientation of the CFRP in 5° steps. Furthermore, it checks after each variation whether the parameters of the new hybrid laminate improved compared to the original steel component. These parameters are global torsion stiffness and mass.

Due to selected FRP pre-products and available material thicknesses, different solutions exist and three alternatives of the hybrid laminate are listed in Table 2. The first variant (A) includes 0.5 mm thick steel top layers and a CFRP core consisting of 5 unidirectional layers. Based on the different material orientations of the CFRP, it is possible to maintain the torsional stiffness while simultaneously reducing the mass by 25 %. In variant B a change from unidirectional CFRP to CFRP-fabric is realized, what reduces the achievable mass reduction by 2 %. At the same time, the torsional stiffness increases slightly (+ 0.37 %). The main difference between variant B and C is the thickness of the steel sheet. In the latter variant C, the thickness is changed from 0.5 mm to 0.6 mm. Therefore, the thickness of the CFRP-fabric can be reduced from 1.05 mm to 0.7 mm (2 layers instead of 3 layers CFRP-fabric) and as a consequence the mass of the component increases. Nevertheless, with this design the mass can be still reduced by 14.6 % compared to the reference. The torsional stiffness remains unaffected.

The influence of these variants on further BIW properties is investigated in section 5.

Table 2: Result list of different hybrid stack variants

	Variant A	Variant B	Variant C
Layer 1	Steel 0.5 mm	Steel 0.5 mm	Steel 0.6 mm
Layer 2	CFRP 0.2 mm 40°	CFRP 0.35 mm +45°/–45°	CFRP 0.35 mm +45°/–45°
Layer 3	CFRP 0.1 mm 145°	CFRP 0.35 mm +45°/–45°	CFRP 0.35 mm +45°/–45°
Layer 4	CFRP 0.3 mm 180°	CFRP 0.35 mm +45°/–45°	Steel 0.6 mm
Layer 5	CFRP 0.2 mm 145°	Steel 0.5 mm	-
Layer 6	CFRP 0.1 mm 140°	-	-
Layer 7	Steel 0.5 mm	-	-
Global torsional stiffness	+0.03 %	+0.37 %	+0.37 %
Component mass	–25 %	–23.2 %	–14.64 %
Lightweight Index	–0.42 %	–0.60 %	–0.49 %

4.2 Geometry design

As reported in [17] the deep-drawing of stacked hybrid laminates can lead to complex failure modes, as for example delamination, buckling or wrinkling, which, however, can be avoided by an adapted part design and special processing techniques. To meet these new material-related manufacturing requirements a redesign of the part geometry is necessary.

Figure 9: Topology optimization of the rear cross member – a) reference design; b) package model of the available space; c) result of the optimized topology

At the outset, an investigation of the optimal part design within the available package is carried out by the aim of SIMP-based topology optimization. As depicted in Figure 9 c, the torsion-optimal structure consists of only few bars forming a truss in the rear underfloor area of the car body.

Based on the optimal component geometry, space restrictions and forming limitations a new part geometry is developed. Additionally, in order to avoid changes in the joining sequence, the quantity and location of connection points to adjacent parts are kept unchanged. As shown in Figure 10 b, the new part geometry uses almost the entire free package and avoids leaps in the drawing ratio. That guarantees a high stiffness and manufacturability of the new hybrid part at the same time. The possibility of a later subdivision into three separate components is also provided by this geometry design.

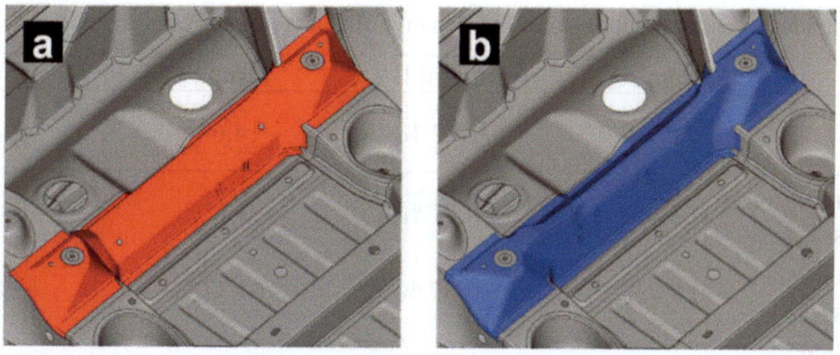

Figure 10: a) the reference rear cross member design; b) new hybrid-optimal rear cross member design

Strictly, the optimization-based material design process introduced in section 2.2 should be repeated for the new part geometry, since the changed geometry can lead to a different optimal layer design. Nevertheless, as a proof of concept the stacking sequence of variant C is applied to the new part geometry (variant D, Table 3). Due to a higher overall surface of the new part geometry the achieved mass reduction is marginal. However, in comparison to the reference geometry, it is possible to increase the torsional stiffness by 5.89 % and decrease the lightweight index by –5.38 %. A further improvement in lightweight potential can be gained by performing a full material design process. This will be the subject of future work.

Table 3: Result list of the optimized new hybrid rear cross member design

Variant D		Change in BIW Properties		
		Torsional stiffness	Mass	Lightweight Index
Layer 1	Steel 0.6 mm	+5.89 %	–0.45 %	–5.38 %
Layer 2	CFRP 0.35 mm +45°/–45°			
Layer 3	CFRP 0.35 mm +45°/–45°			
Layer 4	Steel 0.6 mm			

5 VALIDATION

Finally, the developed materials undergo a series static and dynamic BIW stiffness simulations. The results are compared with the InCar® plus reference to highlight the potentials of the tailored hybrid material design. In all cases, a redesign of the through thickness properties by a tailored hybrid stack leads to a weight reduction while maintaining or even improving the overall vehicle properties, compare Table 4.

Table 4: Comparative assessment of static and dynamic stiffness properties of the tailored hybrid stacks

	Variant A	Variant B	Variant C	Variant D
Mass	–25 %	–23.2 %	–14.64 %	–0.45 %
Bending stiffness	–0.71 %	–0.71 %	–0.72 %	–0.40 %
Torsional stiffness	+0.03 %	+0.37 %	+0.37 %	+5.89 %
Lightweight index	–0.42 %	–0.60 %	–0.49 %	–5.38 %
1st bending mode	unchanged	unchanged	unchanged	+0.02 %
1st torsion mode	+0.27 %	+0.40 %	+0.33 %	+2.32 %
2nd bending mode	+0.07 %	+0.07 %	+0.04 %	+0.02 %

6 CONCLUSIONS AND OUTLOOK

Numerical optimization processes introduced within the scope of this contribution allowed to design novel requirement-optimal hybrid materials directly within a BIW. These materials led to a weight reduction of up to 25 % and hold further light-weight potentials. The developed materials were hitherto investigated only in numerical simulations and coupon based experimental tests. The experimental validations on real component geometries are still pending. The chosen demonstrator component will be subjected to a series of crash, stiffness and durability tests to point out the qualification for automotive applications. Further, the developed approach will be consolidated to a user-friendly numerical tool to provide usability in the serial development of mechanical systems. Distal the automotive field the approach could be used for other disciplines, as for example in the aeronautical or energy sector.

ACKNOWLEDGEMENTS

The authors would like to give thanks to the European Regional development Fund and the State of North Rhine-Westphalia for funding the research project "LHybS" (www.tecup.de/lhybs/) and the lead partner Projektträger Jülich (PTJ), Forschungszentrum Jülich GmbH. Sincere thanks also to all of the industrial partners, especially to the thyssenkrupp Steel Europe AG for providing the InCar® plus model.

LHybS 2014 EFRE.NRW Investitionen in Wachstum und Beschäftigung

EUROPÄISCHE UNION Investition in unsere Zukunft Europäischer Fonds für regionale Entwicklung

This project is founded by the European Union and the State North Rhine-Westphalia.

REFERENCES

[1] A. A. Camberg, T. Tröster, Optimization-based material design of tailored stacked hybrids for further improvement in lightweight car body structures. *Proc Hybrid 2018, 3rd Int Conf Hybrid Mat & Struct (Eds. J. M. Hausmann, M. Siebert, A. von Hehl), Bremen, Germany, 18-19.04.2018,* DGM e.V.

[2] X. Guo, G.-D Cheng, Recent development in structural design and optimization. *Acta Mech Sinica*, **26(6)**, 2010, pp. 807–823.

[3] M.P. Bandsøe, Optimal shape design as a material distribution problem. *Struct Optim*, **1(4)**, 1989, pp. 1–16.

[4] M. Zhou, G.I.N. Rozvany, The COC algorithm, part II: topological, geometry and generalized shape optimization. *Comput Methods Appl Mech Eng*, **89(1)**, 1991, pp. 309–336.

[5] C. Li, I.Y. Kim, Multi-material topology optimization for automotive design problems. *Proc IMechE Part D: J Automobile Engineering*, 2017. **DOI: 10.1177/0954407017737901**.

[6] M.P. Bandsøe, O. Sigmund, Topology optimization: theory, methods and applications, Springer Vieweg, Berlin, Heidelberg, 2013. **DOI: 10.1007/978-3-662-05086-6**.

[7] W. Zuo, K. Saitou, Multi-material topology optimization using ordered SIMP interpolation, *Struct Multidisc Optim*, 2016. **DOI: 10.1007/s00158-016-1513-3**.

[8] J. Stegmann, E. Lund, Discrete material optimization of general composite shell structures, *Int J Numer Meth Engng,* **62**, 2005, pp. 2009–2027. **DOI: 10.1002/nme.1259**.

[9] A. Grüneklee et al., Das Projekt ThyssenKrupp InCar plus. Lösungen für automobile Effizienz. *ATZ Extra,* **Oktober 14**, Springer Vieweg Verlag, Wiesbaden, 2014.

[10] B. Pohl, S. Rützel, The all new Opel Insignia – Body development and manufacturing, *Conf Proc Aachen Body Engineering Days 2017 (Ed. L. Eckstein), Aachen, Germany, 18–19.09.2017,* ika RWTH Aachen, 11, 2017, pp. 205–230.

[11] J. Helsen, L. Cremers, P. Mas, O. Sas, Global static and dynamic car body stiffness based on a single experimental modal analysis test. *Proc Int Conf NVH Eng (ISMA2010) Including USD2010 (Ed. P. Sas), Leuven, Belgium, 20–22.09.2010,* Katholieke Universiteit Leuven, Department of Mechanical Engineering, 2010, pp. 2505–2521.

[12] O. Danielsson, A. Gonzálcz Cocaña, Influence of Body Stiffness on Vehicle Dynamics Characteristics in Passenger Cars, Chalmers University of Technology, Göteborg, 2015.

[13] B. Rediers, B. Yang, V. Juneja, Static and dynamic stiffness – One test both results. *Proc 16th Int Modal Analysis Conf IMAC (Ed. A. L. Wicks), Santa Barbara, Calif. USA, 02–05.02.1998*, Society for Experimental Mechanics, Inc., 1998, pp. 30–35.

[14] D. E. Malen, Fundamentals of Automobile Body Structure Design, SAE International, 2011.

[15] M. Kiani, H. Shiozaki, K. Motoyama, Simulation-based design optimization to develop a lightweight Body-In-White structure focusing on dynamic and static stiffness. *Int J Vehicle Design*, **Vol.67**, No.3, 2015, pp. 219–236.

[16] H.-H. Braess, U. Seiffert (Eds.): Vieweg Handbuch Kraftfahrzeugtechnik, Springer Vieweg, Berlin, Heidelberg, 2013.

[17] A. A. Camberg, K. Engelkemeier, J, Dietrich, T. Heggemann, Top-Down Design of Tailored Fibre-Metal Laminates. *lightweight.design*, **2|2018**, Springer Vieweg Verlag, Wiesbaden, 2018.

A MANUFACTURE CONSTRAINED DESIGN METHODOLOGY APPLICATION FOR A TAILORED FORMING HYBRID COMPONENT

Renan S. Siqueira[1], Roland Lachmayer[2]

[1] Leibniz Universität Hannover, siqueira@ipeg.uni-hannover.de
[2] Leibniz Universität Hannover, lachmayer@ipeg.uni-hannover.de, www.ipeg.uni-hannover.de

Keywords:

Tailored forming, Hybrid components, Topology optimization, Manufacturing restrictions

ABSTRACT

Components with high performance, such as light weight, long life-cycle, good cost-benefit and other specific properties, are a constant goal of the industry. To achieve such performances, a constant progress is required in both manufacturing and design processes. These two areas must develop together and connected, at the same time that one challenges the other. Tailored Forming is a new manufacturing technique in development, which consists in a process chain to create massive hybrid material structures made of two different metals. This new technology presents a new range of manufacturing restrictions and requires a suitable design methodology to deal with the multi-material problem. The objective of this paper is to present an overview of the methodology that is currently being used to generate feasible designs for Tailored Forming. This methodology consists in an optimization tool that searches for an optimal material distribution and generates a first concept for the component, and a parametric analysis that generates a large solution space with a finer ready-to-use design result. Some examples of the applicability of these tools are here showed, with focus on a current Tailored Forming demonstrator, which is a hybrid shaft. Despite the simplicity of this application, it involves a challenging implementation, due to the manufacturing restrictions present. At the end, a final design is presented, which is not only suitable for the manufacturing process but also raises the advantages that this technology provides.

1 INTRODUCTION

With the growing needs of the industry for materials and components with higher performance, new manufacturing techniques and design methodologies arise. Here, a combination between manufacture and design is essential, since one has a direct influence on the other. Thus, although generic design methodologies have been used to achieve optimized geometries, such as minimization of weight, the presence of manufacture restrictions are constantly neglected, ignoring this natural mutual influence. For that reason, it is fundamental that design and manufacture methods stay attached in pairs, where the manufacturing technique has a specifically adapted designing tool.

In this study, we want to present the current research state made in the field of design for hybrid components generated through Tailored Forming [1], theme of a collaborative research at the Leibniz University. Tailored Forming is a process chain with the objective of creating hybrid solid components by using semi-finished work-pieces. One of the advantages that this process offers is the ability of combining different metals, such as Steel and Aluminum. This enables components with stress adjusted profiles and, consequently, parts that can handle the same amount of load with less weight, for example.

A new technology such as Tailored Forming, brings not only new possibilities, but also a series of restrictions that must be considered in the design phase. For instance, the multi-material design brings a big amount of degrees of freedom, requiring a systematic methodology for its construction, as seen

K. Dröder et al (eds), *Technologies for economical and functional lightweight design*, Zukunftstechnologien für den multifunktionalen Leichtbau, https://doi.org/10.1007/978-3-662-58206-0_13

in the works of Kleemann [2] and Brockmöller [3]. The use of construction catalogues [4] and knowledge-based systems [5] has also been served as support for such designs and its use for Tailored Forming has been also investigated [6]. Not only in the methodological point of view, the design parameters must also be carefully studied and analyzed, in order to explore all the advantages that it brings in the most effective way. So, a state-of-the-art research for multi-material design was executed, in order to find methods that take in consideration all the manufacturing restrictions. Based on that, we search for the answer of two main questions:

- How the material should be distributed in the domain space?
- How the joining zone between the two metals should be designed?

For the first question, topology optimization techniques were analyzed and evaluated, in order to serve as solution generator. Here, a new method is proposed based on evolutionary algorithms. This new method allows a direct control of the manufacturing constraints in the final design, which is seen often as one of the disadvantages of topology optimization [7]. Another challenge here is the presence of multi-materials in the optimization task, which enlarges the complexity of the problem. The method presented here makes use of heuristic approaches to generate fast and practical solutions. As result, the concept for some classical problems were generated using different manufacturing constraints and, at the end, the method was used for a real application.

For the second issue, about the joining zone geometry, a finer optimization method is required to generate the final model already in a CAD software, where it can be directly sent to manufacture. For that, it was used a parametric approach [8], where it is made a characterization of the component based on the concept found in the topology optimization step, also considering the manufacture constraints involved. This allows the generation of a large number of design solutions with simulations results. All this information is saved as a databank, where engineers can analyze and select the most suitable solution.

2 METHODOLOGY

2.1 Tailored Forming

Tailored Forming is a new manufacturing technique that allows the construction of hybrid high performance components, being the aim of the Collaborative Research Project (CRC) 1153 [9] established at the Leibniz University of Hannover. This technology consists in a process chain that combines different manufacturing techniques in order to produce functional final parts. The process includes the following sequence: generation of the separated mono-material parts; joining process performed by friction welding, laser welding or compound profile extrusion to generate semi-finished hybrid workpieces; metal forming process through high temperatures, such as cross wedge rolling, forging or impact extrusion; and finally machining and heat treatments are executed to obtain the final piece [10]. Figure 1 shows a diagram with this sequence.

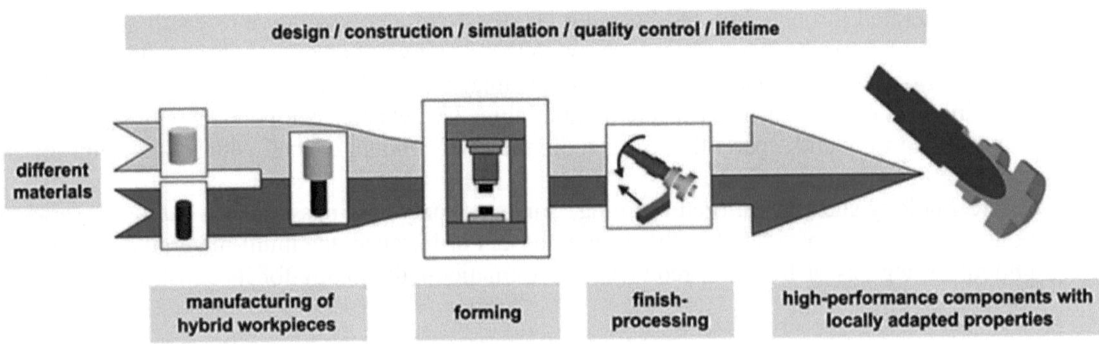

Figure 1: Process chain of Tailored Forming technology (CRC 1153) [9]

So, in this initial phase of research, the design process normally starts with the concept generation for a component based on the available forming technology. However, in order to expand the possibilities, the work of design and manufacture must be parallel, so that new optimized candidates can be delivered. This requires a design process that take into consideration the limitations of the technique, which is the focus of this paper. As an example, we take the design of a shaft, which is one of the demonstrators from the CRC 1153, seen in (Figure 2). This shaft is manufactured through a laser welding of the mono-material workpieces, followed by a process of impact extrusion or cross wedge rolling, finishing with a machining technology step [10].

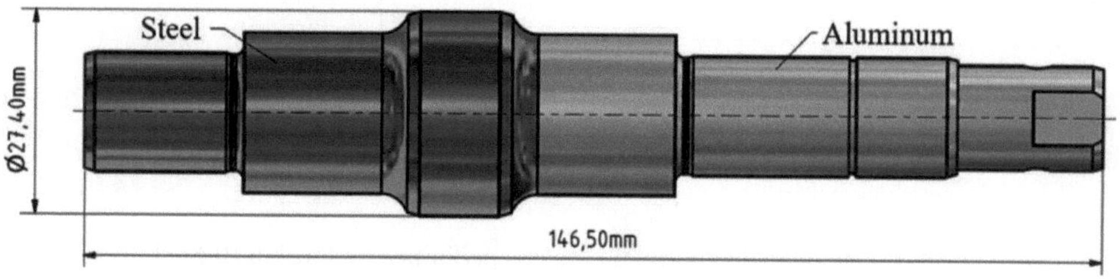

Figure 2: Hybrid shaft example manufactured by Tailored Forming

The shaft presented in Figure 2 is made of steel and aluminum, with the intent of having steel at the region where the mechanical load is more intense (section with higher diameter). Although in this example the material distribution is mostly predetermined, it presents already challenges to find which design method should be used to reach this result, which will be seen in next section. The geometry of the interface between the two materials must be also analyzed and optimized in a way that the manufacturing constraints are obeyed.

2.2 Topology Optimization

Topology Optimization is a wide scientific area that includes a large number of publications, but still offers many challenges in the application field [11]. The pioneer work of Bendsoe [12], where SIMP (Solid Isotropic Material with Penalization) theory is introduced. This theory was deeply investigated in [13] and also expanded in last decades, such as the development of filters [14] and use of stress-based approaches [15]. Many specific applications were researched in meanwhile, such as vibration problems [16], use for additive manufacturing [17] and thermal problems [18]. In our current study, however, a very specific case is under investigation, which is the use of topology optimization for multi-material components and manufacturing restrictions. These are areas where the scientific community hasn't agreed in an ultimate solution yet.

The isolated problem of multi-materials was also largely investigated. The first works in this field also worked as an extension of SIMP [13]. Due to the high number of design variables in these problems, the method becomes complex with the use of more materials. Some other adaptations were made to avoid these problems, as seen in [19, 20, 21]. Other recent approaches can be found in [22] and [23], where different solving techniques are used to deal with the optimization problem. However, although these studies give an essential ground research in the field, it is seen that they don't focus on a specific manufacturing technique, but in a general formulation. This approach was implemented in [24] for a 3-D printer, but with no manufacturing constraints. With that, multi-material designs achieved in topology optimization methods are not so practical for other current manufacturing techniques available, such as Tailored Forming.

In the field of manufacturing constraints, a similar scenario is found in literature [25]. Some implementations for specific manufacturing restrictions can be found, as seen in [26]. There are also applications with different optimization methodologies, such as evolutionary algorithms, as proposed in [15, 27], or the use of level-set method, as seen in [28]. In [29], a unified method for implementation was presented, where many traditional manufacturing methods are related to specific geometric constraints. In the current study, a very specific method is in question, which is Tailored Forming. Since it is a technique still in development, the manufacturing constraints are not quite established. However, some of them can be already translated to the geometric field. In this point of research, the generation of optimized solutions that can't be manufactured (but are close enough to possible solutions) bring also valuable information to the community.

In order to create an optimization tool that can deal with all the difficulties discussed above, a new method was proposed in [7], called Interfacial Zone Evolutionary Optimization (IZEO). This method is based on the evolutionary algorithms theory, which is an optimization methodology alternative to SIMP that has gained more and more space in research. The state-of-the-art of this technique is called Bi-directional Evolutionary Structure Optimization (BESO), as presented in [30]. Evolutionary Algorithms consist in an iterative process, where the structure changes gradually according to a sensibility field, as an evolution (Figure 3).

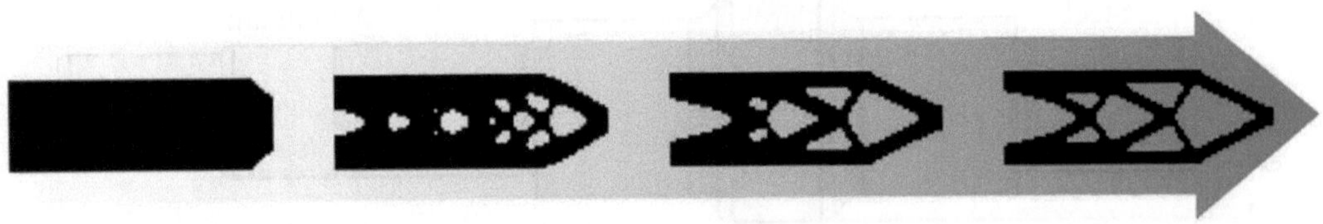

Figure 3: Evolutionary Optimization Process performed by soft-kill BESO [30]

The objective function here is maximization of stiffness, with a volume constraint, as described in Equations (1) and (2).

$$\min \mathbf{f}^T\mathbf{u} \tag{1}$$

$$\sum_{i=1}^{N} V_i x_i = V^* \tag{2}$$

Where **f** is the applied load vector; **u** is the nodal displacement vector; N is the number of elements in the mesh; V_i is the volume of an individual element; x_i is the density of the element and the design variable; and V^* is the prescribed volume.

Differently from SIMP, that generates the so-called "grey zones" with intermediate densities values, Evolutionary Algorithms have final result presented as a 1-0 design, since the algorithm completely removes material during the evolution. The actual standard model of BESO is called soft-kill BESO, which consists on using a very small density for the void spaces, in order to calculate sensitivity values for this region. The use of this technique with multi-materials was also investigated by the same authors in [30] and other recent improvements can also be found in literature, such as in [22], where a new filtering scheme and a continuation approach is used. Separately, some approaches of the method with manufacturing constraints can be also found, as in [27, 31].

In the case of our proposed method, IZEO, a set of restrictions for the evolution is implemented, so that the design doesn't take undesired forms. The basic concept is that in the beginning of our process, the whole design domain is set to the first material; then, an initial point of growth is set by the user; and from this point on, the second material grows over the first one, interactively. So, the first restriction implemented is that the evolution can happen only at the interfacial zone between materials (Figure 4).

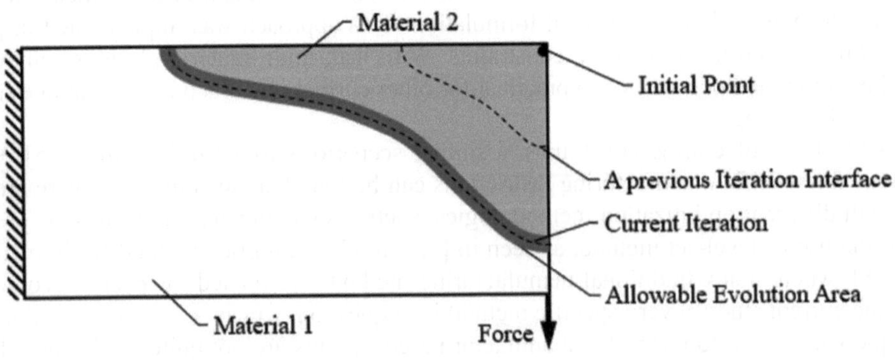

Figure 4: Scheme showing the evolution process of IZEO

The optimization restriction is set to be the proportion of each material in the domain space. The sensitivity analysis can be calculated as following:

$$\alpha_i = \frac{1}{2}\mathbf{u}_i^{\mathrm{T}}\mathbf{K}_i\mathbf{u}_i \quad \forall \quad i \in \Lambda \tag{3}$$

Where α_i is the sensitivity of the element; $\mathbf{u}_i$ is the elemental nodal displacement vector; $\mathbf{K}_i$ is the stiffness matrix of the element; and Λ is the set of elements that belongs to the allowable evolution area. With these implementations, the addition of manufacturing constraints don't have to deal with the whole domain, but only with the interfacial area to control the design geometry. As well as in [29], some geometric constraints were here implemented and, with the combination of them, many different restrictions can be described. They are the following:

- Minimal Member Size: related to how small the structures in the design are allowed to be. In mono-material designs, a minimal hole size restriction is also necessary to limit the size of the holes. It is implemented through the use of density filters, which are also important to avoid mesh-dependency and checkboard problems.
- Symmetry: restriction that generates symmetric components. It is implemented through a mirroring of the sensibilities in the optimization process.
- Unidirectional Growth: this restriction works like an "extrusion" restriction on mono-materials. The evolution here is restricted to only one direction specified by the user.
- Single Phases: this restriction avoids the "closure" of phases during the evolution. With that, both material phases remain continuous and unified during the whole evolutionary process.
- Outside Surface: here the evolution can happens at any outside surface of the material, in any point.

With all the constraints implemented, different solutions can be achieved for the same problem, according to which ones are active. Figure 5 presents some results achieved with IZEO for the classical problem of a MBB beam with different manufacturing restrictions. These results present also different bi-material and whole component restriction, which means that one kind of restriction was applied to form the bi-material component, and a different kind of restriction was applied to the whole component.

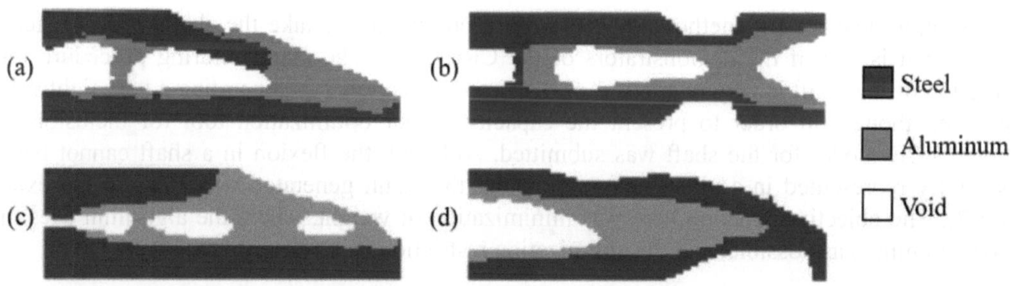

Figure 5: Steel/Aluminum/Void designs with different bi-material restrictions: (a) no restriction; (b) Symmetry and Unidirectional Growth from the right; (c) Unidirectional Growth form the right; (d) Unidirectional Growth from the left and Single Phases. All of them have also Minimal Element Size and Outside Surface restrictions at the component level

2.3 Parametric Analysis

As mentioned earlier, the topology optimization technique for Tailored Forming is a powerful tool for concept generation. The implementation of the manufacturing constraints contributes for the robustness of the tool, generating significate designs. However, the model implemented cannot address fine problems, such as very specific manufacturing constraints. Another limiting factor, is that the final result is discretized in finite elements, which is obviously not reasonable for manufacturability.

So, after a concept is generated, the model must be constructed in a CAD environment. Here, the final shape of the component can be drawn and specific parametric analysis can be executed. This analysis are made in points of interest, which for us is the connection zone between the two materials. Here, the software Inventor [32] was used for the construction of the parametric model and an interface with a finite element software, in our case Abaqus CAE [33], was constructed. All the parametric data and user-defined results are than saved in a database (Figure 6).

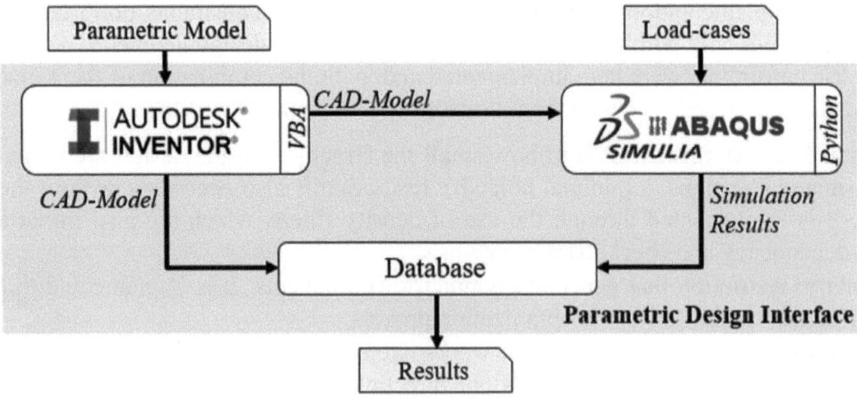

Figure 6: Parametric design process

As shown in the given diagram, the parametric models are generated with a specific range of parameters and its combinations. With the finite element analysis and database generation, the user can identify how each parameter influences the design and choose the best option. Naturally, the solutions that present not reproducible designs can be filtered, in order to see only those that fit the manufacturing process. The choice is then defined as the best design in our solution space and the model for this solution is already saved as a CAD-model in our database.

3 APPLICATION FOR A TAILORED FORMING SHAFT

As an example here for the methodology procedure presented, we take the shaft made of steel and aluminum that is one of the demonstrators of the CRC 1153. The manufacturing procedure for the construction of this shaft is very restrictive, as discussed in section 2.2, generating a very tight solution space. Even though, in order to present the capacity of our optimization tool for inclusion of restrictions, a 2D model for the shaft was submitted. Although the flexion in a shaft cannot have the stress rightly represented in a plane-state simulation, the result generated was close to the expected (Figure 7). The objective function here was minimization of weight, where the algorithm tries to use the most aluminum as possible, and the optimization restriction was a determined safety factor.

Figure 7: Design result for the demonstrator shaft using IZEO with Tailored Forming restrictions

In order to achieve the result seen in Figure 7, many constraints had to be implemented. Symmetry and Growth Direction are naturally present, since it is a symmetric component and the manufacturing process consists in joining two parts together through laser or friction welding. A third and more complicated restriction had also to be implemented, which is related to the fact that the steel must be inside the aluminum at the joining zone. This is a very specific restriction inherent to the process and it was implemented through the intersection of Growth Direction restriction in three different directions.

With that, we achieved the expected concept geometry. As discussed, this result contains no relevant solution for the manufacturing side, where a detailed representation is needed, but it showed the adaptability of the tool. In the next step, we create the parametric representation of the joining zone based on two parameters: length and volume. Figure 8 shows the CAD representation for the connection zone using a function *f(x)*.

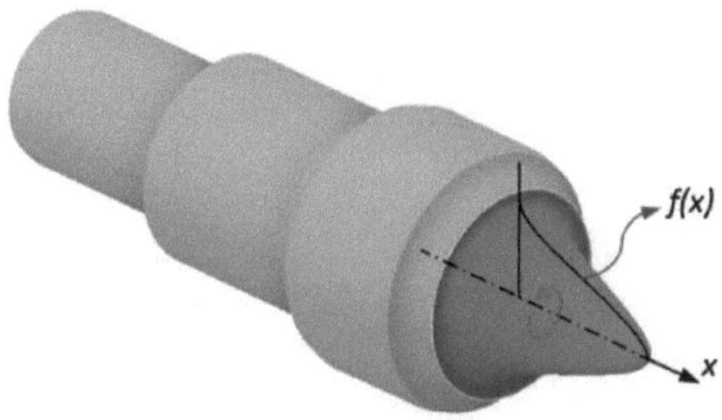

Figure 8: CAD model of the Steel part of the shaft, showing the connection zone surface

To make this description, we used the following equation:

$$f(x) = R\sqrt{1 - \left(\frac{x}{L}\right)^{\left(\frac{V}{\pi R^2 L - V}\right)}}; \qquad L > \frac{V}{\pi R^2} \tag{4}$$

Where R is the radius of the connection zone; L is the length; V is the volume of the interfacial region; and x is the axial axis to the shaft. This equation was chosen in order to create a smooth interface, where length and volume can act as independent parameters. It is important to remember that this function must also take into consideration the manufacturing constraints.

Finally, the model was submitted to our parametric design process, between Inventor and Abaqus, with a large range of parameters. This range must also consider the manufacture restrictions, but some parameter combinations might still generate unfeasible designs that must be later excluded. However, design solutions that are close enough to a manufacturable solution are also valuable in this point, since they can set a target to the manufacturing research. The graph presented in Figure 9 shows one of the results of the finite element analysis, which is the minimal stress found at the joining region for a variety of parameters L and V.

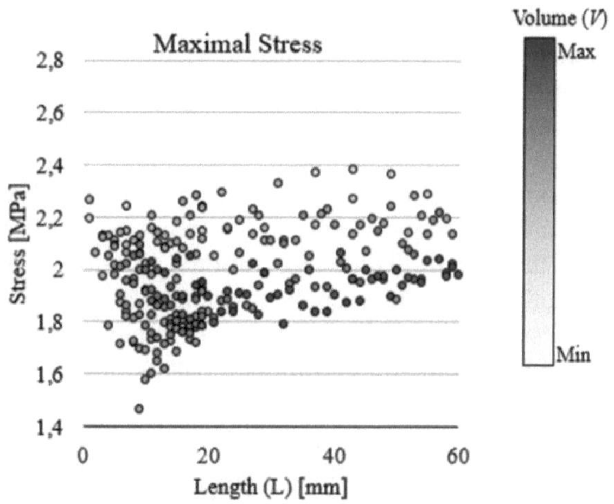

Figure 9: Finite Analysis result for Maximum Stress found at the joining zone for the solution space generated

Since only linear elastic cases were evaluated, the magnitude of the stresses are not a concern, but the difference between them. Taking this maximum stress as a decision parameter for the joining zone geometry, the best solution should be the one that presents minimum stress. This criterion presents a very distinct value from the others, since in Figure 9 it is seen the clear presence of a minimum value. This result is presented in Figure 10, where it is possible to identify the geometry of the joining zone.

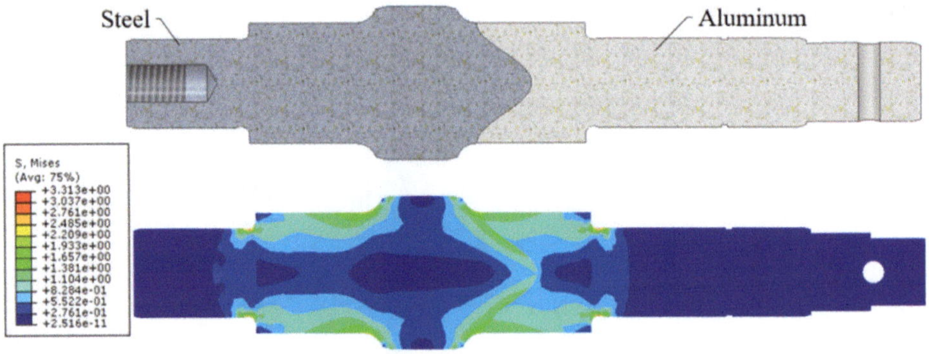

Figure 10: Design solution with smallest maximum stress value at the joining zone and its stress field

Taking a deeper look in the results and analyzing the influence of the parameters, some conclusions can be made about the design of this shaft. After observing different kinds of designs, it is seen that the angle formed between the shaft surface and the interfacial zone play an important role on the homogenization of the stress. The result presented in Figure 10 also has a uniform stress distribution over the joining zone.

As seen, these conclusions are based in different parameters from the used ones. Although these parameters are also controllable with the description made, this shows how the right choice of parameters can impact the result, even though here just an explorative method was the focus. However, these conclusions are enough to serve in future as guideline rules for this type of construction, facilitating the work of design engineers.

4 CONCLUSIONS

This work presented an overview of the design approach used for the design of Tailored Forming hybrid components. Two different techniques were used to address two main objectives of this design: evolutionary topology optimization to handle material distribution and parametric analysis to establish a connection zone geometry. Both approaches had to include the manufacturing process restriction in order to generate reasonable results.

For the first step, it was seen that IZEO has a good capability to include geometric constraints and generate reasonable designs. Although the process is still highly heuristic, a large solution space can be generated, providing support to engineers that deals with such complex tasks. For that, the strong limitations present in the manufacturing process can be translated into restrictions in the algorithm, increasing the potential of the use of this tool for Tailored Forming. The implementations in 2D can still be expanded, where other restrictions can be implemented and different objective functions can be tested. Even though the results are already consistent, an implementation in a 3D environment enlarges the difficulty to set the same geometric constraints, but the implementations of it would follow the same logical procedure.

In the parametric analysis, the approach is quite straightforward. However, the task of making the parametric description plays a fundamental role, since it contains the desired geometric restrictions. Preferably, the solution space should contain the highest number of solutions as possible, giving the possibility of a filtering stage after all, where the user can select only the designs that are suitable for the manufacturing process. The design guidelines reached at the end showed the efficacy of such a method, transforming a large number of complex data in simple standard practices.

At the end, a suitable optimal design for the proposed shaft was reached. Moreover, a suitable design methodology for Tailored Forming is in development and the area still opens space for a large research field. The same process here described can be used equally for different components. Furthermore, the use of such techniques makes possible the achievement of higher performance machine components.

ACKNOWLEDGEMENTS

The results presented in this paper were obtained under the umbrella of Collaborative Research Centre 1153 "Process Chain for Manufacturing Hybrid High Performance Components by Tailored Forming", preliminary inspection project C2. The authors would like to thank the German Research Foundation (DFG) and the CRC 1153 for its financial and organizational support.

REFERENCES

[1] Behrens, B.-A., Overmeyer, L., Barroi, A., Frischkorn, C., Hermsdorf, J., Kaierle, S., Stonis, M., Huskic, A., "Basic study on the process combination of deposition welding and subsequent hot bulk forming", *Production Engineering*, Vol. 7 No. 6, pp. 585–591. 2013.

[2] Kleemann, S., Fröhlich, T., Türck, E., Vietor, T., "A Methodological Approach Towards Multi-Material Design of Automotive Components", Procedia CIRP, Vol. 60, pp. 68–73, 2017.

[3] Brockmöller, T., Mozgova, I., Lachmayer, R., "An Approach to Analyse the Potential of Tailored Forming by TRIZ Reverse", *Proccedings of the 21st International Conference on Engineering Design (ICED 17) Vol 4: Design Methods and Tools, Vancouver, Canada, 21–25 August*, pp. 445–452, 2017.

[4] Bibani, M., Gembarski, P. C., Lachmayer, R., "Ein wissensbasiertes System zur Konstruktion von Staubabscheidern", *DFX 2017: Proceedings of the 28th Symposium Design for X, Oct 4–5, Bamburg, Germany*, 2017.

[5] Gembarski, P.C., Sauthoff, B., Brockmöller, T., Lachmayer, R., "Operationalization of Manufacturing Restrictions for CAD and KBE-Systems", *Proceedings of the DESIGN 2016 14th International Design Conference, Cavtat-Dubrovnik, Croatia, 16–19 May*, 2016.

[6] Brockmöller, T., Gembarski, P.C., Mozgova, I. and Lachmayer, R., "Design Catalogue in a CAE Environment for the Illustration of Tailored Forming", *59th Ilmenau Scientific Colloquium, Ilmenau, Germany, Sep 11 – 15, 2017, Technische Universität Ilmenau*, 2017.

[7] Siqueira, R., Mozgova, I., Lachmayer, R., "Development of a topology optimization method for tailored forming multi-material design", *24th ABCM International Congress of Mechanical Engineering, December 3–8, Curitiba, PR, Brazil*, 2017.

[8] Harzheim, L., "Strukturoptimierung: Grundlagen und Anwendungen", Europa Lehrmittel, Germany, ISBN 9783808556580, 2008.

[9] SFB 1153 – *Prozesskette zur Herstellung hybrider Hochleistungsbauteile durch Tailored Forming,* [online] Leibniz Universität Hannover. Available at: https://www.sfb1153.uni-hannover.de/sfb1153.html (accessed 08.11.2017).

[10] Behrens, B.-A., Bouguecha, A., Moritz, J., Bonk, C., Stonis, M., Klose, C., Blohm, T., Chugreeva, A., Duran, D., Matthias, T., Golovko, O., Thürer, S. E., Uhe, J., "Aktuelle Forschungsschwerpunkte in der Massivumformung", *22. Umformtechnisches Kolloquium, Hannover, Germany, Mar 15–16, Hannoversche Forschungsinstitut für Fertigungsfragen*, 2017.

[11] Fiebig, S., Sellschopp, J., Manz, H., Vietor, T., Axmann, K., Schumacher, A., "Future challenges for topology optimization for the usage in automotive lightweight design technologies". *In Proc. of 11th World Congress on Structural and Multidisciplinary Optimization, Sydney, Australia*, 2015.

[12] Bendsoe, M.P. and Kikuchi, N., "Generating optimal topologies in structural design using a homogenization method", *Comput. Methods Appl. Mech. Eng.*, Vol. 71, No. 2, pp. 197–224, 1988.

[13] Bendsoe, M.P., Sigmund, O., "Topology Optimization: Theory, Methods and Applications". Springer. ISBN 9783540429920, 2004.

[14] Zhou, M., Shyy, Y., Thomas, H., "Checkerboard and minimum member size control in topology optimization", *Structural and Multidisciplinary Optimization*, Vol. 21, No. 2, pp. 152–158, 2001.

[15] Fiebig, S., Axmann, J.K., "Using a binary material model for stress constraints and nonlinearities up to crash in topology optimization", *In 10th World Congress on Structural and Multidisciplinary Optimization*, pp. 19–24, 2013.

[16] Zargham, S., Ward, T.A., Ramli, R., Badruddin, I.A., "Topology optimization: a review for structural designs under vibration problems", *Structural and Multidisciplinary Optimization*, Vol. 53, No. 6, pp. 1157–1177, 2016.

[17] Ferguson, I., Frecker, M., Simpson, T.W., Dickman, C.J. *et al.*, "Topology optimization software for additive manufacturing: A review of current capabilities and a real-world example". *In ASME 2016 International Design Engineering Technical Conferences and Computers and Information in Engineering Conference. American Society of Mechanical Engineers*, pp. V02AT03A029–V02AT03A029. 2016.

[18] Li, T., Wu, T., Ding, X., Chen, H., Wang, L., "Design of an internally cooled turning tool based on topology optimization and cfd simulation". *The International Journal of Advanced Manufacturing Technology*, Vol. 91, No. 1, pp. 1327–1337, doi:10.1007/s00170-016-9804-9, 2017.

[19] Zuo, W., Saitou, K., "Multi-material topology optimization using ordered SIMP interpolation", *Structural and Multidisciplinary Optimization*, Vol. 55, No. 2, pp. 477–491, 2017.

[20] Tavakoli, R., Mohseni, S.M., "Alternating active-phase algorithm for multimaterial topology optimization problems: a 115-line matlab implementation". *Structural and Multidisciplinary Optimization*, Vol. 49, pp. 621–642, 2014.

[21] Zhou, S., Wang, M.Y., "Multimaterial structural topology optimization with a generalized cahn–hilliard model of multiphase transition", *Structural and Multidisciplinary Optimization*, Vol. 33, No. 2, pp. 89–111, 2007.

[22] Ghabraie, K., "An improved soft-kill BESO algorithm for optimal distribution of single or multiple material phases", *Structural and Multidisciplinary Optimization,* Vol 52, pp.773–790, https://doi.org/10.1007/s00158-015-1268-2, 2015.

[23] Mirzendehdel, A.M., Suresh, K., "A pareto-optimal approach to multimaterial topology optimization", *Journal of Mechanical Design*, Vol. 137, No. 10, p. 101701, 2015.

[24] Meisel, N., Gaynor, A., Williams, C., Guest, J., "Multiple-material topology optimization of compliant mechanisms created via polyjet 3d printing", *In 24th Annual international solid freeform fabrication symposium an additive manufacturing conference*, p. 28., 2013.

[25] Zhou, M., Fleury, R., Shyy, Y.K., Thomas, H., Brennan, J., "Progress in topology optimization with manufacturing constraints", *Proceedings of the 9th AIAA MDO conference, Atlanta, USA, 4–6 September,* 2002.

[26] Aremu, A., Ashcroft, I., Hague, R., Wildman, R., Tuck, C., "Suitability of SIMP and BESO topology optimization algorithms for additive manufacture", *In Twenty First Annual International Solid Freeform Fabrication Symposium*, Austin (pp. 679–692), 2010.

[27] Fiebig, S., Axmann, J. K., "Combining nonlinear FEA simulations and manufacturing restrictions in a new discrete Topology Optimization method", *9th World Congress on Structural and Multidisciplinary Optimization*, June 13–17, Shizuoka, Japan, 2011.

[28] Michailidis, G., "Manufacturing Constraints and Multi-Phase Shape and Topology Optimization via a Level-Set Method", *Ph.D. thesis*, Ecole Polytechnique, Palaiseau, 2014.

[29] Vatanabe, S.L., Lippi, T.N., de Lima, C.R., Paulino, G.H., Silva, E.C., "Topology optimization with manufacturing constraints: A unified projection-based approach", *Advances in Engineering Software*, Vol. 100, pp. 97–112., 2011.

[30] Huang, X., Xie, M., "Evolutionary topology optimization of continuum structures: methods and applications", John Wiley & Sons, 2010.

[31] Aremu, A., Ashcroft, I., Hague, R., Wildman, R., Tuck, C., "Suitability of SIMP and BESO topology optimization algorithms for additive manufacture", *In 21st Annual International Solid Freeform Fabrication Symposium (SFF)*, An Additive Manufacturing Conference. pp. 679–692, 2010.

[32] Inventor Professional (2016). Version 2017 RTM, Autodesk Inc. ©.

[33] Abaqus CAE (2014). Version 6.14-2. Dassault Systemes ©.

New Production Technologies

MATERIAL- AND PROCESS CHARACTERIZATION OF FIBRE-METAL-ELASTOMER LAMINATE COMPONENTS WITH HIGH FORMING DEGREES

Sven Roth[1], Sven Coutandin[2], Jürgen Fleischer[3]

[1] Karlsruhe Institute of Technology, wbk Institute of Production Science, Kaiserstraße 12, 76131 Karlsruhe, Germany, sven.roth@kit.edu, www.wbk.kit.edu
[2] Karlsruhe Institute of Technology, wbk Institute of Production Science, Kaiserstraße 12, 76131 Karlsruhe, Germany, sven.coutandin@kit.edu, www.wbk.kit.edu
[3] Karlsruhe Institute of Technology, wbk Institute of Production Science, Kaiserstraße 12, 76131 Karlsruhe, Germany, juergen.fleischer@kit.edu, www.wbk.kit.edu

Keywords:

Hybrid Laminates, Fibre-Metal-Elastomer Laminates, FMEL, Damping behaviour, Innovative manufacturing processes

ABSTRACT

Hybrid material concepts provide a high variability in the resulting part properties, and thus are often applied to satisfy multiple component demands. Fibre-metal laminates (FML) are widely spread in aerospace applications and are being used for decades as they show a high lightweight potential and a good fatigue behaviour. However, a broad conventional use of hybrid laminates in the automotive sector is not existing until today. The high manufacturing costs, caused by the surface pre-treatment of the metal layer, as well as long process cycles and a limited formability of current laminates are not suitable for automotive applications.

This paper presents an approach, which allows the processing of hybrid laminates for high-volume applications and enables high forming degrees of the manufactured parts. As an additional elastomer layer is used to separate the metal from the fibre reinforced layer, carbon fibre reinforced polymers (CFRP) can be used instead of conventional glass fibres, preventing a galvanic corrosion between carbon and the metal. In addition to the manufacturing process itself, the influence of the formability will be discussed with regards to the distribution of the laminate layers, determining achievable forming degrees of the manufactured fibre-metal-elastomer laminate (FMEL) specimen. The laminate behaviour during the forming of the uncured laminate will be described by analysing micro sections. Furthermore, the results of an experimental modal analysis will be presented in order to determine the damping properties of the investigated hybrid laminates.

1 INTRODUCTION

The increasing demands of the environmental and climate protection require an economical implementation of lightweight design in the production of new vehicles [1]. Fibre-metal laminates are being used in aviation for more than 30 years as they provide good mechanical properties, a high lightweight potential and a good fatigue behaviour [2, 3]. By adding an additional elastomer interlayer between the fibre reinforced polymer and the metal layer, the damping behaviour can be increased and carbon fibres can be used instead of conventional glass fibres without inducing a corrosion between the carbon and the metal layer [4, 5]. However, a wide implementation of fibre-metal laminates into the automotive sector is currently not rewarding due to the high cycle times and manufacturing costs [6]. Furthermore, current manufacturing processes are limiting to the geometry, especially to the achievable bending radii of the manufactured part [3, 7].

K. Dröder et al (eds), *Technologies for economical and functional lightweight design*, Zukunftstechnologien für den multifunktionalen Leichtbau, https://doi.org/10.1007/978-3-662-58206-0_14

For this purpose, a novel manufacturing process was developed by Fleischer et al., capable to produce FMEL components with a high forming degree at low cycle times. The achievable bending radii of 1.2 mm at a bending angle of 90° are promising and enable new possibilities for the design of hybrid laminates. However, the increased forming degree leads to a redistribution of the laminate constituents in the formed area, which is not fully understood. [8]

Therefore, the influence of the manufacturing process on the proportion of the laminate constituents of highly formed FML and FMEL will be investigated in this study. Different laminate variants will be compared in order to determine the influencing factors on the distribution of each laminate component in the formed region. Furthermore, the influence of an elastomer interlayer on the damping behaviour will be characterized by the use of an experimental modal analysis.

2 MATERIALS

Fibre-metal laminates consist of alternating CFRP & metal layers, whereas fibre-metal-elastomer laminates have an additional elastomer interlayer, separating the metal from the CFRP. In this study, aluminium was chosen as metal sheet. Four different layups have been realized, which distinguished in the use of aluminium or CFRP as a face layer as well as the presence of elastomer as an interlayer, see Figure 1. The shown stacking sequence of the CFRP layer was chosen to obtain a comparable weight to stiffness ratio between variant 1-0 and variant 1-1 as well as a comparable laminate thickness of variant 1-0, 1-2 and 2-0. Additionally, a symmetrical layup was intended.

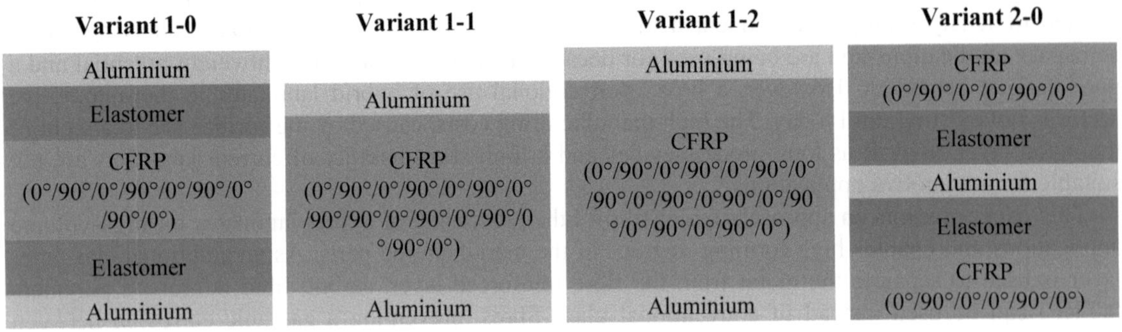

Figure 1: Laminate structure of investigated layups

2.1 CFRP

The selected CFRP was a Hexcel prepreg “M77/42 %/UD90/CHS”. The fast curing epoxy resin and a nominal sheet thickness of 0.1 mm provide low curing cycles and a variable layup of the CFRP layer. Different laminates with a varying stacking sequence of the CFRP layer have been investigated in this study, whereas the unidirectional sheets have been arranged to biaxial layer with a symmetrical succession. For the chosen prepreg, the recommended curing time at a temperature of 150° C and a minimum pressure of 5 bar is 120 s.

2.2 Aluminium

The metal chosen in this study was an aluminium 2024-T3 with a commercial available sheet thickness of 0.3 mm. The selected alloy is common in FML applications and is used in many previous investigations and therefore ensures a good comparison to this study [2, 3, 9]. In the available condition T3, a minimal bending radii of 1.2 mm at 90° is expected [10].

2.3 Elastomer

The elastomer selected for this study was a Kraiburg mixture “SAA-9579/52” in a sheet thickness of 0.5 mm, developed for applications with high damping requirements. Optimized for CFRP & metal applications, it provides a good adhesion to CFRP with epoxy matrices and to aluminium. Implemented as an interlayer between CFRP and aluminium, the elastomer mixture prevents from galvanic corrosion and compensates the different thermal expansion coefficients in the laminate. For the non-cross-linked elastomer mixture, the recommended curing time at a temperature of 150° C and a pressure of 23 bar is 300 s.

3 EXPERIMENTAL

In order to investigate the process-property relationship on the manufacturing of hybrid laminates, the manufacturing process itself is described in the following. The hereby manufactured specimen are being investigated by a microscopic analysis to characterise the influence of the forming process and the tooling concept on the local layer distribution. Furthermore, the influence of the elastomer interlayer on the damping behaviour is investigated by the use of an experimental modal analysis.

3.1 Manufacturing process

The basic manufacturing steps for the manufacturing of FMEL parts are shown in Figure 2. During the stacking of each laminate component in the beginning, the desired layup is realised. In the second step, the two dimensional stack with its non-cross-linked CFRP and elastomer layers is placed in a pressing mould.

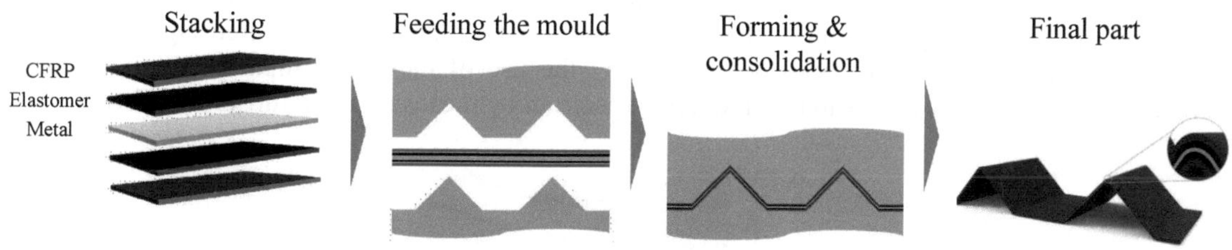

Figure 2: Process scheme for the manufacturing of formed hybrid laminate

In this study, two different mould geometries have been used. For the formed laminates, a double v-shaped mould design was chosen as it implements two different bending angles of 45° & 90° (see Figure 3 a)). By the use of different mould variants, an additional flexibility of the bending radii (R_i) was implemented. The bending radii investigated in this study was set to 1.2 mm. The sample width of formed laminates was 100 mm. Furthermore, an additional flat shaped tool was used for the manufacturing of two dimensional specimen in the size of 300 × 300 mm^2.

After the laminate constituents are placed in the mould, the forming of the laminate takes place in one single stroke. The forming step itself is similar to the bottom bending used in the sheet forming of metal. As soon as the mould is being closed, the forming of the stack is followed by the consolidation of the laminate at 23 bar pressure. After a curing time of 300 s at 150° C, the part can be removed in a final step. Figure 3 b) exemplary shows two manufactured FMEL specimen of variant 1-0 and 2-0.

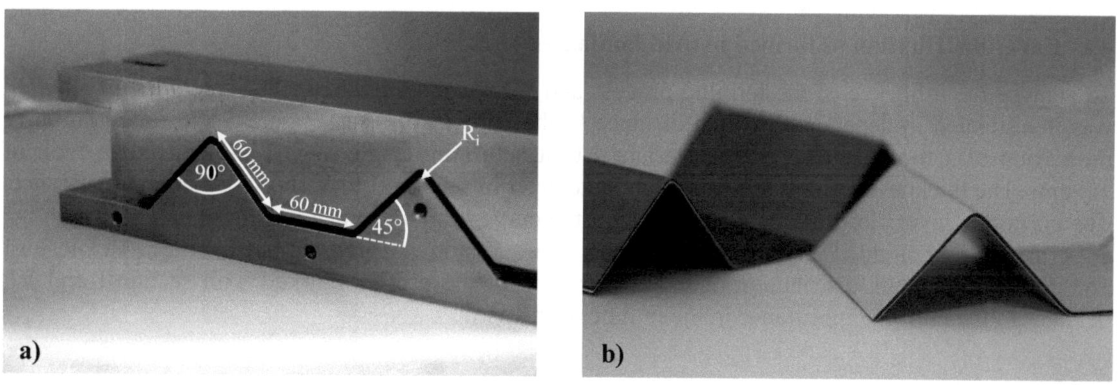

Figure 3: a) Tool for the manufacturing of formed laminate; b) Manufactured FMEL profiles of the variant 1-0 (right) & 2-0 (left)

3.2 Optical analysis

Different micro sections of formed FML & FMEL specimen have been investigated in order to analyse the influence of the manufacturing process and the laminate structure on the layer distribution. The sections have been cut by a water-cooled cutting off machine from Struers GmbH type Accutom-5 and have been embedded by the use of Varidur 200, a castable mounting compound. The polishing up to a grit size of 4000 of the materialgraphic specimen was done by using a grinding polisher type EcoMet 250 by ITW Test & Measurement GmbH European Headquarters. The optical analysis was done by the use of a microscope of the type Leitz Aristomet and the software Olympus Stream by Olympus Industrial.

3.3 Modal analysis

The influence of an elastomer interlayer on the damping behaviour of two dimensional laminates was characterized by the use of a modal analysis. As shown in Figure 4 a), the free-mounted specimen with the dimensions of 300 × 300 × 2.5 mm³ were suspended by the use of a thin nylon thread. Using an impact hammer of the type Kistler 9726A20000, the structure-borne noise was generated (see Figure 4 c)). Five acceleration sensors of the type Kistler 8692C50 (see Figure 4 b)) have been attached to the two dimensional laminate specimen in order to record the vibration. Figure 4 c) shows the frontend of the test setup, transferring the data of the attached sensors to a laptop. The desired eigenfrequencies and damping coefficients have been determined by the use of the LMS Impact Testing Software, transferring the vibration signals to the frequency spectrum by means of a fast furrier transformation.

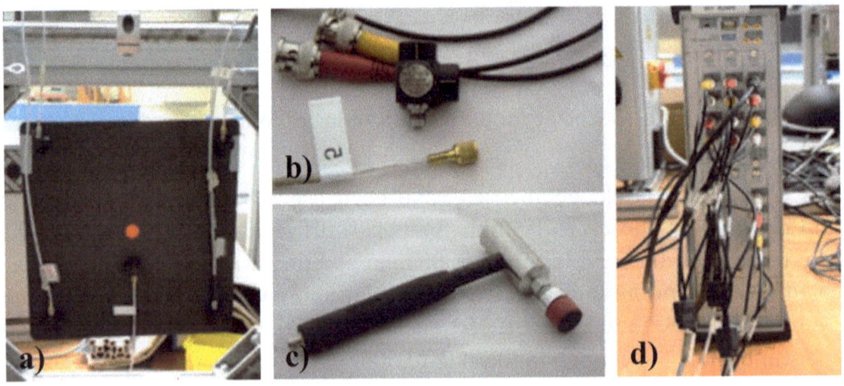

Figure 4: Modal test setup: a) Free-mounted specimen; b) Acceleration sensor; c) Impact hammer; d) Frontend

4 RESULTS

4.1 Layer distribution of formed hybrid laminates

To analyse the influence of the forming process and the chosen mould concept, the layer distribution was investigated in dependence of the distance to the edge of the part. Figure 5 shows the laminate proportion of a formed FMEL of the variant 1-0 at a forming angle of 90° and a bending radii of 1.2 mm. The thickness distribution of the laminate layer in section II to V shows a similar appearance. The elastomer layer closer to the convex side of the curve has an average thickness of 777 µm on the apex in the four middle sections, whereas the inner elastomer layer has an average thickness of 165 µm. The laminate distribution of the edge region of the profile, marked with section I and VI, deviates to the uniform appearance in the middle of the part.

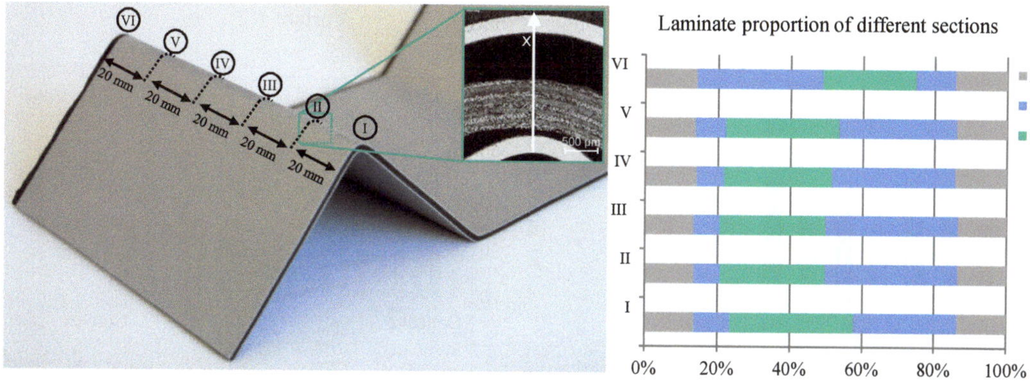

Figure 5: Layer distribution of formed FMEL in dependence of the distance to the part

To analyse the influence of the elastomer on the homogeneity of the CFRP layer, the formed area of FML & FMEL was compared. The sections have been cut out of the center of the part, to exclude any superpositioning of the layer distribution as a result of a near-edge position. Figure 6 shows the micro section of a formed variant 1-0 and 1-2 with a bending angle of 90° and a bending radii of 1.2 mm. The CFRP sheets in variant 1-0 show a more homogeneous appearance in comparison to variant 1-2, whereas in the latter case, the sheets oriented parallel to the bending axis tend to gain in thickness with a growing distance to the concave side.

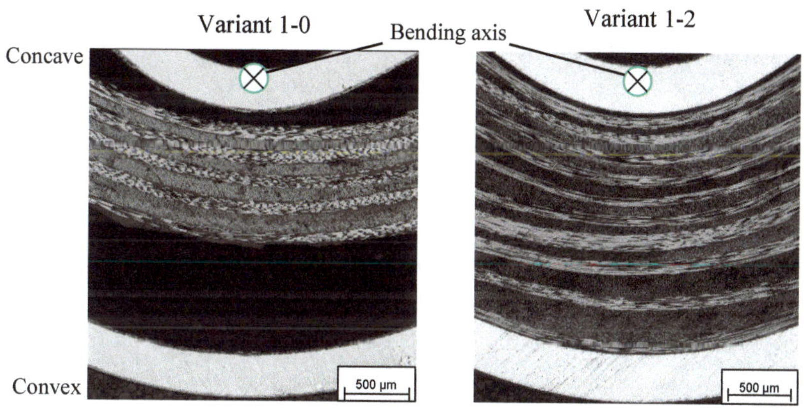

Figure 6: Micro section of formed laminates with a bending angle of 90° and a bending radii of 1.2 mm

To quantify the gain in thickness of the sheets with a fibre orientation in parallel to the bending axis, each CFRP layer has been measured on the apex of the formed region. Figure 7 exemplary shows the results of the microscopic measurement. The FMEL with an elastomer interlayer show a certain variance in the thickness of each CFRP sheet. The average thickness of the sheets orthogonal to the bending axis is 95 µm, whereas the average thickness of the parallel oriented sheets is 100 µm. In comparison to variant 1-0, the FML show a similar variance in the thickness of each CFRP sheet on the concave side, whereas the average sheet thickness is on a lower level. For example, the first six orthogonal oriented layer on the concave side show an average thickness of 67 µm, the first six parallel oriented sheets are in average 62 µm thick. However, an increase of the thickness of parallel oriented sheets on the convex side of the formed profile up to 202 µm can be observed.

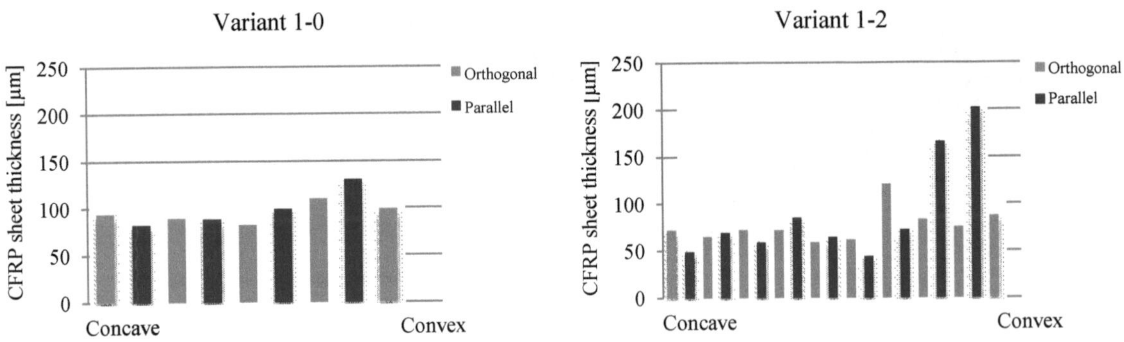

Figure 7: CFRP sheet thickness laminates with a bending angle of 90° and a bending radii of 1.2 mm

4.2 Modal analysis of FMEL

The damping behaviour of hybrid laminates was investigated on the laminate variant 1-0 and 1-1 by the use of a modal analysis. Three specimen each variant have been analysed, whereas each specimen was evaluated by the use of the mean value out of 5 measuring strokes. The available test-setup showed a coherence in the frequency range between 50 Hz – 400 Hz, and thus allowed the comparison of the first two eigenfrequencies. As shown in Table 2, the first two eigenfrequencies of variant 1-0 and 1-1 have an average deviation of 4 and 20 Hz, allowing a qualified comparison of the damping rates on both laminates. On the first eigenfrequency, the FML shows an average damping rate of 0.55 %, whereas the damping value of the FMEL is on average 0.94 %. On the second eigenfrequency, the increased damping rate of the laminate with an elastomer interlayer is even more significant and exceeds the value of the FML by ~490 %.

Table 2: Eigenfrequencies and damping rates of laminate variant 1-0 and 1-1

	Eigenfrequency [Hz]	σ [Hz]	Damping rate [%]	σ [%]
Variant 1-0	134	0.45	0.94	0.13
	339	9.93	2.54	0.26
Variant 1-1	138	2.38	0.55	0.09
	319	2.71	0.52	0.04

5 DISCUSSION

The introduced manufacturing process enables the manufacturing of fibre-metal-elastomer laminates with high forming degrees and a uniform appearance of the laminate constituents. However, an influence of the tooling concept on the near-edge layer distribution of the laminate within the first 20 mm, starting from the edge of the component, can be observed. The non-uniform laminate proportion, shown in Figure 5, appears as no vertical flash face was implemented into the pressing mold. Hence, a displacement of the viscous resin and elastomer within the first 20 mm of the component edge takes place, resulting in an inhomogeneous near-edge laminate structure.

Despite this peripheral impact, an influence of the forming process on the layer distribution can be observed in the formed region. Induced by the different forming behaviour of the laminate constituents, the elastomer interlayer on the concave side is suppressed during the forming, whereas the layer on the convex side increases in thickness. This effect is not typical in common forming processes of fibre-metal laminates as the hybrid laminate is cured before the forming step takes place. As a result to this, the formability of common fibre-metal laminates is limited. The bending radii of a comparable FML component with a sheet thickness of 2.0 mm for instance is of the same order of magnitude as the metal used in a sheet thickness of 2.0 mm. Higher forming degrees lead to a damage of the material, for example by delamination or a buckling of the inner layer. In comparison to this, the bending radii of FMEL is limited by the used metal sheet in the sheet thickness of 0.3 mm, allowing a bending radii of 1.2 mm instead of 10.0 mm, as applied for a comparable FML. As the forming of the laminate takes place in an uncured state, the deformation induced stress into the laminate, especially of the CFRP layers, is expected to be lower compared to laminates which are being formed in a cured state. However, an analysis of the influence of the forming degree on the mechanical properties using the presented FMEL manufacturing process is pending. [7, 10]

The thinnest elastomer layer observed was measured with 90 µm, sufficient to prevent a galvanic corrosion. An interruption of the elastomer sheet was not ascertained. The CFRP sheets of FMEL specimen showed a certain variance in its thickness oriented parallel and orthogonal to the bending axis, which is returned to the manual manufacturing process and a dispersion of the thickness in the prepreg raw material. The average sheet thickness of both orientations is on a similar level, resulting in a homogeneous appearance of the CFRP layer. In absence of the elastomer interlayer, formed specimen of variant 1-2 reveal thinner CFRP layer on the concave side and a thicker layer in parallel to the bending axis on the convex side of the profile. An increase in thickness of the first and the last parallel oriented sheet in the CFRP stack of 412 % was observed. This effect is affiliated to the higher stress induced to the orthogonal oriented sheets as a result of the missing soft elastomer during the forming of the FML, leading to a displacement of resin and fibres to the parallel oriented layer in the convex area.

Investigated by the use of a modal analysis, hybrid laminates with an additional elastomer interlayer show a significant increase in the damping rate, especially in the frequency range of the second eigenfrequency between 319 Hz and 339 Hz. The frequency dependence of the increase in the damping rate of FMEL is ascribed to the frequency dependence damping behaviour of the used elastomer interlayer. The eigenfrequencies of the tested laminates distinguish in average of 4 Hz respectively 20 Hz, allowing a comparison of the damping rate in a good approximation. It should be noted, that the used elastomer mixture was not optimized for the present frequencies. The maximum damping potential of an optimized elastomer mixture is currently unknown and supposedly unexhausted. Furthermore, the laminate layup was not optimized regarding to a maximum damping coefficient, hence the damping capabilities especially of an increased laminate thickness remain unexploited.

6 CONCLUSION

In this study, the influence of the forming process on the distribution of the laminate constituents was investigated. Two influencing factors could be identified. On one hand, the mould concept led to an inhomogeneous edge region of the manufactured specimen. On the other hand, the forming process itself influences the distribution of the constituents in the formed area as a result of the individual formability of the different layer. Here, the elastomer interlayer reduces the stress of the CFRP induced by the forming process, leading to a more homogeneous thickness of the CFRP sheets. Furthermore, a significant influence of the elastomer interlayer on the damping properties of the hybrid laminate could be observed, in particular in the frequency range of the second eigenfrequency between 319 Hz and 339 Hz.

ACKNOWLEDGEMENTS

The research of this paper was kindly financed by the Baden-Württemberg Stiftungs project "Faser-Metall-Gummi-Hybridlaminate (FMGL) ein neuartiges, nachhaltiges Werkstoffkonzept für den Fahrzeugleichtbau", support code MAT0012 of the research program "Rohstoff- und Materialeffzienz in der Produktion".

The authors kindly thank the Kraiburg Holding GmbH und Co. KG, providing the elastomer raw material for this study.

REFERENCES

[1] S. Pischinger, U. Seiffert, Vieweg Handbuch Kraftfahrzeugtechnik, 8th edn. Springer Vieweg, Wiesbaden, 2016.

[2] L.B. Vogelesang, A. Vlot, Development of fibre metal laminates for advanced aerospace structures, *Journal of Materials Processing Technology*, **103**, 2000, pp. 1–5.

[3] A. Vlot, J.W. Gunnink, Fibre Metal Laminates – An Introduction. Springer Netherlands, Dordrecht, 2001.

[4] M. Stoll, F. Stemmer, S. Ilinzeer, K.A. Weidenmann, Optimization of Corrosive Properties of Carbon Fiber Reinforced Aluminum Laminates due to Integration of an Elastomer Interlayer, *Key Engineering Materials*, **742**, 2017, pp. 287–293.

[5] V. Sessner, M. Stoll, A. Feuvrier, K.A. Weidenmann, Determination of the Damping Characteristics of Fiber-Metal-Elastomer Laminates Using Piezo-Indicated-Loss-Factor Experiments, *Key Engineering Materials*, **742**, 2017, pp. 325–332.

[6] T. Sinmazçelik, E. Avcu, M.Ö. Bora, O. Çoban, A review: Fibre metal laminates, background, bonding types and applied test methods, *Materials & Design*, **32**, 2011, pp. 3671–3685.

[7] J. Sinke, Manufacturing of GLARE Parts and Structure, *Applied Composite Materials*, **10**, 2003, pp. 293–305.

[8] J. Fleischer, S. Roth, C. Sommer, Faser-Metall-Gummi-Hybridlaminate – Produktionsprozessentwicklung für ein neuartiges Materialsystem, *ZWF Zeitschrift für wirtschaftlichen Fabrikbetrieb*, **111**, 2016, pp. 483–486.

[9] S. Hinz, T. Omoori, M. Hojo, K. Schulte, Damage characterisation of fibre metal laminates under interlaminar shear load, *Composites Part A: Applied Science and Manufacturing*, **40**, 2009, pp. 925–931.

[10] J. Datta, Aluminium-Werkstoff-Datenblätter, 3rd edn. Aluminium-Verlag, Düsseldorf, 2001.

Material Concepts

FILM-ADHESIVES FOR POLYMER-METAL HYBRID STRUCTURES FROM LABORATORY TO CLOSE-TO-PRODUCTION

Thorsten Schnettker[1], Philipp Dreessen[1,2], Klaus Dröder[2], Christian Vogler[3], David Koch[3], Thomas Frey[3], Benjamin Poller[4], Andreas Wedemeier[4], Christian Vree[4], Jörg Kosowski[5], Daniel Riss[2]

[1] SITECH Sitztechnik GmbH, Wolfsburg, Germany
[2] Technische Universität Braunschweig, Institute of Machine Tools and Production Technology, Braunschweig, Germany
[3] nolax AG, Sempach Station, Switzerland
[4] Salzgitter Mannesmann Forschung GmbH, Eisenhüttenstraße 99, 38239 Salzgitter, Germany
a.wedemeier@sz.szmf.de
[5] Salzgitter Flachstahl GmbH, Salzgitter, Germany

Keywords:

Film-Adhesives, Thermoplastic, Steel, Hybrid material concepts, Seating structures

ABSTRACT

For achieving efficient lightweight materials, automotive development focusses more and more on hybrid plastic structures. In this context target was to combine plastic and steel to develop a cost and weight optimized hybrid structure. Their performance strongly depends on the adhesion between the combined materials. Mechanical interlocking is commonly used for the joining of the two materials. However, an adhesive bonding of the polymer and metallic surface is more effective. Therefore, different kinds of adhesion promotion primers are available. Several of these were tested by SITECH Sitztechnik GmbH and within the project "TRoPHy2" by Salzgitter Mannesmann Forschung GmbH and the project partners. Regarding the tested primer systems, the multilayer film-adhesives by nolax AG showed the best performance. The film-adhesives were developed for hybrid structures for interior parts with decorative applications, for example metal decor parts out of aluminium with a thermoplastic back injection moulding layer.

The suitability of these film-adhesives on steel surfaces was proved within a large test program including adhesion, corrosion, forming and other automotive processing steps. Among these tests suitability for large-scale production of film coated steel was also estimated. All tests showed good results.

At the opening of the Open Hybrid LabFactory e. V. (OHLF) a first hybrid seat structure, produced by injection moulding, was presented. This structure was validated regarding to function and safety requirements. In view of a large-scale production, first lamination tests were done on the coil coating line of Salzgitter Flachstahl GmbH. A series launch is possible and economic lightweight structures are realizable.

1 INTRODUCTION

Hybrid material concepts are increasingly focused by the development departments of the automotive industry. Especially injection moulded polymer-metal hybrid structures show a promising approach for cost-efficient lightweight design in a high-volume production. The load-optimized application of the two material components allows the utilization of their specific advantageous mechanical properties (Figure 1). The joining of the two components is crucial for the overall performance of the load-bearing part. Investigations on different bonding agents show a significant increase of the stiffness and energy absorption of hybrid components due to the use of the film-adhesive nolax Cox 490-1 [1]. So

K. Dröder et al (eds), *Technologies for economical and functional lightweight design*, Zukunftstechnologien für den multifunktionalen Leichtbau, https://doi.org/10.1007/978-3-662-58206-0_15

far, there is no large-scale production process for the application of the film-adhesive and it is flatbed laminated onto the metal sheet in laboratory scale.

Furthermore, the film-adhesive shows streaks and scratches after flatbed lamination with the state-of-the-art process (Figure 2). For these reasons, a coil coating process is being investigated and established for the integration of the film-application into serial production.

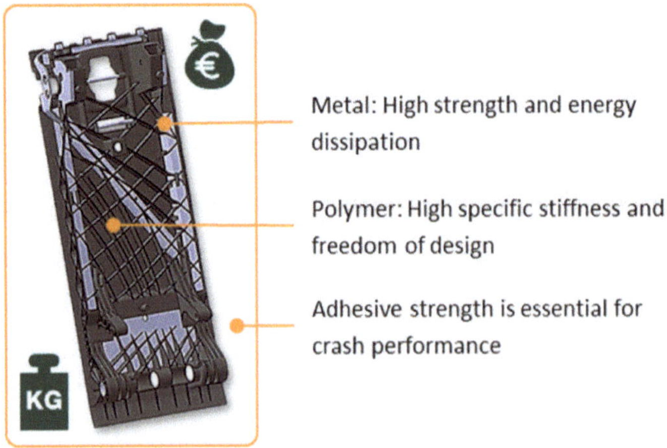

Figure 1: Polymer-metal hybrid seat structure of SITECH Sitztechnik GmbH

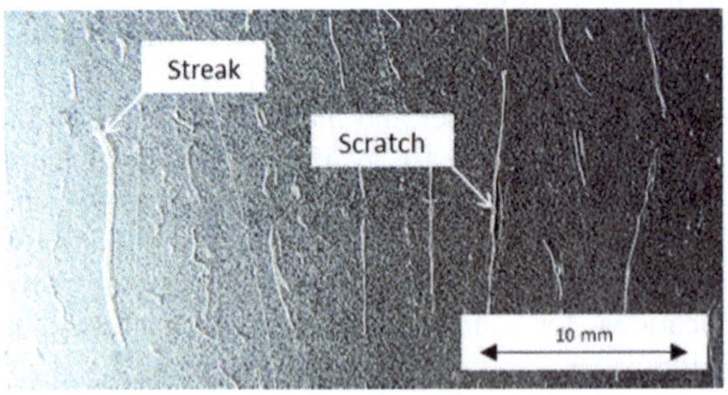

Figure 2: Surface defects in the film-adhesive after flatbed lamination on DX57+Z

2 STATE OF THE ART

Film-adhesives can be produced amongst other methods such as blow extrusion, which can be carried out as multilayer blow extrusion (Figure 3). The multilayer design allows a better adaption of the layers to the properties of the bonding substrates. An optimised film-adhesive for polyamide-metal bonding is nolax Cox 490-1.

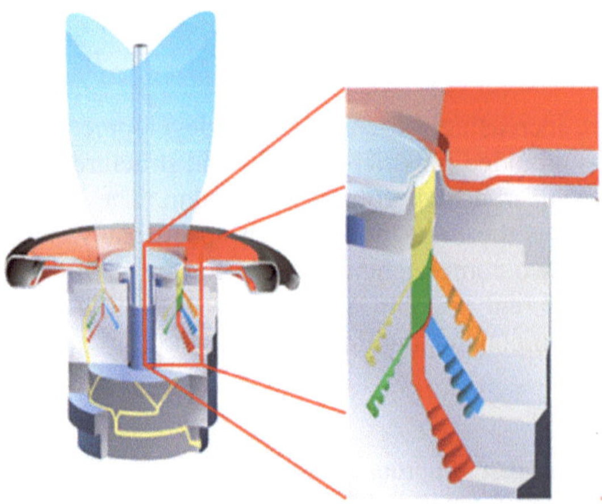

Figure 3: Multilayer blow extrusion

The application of the film-adhesive onto a metal sheet in laboratory scale is performed by means of flatbed lamination technique. The flatbed laminator consists of a heating zone for the melting and lamination of the film-adhesive and a cooling zone for the solidification. Pressure application is possible via a nip roll after the heating zone (Figure 4). However, the lamination quality is not always satisfying, as surface defects like streaks and scratches can occur (Figure 2).

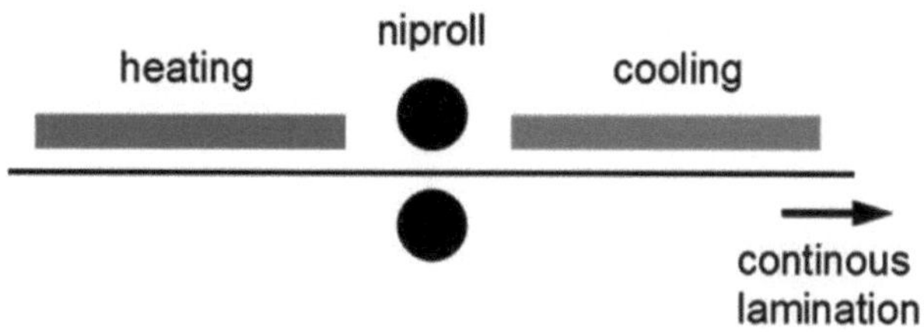

Figure 4: Schematic view of flatbed lamination

2.1 Investigations of the influence of flat steel metallic and organic coatings on bonding properties

In laboratory tests, the adhesive strength of the nolax Cox 490-1 for bonding metal and polyamide 6 (PA6) was evaluated. The tests were carried out with single lap shear samples according to DIN EN 1465 (Figure 5). Several material combinations of large-scale produced steel surfaces with the nolax film-adhesives and PA6 were fabricated and tested.

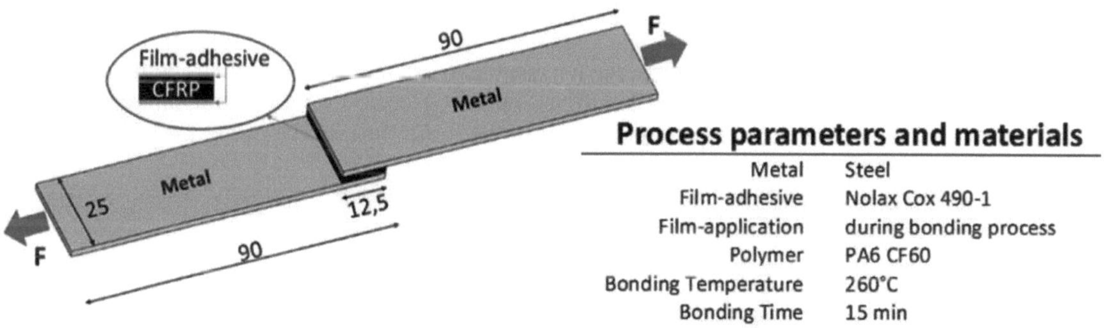

Figure 5: Metal-polymer-metal single-lap shear specimen for tensile test based on DIN EN 1465

In Figure 6, the results of the tensile tests of steel-polyamide 6 composites with nolax Cox 490-1 are shown. The investigation points out the adhesion promoting effect of the nolax film in the direct adhesion bonding of steel with CFRP. However, the combination of electro galvanized steel sheet with weldable corrosion protection primer (EG + CCP) shows a higher tensile strength without the film-adhesive. The results in material screening demonstrate the adhesion promotion for several steel surfaces by the nolax films. Most of the material combinations have tensile shear strengths between 11 and 13 MPa. On the matter of costs, an additional coating upon the metallic surface is not necessary, in order to achieve a high strength. Consequently, it seems promising to laminate the nolax films directly onto the metallic coating.

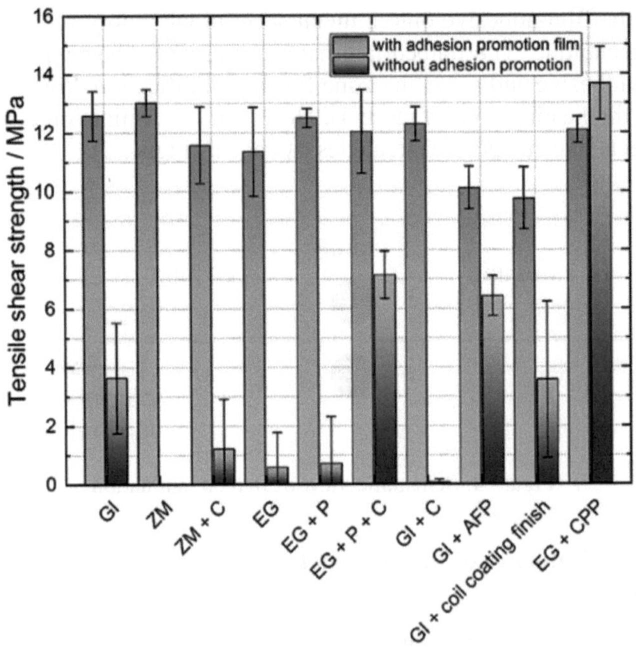

Figure 6: Tensile shear strength for several combinations of steel surfaces and polyamide 6 (CFRP). As a reference, analogue combinations of steel and PA6 without adhesion promotion films were tested. (GI: hot dip zinc coating; ZM: hot dip coating with zinc-magnesium coating; EG: electrogalvanized coating; P: pre-phosphated; C: chemically passivated; AFP: anti finger print; CPP: weldable corrosion protection primer)

In addition to the tests at room temperature, the tensile shear strength and T-peel strength were examined at 50° and 80 °C. These samples were combined out of GI coated steel and polyamide 6 (CFRP). nolax Cox 490-1 was used as adhesion promotion film. The results are shown in Figure 7.

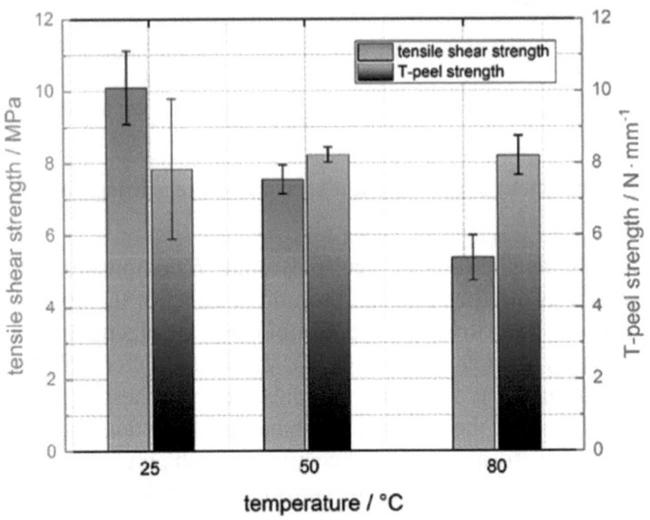

Figure 7: Tensile shear strength and T-peel strength as a function of temperature. Samples were fabricated out of GI coated steel in combination with nolax Cox 490-1 and polyamide 6 (CFRP)

With increasing test temperature, the tensile shear strength decreases. This behaviour is also known by structural adhesive bondings. The T-Peel strength seems to be not influenced by the test temperature. Values of about 8 N/mm were reached. Based on these results, laminating experiments were started in laboratory. It was shown that the nolax film-adhesives could be easily laminated on steel surfaces in pilot plant scale. According to these results, the upscaling on the production line could be realised.

3.2 Investigations on the process-integrated manufactured bonding

Typically, the strength of adhesive joints is assessed by single-lap shear testing according to the norm DIN EN 1465 [2]. Based on the norm, the single-lap shear test is also used to assess the performance of the joint in metal-polymer hybrid structures [1]. In order to investigate the metal-polymer compound in a close-to-reality load case, torsion tests of hybrid technology carriers are conducted. The complex design allows the assessment of the thermal mismatch due to environmental influences. Figure 8 shows the injection moulded hybrid specimen for the investigation of the adhesion in the hybrid three-dimensional compound and summarizes the process parameters and materials.

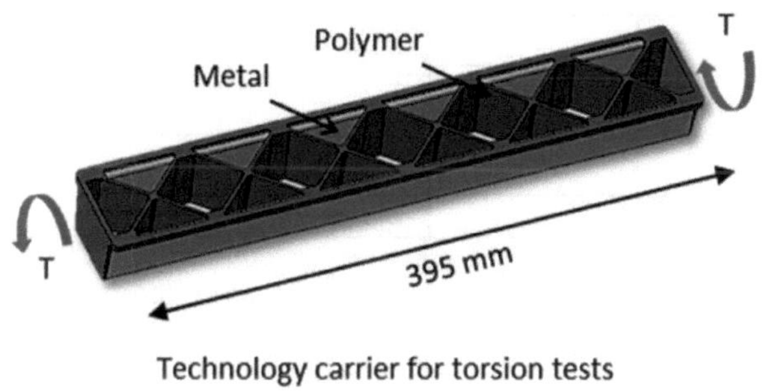

Process parameters and materials	
Metal	DX57+Z
Film-adhesive	nolax Cox 490-1
Film-application	Flatbed lamination
Polymer	PA6 GF30
Injection speed	69 cm³/s
Injection delay	30 s
Mould temperature	90 °C
Mass temperature	280 °C

Figure 8: Injection moulded metal-polymer technology carrier for close-to-reality tests

For corrosion prevention, the steel component is zinc-coated. In the case of the three-dimensional technology carrier, the soft steel DX57+Z with good deep drawing abilities is used. The specimen has a polyamide 6 polymer component with fibre reinforcement (PA6 GF30). Injection moulding of the technology carrier is conducted according to [1]. It is necessary to assess the impact of the presented surface defects (Figure 2) on the corrosion resistance and mechanical properties of the hybrid compound. Therefore, the technology carrier with flatbed lamination was exposed to environmental influences according to the internal test specification PV 1210 of the Volkswagen AG [4]. PV 1210 involves salt spray tests and varying climate conditions with different humidity and temperature. The complex design of the specimen and the thermal mismatch of steel and polyamide 6 lead to a mechanical load in the joining zone. Figure 9 shows the technology carrier after 15 cycles of PV 1210. The sample shows several corrosive spots at the cutting edges as well as amidst the surface. Besides the ones that result from the previously introduced surface defects, the edges at overmoulded areas do not show any significant infiltration.

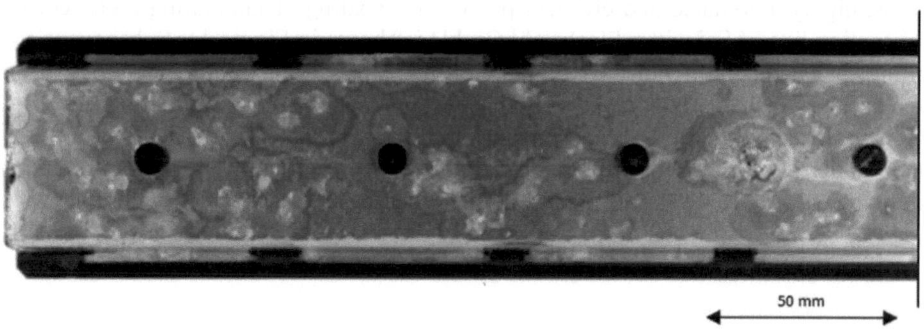

Figure 9: Technology carrier with flatbed lamination after 15 cycles of PV 1210

In addition to the visual inspection of the metal-polymer hybrid compound, destructive tests were conducted after corrosive ageing. Prior to the testing at 23 °C and 50 % rel. humidity, all samples are conditioned according to ISO 1110 at 70 °C at 62 % rel. humidity. The quasistatic torsion tests of the technology carrier at an angle speed of 1,5 °/s as well as the evaluation are carried out analogous to [1]. Figure 10 shows the torsion moment plotted against the torsion angle of the quasistatic torsion tests on the technology carrier. The reference without ageing is compared to samples that were exposed to 15 cycles of PV 1210. Determined are the torsion stiffness (E_T) between 5 and 50 Nm and absorbed energy (W_{T13}) from 0 to 13 °. The torsion stiffness decreases from 14,6 to 13,2 Nm/° (–10 %) and the absorption of energy from 950 to 880 Nm (–7,4 %). Possible reasons for the decreased properties are the corrosive infiltration of the film in load bearing areas or the ageing of the polymers themselves.

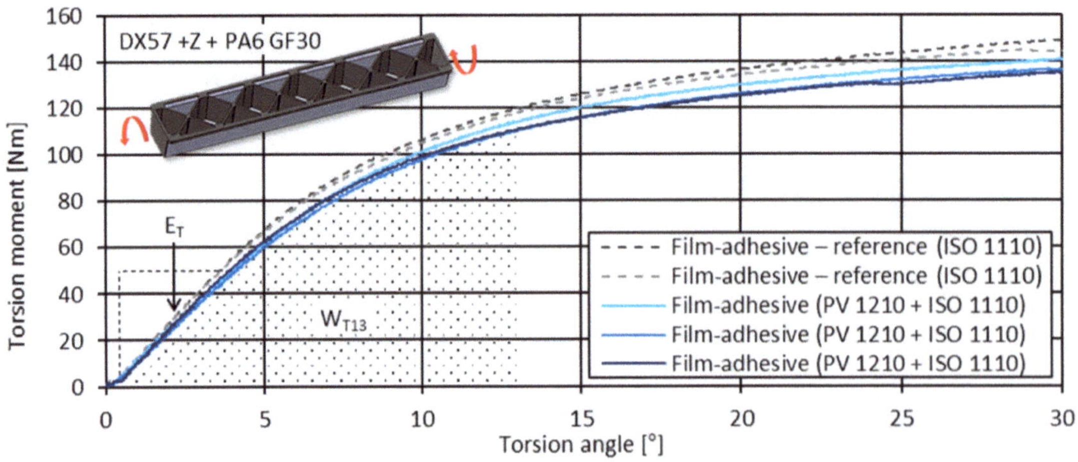

Figure 10: Torsion tests of technology carrier with flatbed lamination after 15 cycles of PV 1210

Good lamination quality is crucial for ageing performance and corrosion performance. The investigations show, that the film-adhesive could possibly function as an anticorrosive, if all surface defects are avoided and open cutting edges overmoulded. A suitable production process for the film-application including efficient quality assurance methods are highly indispensable, for a successful implementation of the film-adhesive in the automotive industry. Preliminary tests on laboratory scale proved that rolling the adhesive film directly onto a pre-heated metal sheet is feasible. Therefore, the industrial scale coil coating was initiated.

3.3 Development of a large-scale application process for the film-adhesive

Coil coating is particularly suitable for large-scale application of laminates on flat steel substrates. It is a continuous, highly automated and efficient process for coating or laminating metal coils. A scheme of the coil coating line of Salzgitter Flachstahl GmbH is shown in Figure 11. Subsequent process steps are decoiling, mechanical stitching, cleaning, pre-treatment, drying, primer coating on one or both sides, oven curing, finish coating on one or both sides, oven curing, film laminating, recoiling.

In a first trial, the film-adhesive nolax Cox 490-1 was laminated on GI steel strip. About 250 m were coated under testing several production parameters like laminating temperature and speed. The upper side was coated with the nolax film while the other surface was coated with a black coil coating finish.

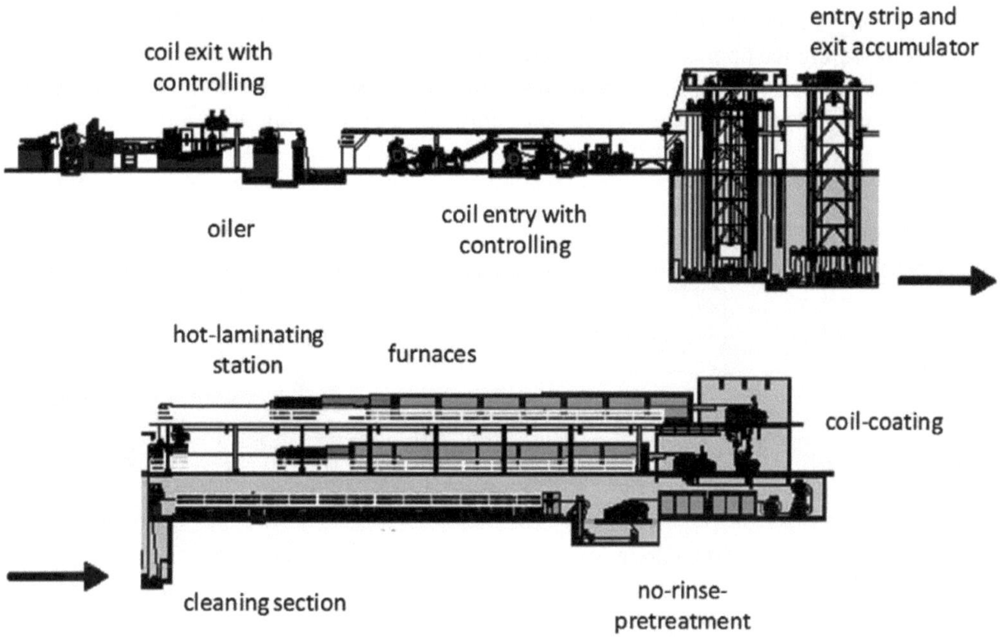

Figure 11: Scheme of coil coating line of Salzgitter Flachstahl GmbH [3]

The coating quality is crucial for a good performance of the subsequent produced parts and therefore an immediate quality test is needed. Different types of test methods were looked at in order to find a suitable quick test such as cupping, pressure water-jetting (ISO 16925:2014), cross cut tests etc. However, these test methods were not able to provide definite differences in quality in the required limited time. Proper rapid testing methods for these films have to be developed and are being investigated in current research.

3.4 Enhanced polymer-metal hybrid structures with coil coating

The single-lap shear specimen (Figure 12) is used to assess the influence of the application process of the film-adhesive nolax Cox 490-1 on the tensile bonding strength. Like the technology carrier, the injection moulding of the single-lap shear specimen takes place according to [1]. In Figure 12, the process parameters and materials are summarized.

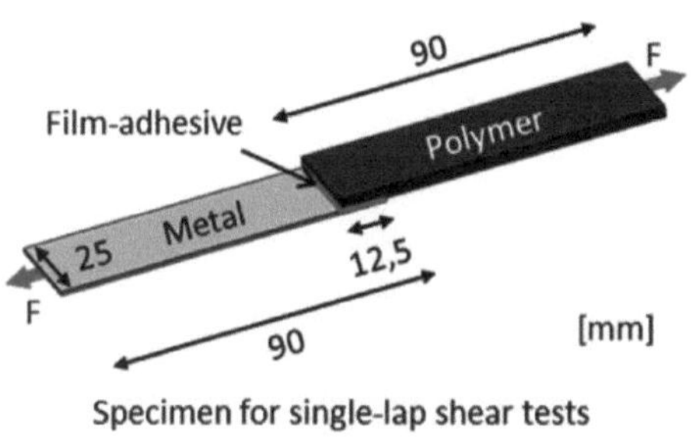

Process parameters and materials	
Metal	HX340 LAD+Z (t = 1 mm)
Film-adhesive	nolax Cox 490-1
Film-application	Coil coating / flatbed lamination
Polymer	PA6 GF30 (t = 3 mm)
Injection speed	30 cm³/s
Injection delay	30 s
Mould temperature	90 °C
Mass temperature	280 °C

Figure 12: Injection moulded metal-polymer single-lap shear specimen for tensile test based on DIN EN 1465

Figure 13 illustrates the achievable average tensile strengths and standard deviations in single-lap shear tests at 5 mm/min. The reference with flatbed lamination shows a tensile strength of 12,3 ± 1,0 MPa. For the coil coating process, a correlation can be drawn between tensile strength and coil speed as well as the peak metal temperature (PMT). Especially higher coil speeds implicate a decreased tensile strength. The combination of low coil speeds and a PMT above 250 °C lead to comparable results as the reference applicated by flatbed lamination. At the lowest coil speed and the highest PMT, up to 14,4 ± 0,2 MPa (+ 17 %) can be reached with coil coating. Furthermore, the overall standard deviation in the coil coating process amounts to only 20 % of the reference.

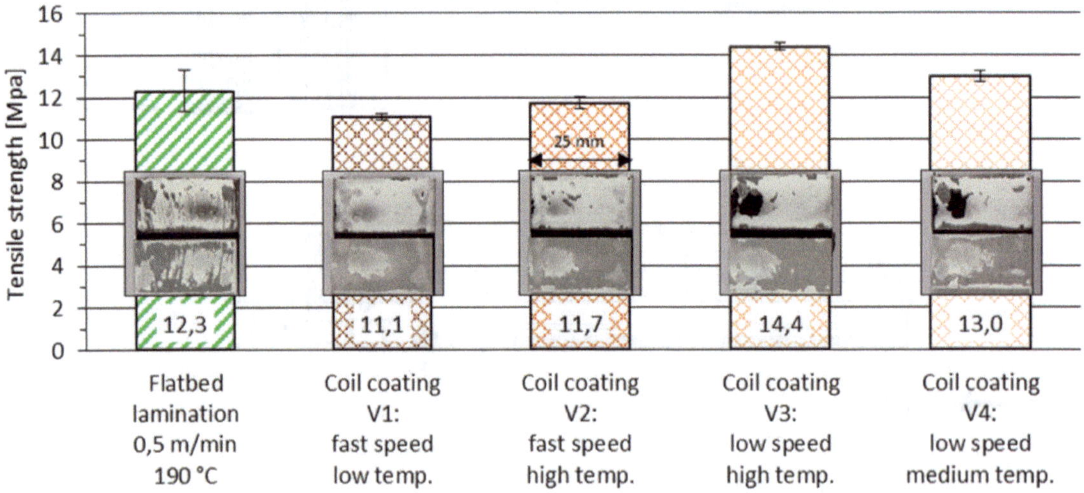

Figure 13: Tensile strength and fracture patterns of flatbed lamination and coil coating in single-lap shear testing

The corresponding fracture patterns provide additional information about the process quality. In the case of the flatbed lamination, a mixed fracture with cohesive parts in the film occurs [cf. 2]. However, it shows streaks and scratches in the surface that were already visible right after lamination (cf. Figure 2). For the coil coating process, the failure occurs more homogeneous and with less surface defects. It shows larger parts of adhesive failure that appear mostly close to the metal substrate. It is observed that the cohesive failure fraction in the film increases at higher tensile strengths.

With the coil coating process, the poor surface quality of the flatbed lamination process could be significantly improved. Figure 14 shows the comparison of the surface topography of the laboratory-scale flatbed lamination process and the developed coil coating process. The topography of the surface (6,5 × 6,7 mm) is visualized with a confocal microscope. Surface damages like linear scratches that are caused by the flatbed lamination, are detected. These partly reach through the entire film down to the zinc coating. Furthermore, irregularities of the film thickness can be observed. The coil coating process on the other hand shows a mostly undamaged surface with a negligible variation of film thickness.

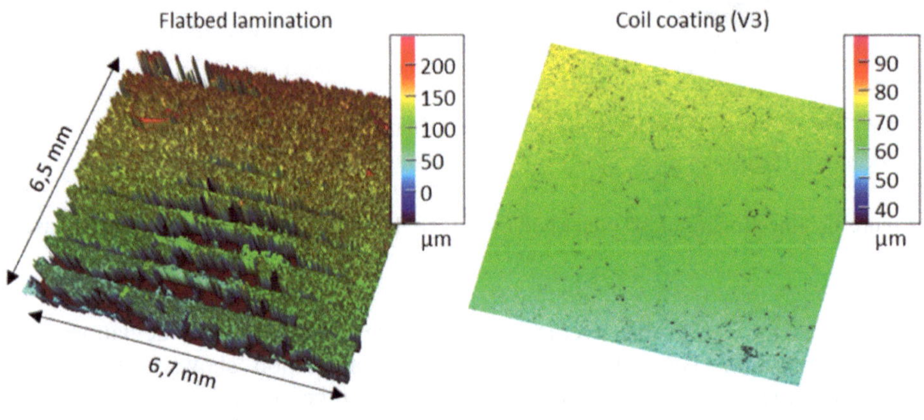

Figure 14: Visual comparison of the achievable surface quality with a confocal microscope

In order to examine the improvement in corrosive resistance, metal sheets (200 x 180 mm) with film-adhesive applied by flatbed lamination as well as coil coating V1 and V3 were exposed to 14 cycles of PV 1210. For a defined damage, scratches down to the metal substrate (100 mm) were preinduced into the film to assess the resulting infiltration or separation of metal and film (Figure 15). All three versions show corrosive infiltrations at the open cutting edges. Surface defects in the flatbed laminated film also lead to corrosive infiltration in several spots. The coil coated film performs significantly better and the infiltration solely takes place where the scratch was preinduced. However, the coil coating version with the lowest tensile strength (V1) shows slightly less distinctive corrosive marks compared to the highest strength (V3).

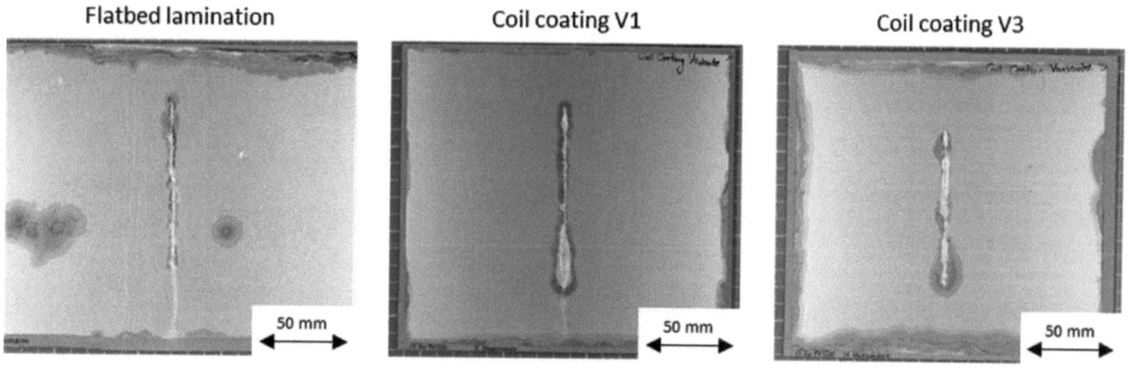

Figure 15: Corrosive infiltration of film-adhesive for different application processes after 14 cycles of PV 1210

3 CONCLUSION AND OUTLOOK

The polymer-metal hybrid technology has been transferred onto a highly safety-relevant component in the automotive industry [1]. Load-optimized design and the film-adhesive nolax Cox 490-1, helped achieving the target of a cost- and weight-efficient polymer-metal hybrid seat structure that meets the safety requirements. The consistent process chain for a close-to-production maturity level has now been established by developing a full-scale and highly efficient coil coating process for the film-adhesive. Compared to the reference process, the tensile strength of the metal-polymer hybrid compound could be improved by 17 % with an increased process stability. Furthermore, the surface quality and corrosion resistance could be significantly improved due to the coil coating process.

Further research has to be done on the investigation of additional load cases. This includes static modes I (tension stress), II (shear stress) and mixed (mixed stress), dynamic loads and more complex close-to-reality load cases. In order to identify the ideal process parameters for the coil coating, an optimization using DoE-methods has to be conducted in a smaller scale. Since most steel components in the automotive industry require corrosive protection, a double-sided film lamination should be in the focus of future research and plant design, in order to seal open edges by overmoulding.

For a holistic estimation in the next step the economic and environmental performance must be clarified and evaluated.

ACKNOWLEDGEMENTS

The project TRoPHy2 (02PQ5134) is funded by the German Federal Ministry of Education and Research (BMBF) within the Forschungscampus “Open Hybrid LabFactory” and managed by the Project Management Agency Karlsruhe (PTKA). The authors would like to thank the partners of the participating research institute of the German Aerospace Center as well as the project partner data M Sheet Metal Solutions GmbH, EDAG GmbH & Co. KGaA, Institute of Joining and Welding of the Technische Universität Braunschweig and Volkswagen AG Nutzfahrzeuge.

Furthermore, we will thank all other direct or indirect involved companies, institute, project teams and appropriation donors who supported this work.

REFERENCES

[1] Dreessen, P.; Dröder, K.: Großserientauglicher Leichtbau durch die prozessintegrierte Herstellung von Kunststoff-Metall-Hybridstrukturen, Leipzig, 2017.

[2] Habenicht, G.: Kleben. Grundlagen, Technologien, Anwendungen. Springer-Verlag, Berlin, Heidelberg, 2009.

[3] Meuthen, B.; Jandel, A.-S.: Coil Coating – Bandbeschichtung: Verfahren, Produkte und Märkte. JOT Fachbuch, Friedr. Vieweg & Sohn Verlag, Wiesbaden, 2008.

[4] Volkswagen AG: Konzernnorm PV1210. Karosserie und Anbauteile – Korrosionsprüfung. Wolfsburg, 2004.

TESTING OF METAL CONNECTIONS USING ADHESIVE BONDING COMBINED WITH SELF-PIERCING RIVETING

M. Reil[1,2,3], O. Knoll[1], D. Morin[2,3], M. Langseth[2,3]

[1] BMW Group, Knorrstrasse 147, 80937 Munich, Germany
[2] Centre for Advanced Structural Analysis (CASA), NTNU, NO-7491 Trondheim, Norway
[3] Structural Impact Laboratory (SIMLab), Department of Structural Engineering, NTNU, NO-7491 Trondheim, Norway

Keywords:

Steel-aluminium connections, Self-piercing riveting, Adhesive bonding

ABSTRACT

A modern car body structure is driven by the multi-material design approach. The best material at the correct place is the objective. Consequently, components made of dissimilar materials need to be joined and the applied joining technology becomes a key issue for crashworthiness and structural reliability. In the present work, connections between steel and aluminium using adhesive bonding combined with self-piercing riveting and their structural design are investigated. The safe application of this hybrid joining technology in a vehicle structure requires detailed knowledge about its mechanical behaviour. For that purpose, riveted, bonded and hybrid connections need to be characterized under tension, shear and mixed mode loading. In the present work, results obtained from a novel test setup are presented. The response of self-piercing riveting and adhesive connections is discussed separately as well as the interaction between both joining technologies. Furthermore, a new test setup for adhesively bonded and point-wise connected components is presented. Here, load combinations comparable to a vehicle crash are introduced into the connections. The developed setup facilitates successive failure of multiple connections and enables a broad validation of numerical connection models.

1 INTRODUCTION

Stricter legal and social requirements on pollutant emissions and energy consumption of passenger cars increase the demand for lightweight vehicle structures. Therefore, a multi-material design approach becomes essential in the design of a modern car body. However, joining of dissimilar materials is a challenging task and requires the application of new joining technologies. Self-Piercing Riveting (SPR) in combination with adhesive bonding is widely used in modern car bodies for joining steel and aluminium components. Virtual tests using numerical methods are applied to improve the crashworthiness and to ensure passenger safety. Consequently, accurate and efficient numerical models of the applied joining technologies are required. These models are calibrated from the mechanical behaviour of the connection. The literature provides a variety of experimental setups to characterize connections under tension, shear, peel and mixed-mode loadings. Typically, these studies are performed separately on SPR connections [1–4] and adhesively bonded connections [5–7]. However, as both joining technologies are applied in combination, suitable experimental setups for this hybrid joining technology are required. The validation of numerical connection models is another important aspect in their development. Tests on component level are applied for that purpose. Here, loading conditions comparable to a vehicle crash are introduced while cost and complexity are reduced significantly. The literature provides information on component tests of SPR connections [8,9] as well as adhesively bonded connections [10,11]. However, modern steel-aluminium car bodies experience combined failure of adhe-

K. Dröder et al (eds), *Technologies for economical and functional lightweight design*, Zukunftstechnologien für den multifunktionalen Leichtbau, https://doi.org/10.1007/978-3-662-58206-0_16

sively bonded and SPR connections under crash loading. Consequently, a suitable component test setup for hybrid connections is needed.

In this study, a multi-scale testing approach for bonded SPR connections is presented. A new test setup is applied to characterize connections under tension, shear and mixed-mode loading. Here, two rectangular plates are joined and assembled to form a cross shaped specimen and loaded up to failure of the connection. Riveted, bonded and riveted bonded (hybrid) cross specimens are tested. Furthermore, a novel component test for joined steel-aluminium structures is presented. Here, two Hat Profile (HP) sections are joined using adhesive bonding combined with SPR and tested in a 3-point bending setup. The setup is optimized to provide a broad validation of numerical connection models. Experimental studies are performed on a steel-aluminium material combination representative for modern car bodies. The top sheet of the connection is a cold rolled micro-alloyed steel. The bottom sheet is a 6000 series aluminium alloy. The sheets are joined using a crash modified, hot curing epoxy adhesive and self-piercing rivets.

2 CHARACTERIZATION OF BONDED SPR CONNECTIONS

2.1 Experimental setup

The mechanical behaviour of riveted, bonded and hybrid connections was investigated. Here, two rectangular plates were overlapped to from a cross shaped specimen. The plates were joined in the center by either SPR, adhesive bonding or a combination of both (Figure 1).

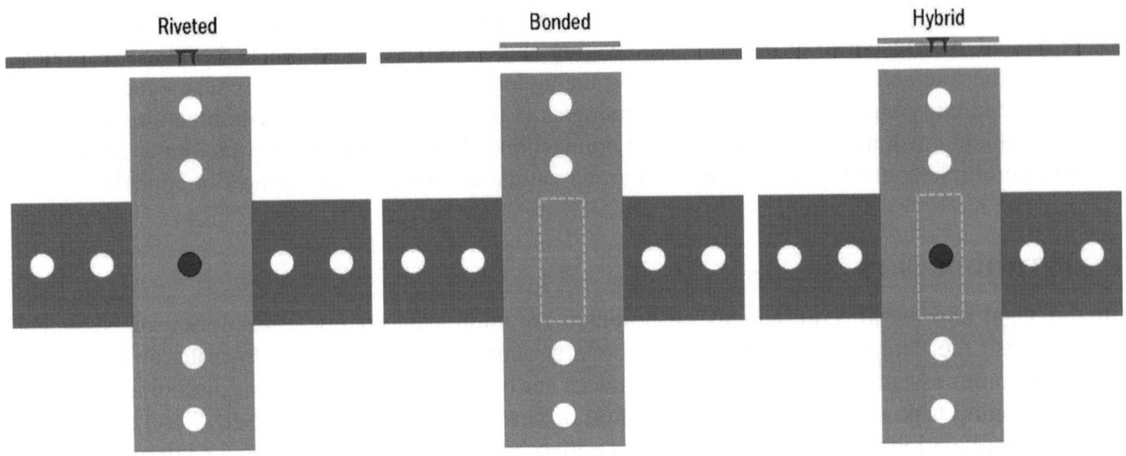

Figure 1: Cross specimen joint configurations

The adhesive of the bonded and hybrid specimens was cured in the ofen at 180° for 20 minutes. The heat treatment might influence the mechanical behaviour of the steel and aluminium plates. Therefore, the riveted specimens were subjected to the same cure cycle.

A new test rig introduced by Sønstabø et al. [12] was applied to characterize connections under tension, shear and mixed mode loading (Figure 2). Cross specimens were bolted to a fixed lower part and a moveable upper part. The upper part was guided inside a stiff steel casing by low friction roller bearings. Hence, the upper part could only move in the direction of the applied force. This was especially important under shear and mixed-mode loading. Here, the asymmetric configuration could cause unwanted rotations and displacements transverse to the loading direction. The rigid design of the new test rig ensured good control over the loading mode and boundary conditions. Cross tests were performed under quasi-static loading conditions

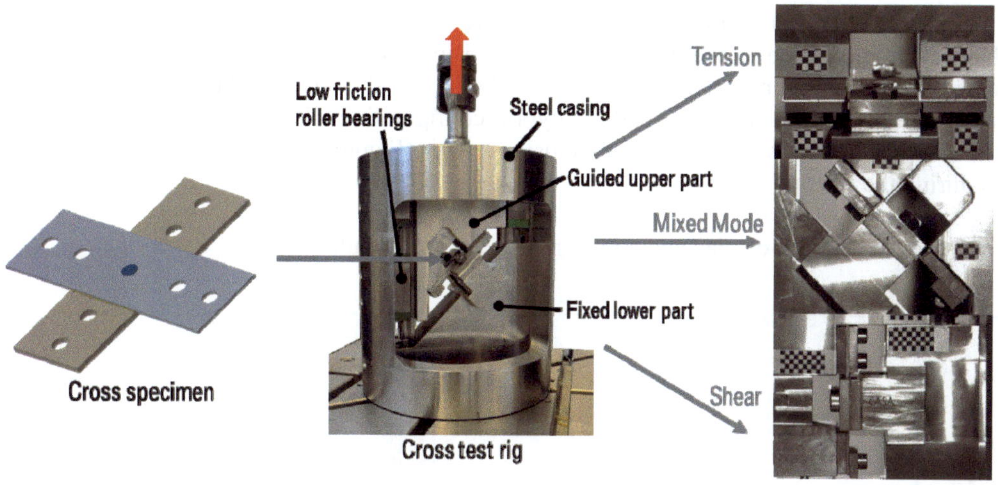

Figure 2: New cross test rig design

2.2 Experimental results

Test results of the riveted, bonded and hybrid cross specimens are discussed in the following. Figure 3 shows the force-displacement response under tension loading.

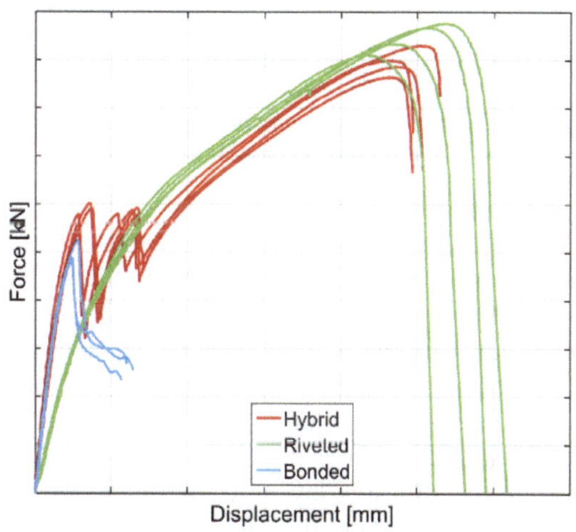

Figure 3: Force-displacement response under tension loading

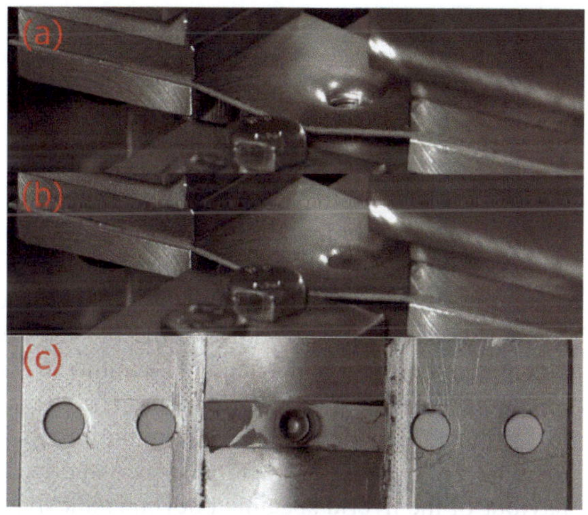

Figure 4: Deformation behaviour under tension loading

The highest joint strength was achieved for the riveted configuration. Here, failure occurred when the top sheet was pulled over the rivet head (Figure 4a). The initial stiffness of the bonded specimen was significantly higher compared to the riveted configuration. The maximum force was reached once a crack was initiated at one side of the adhesive specimen. The maximum force and the failure displacement of the adhesive configuration were significantly lower compared to the riveted configuration. The behaviour of the hybrid specimen was defined by an almost sequential loading of the adhesive and riveted connection. Initially, the response was dominated by the adhesive connection. Failure of the adhesive layer was initiated from one side. However, the crack propagation was stopped by the SPR connection at the center of the specimen. Subsequently, a crack was initiated and propagated from the other side. Two force peaks can be observed as a result of that failure sequence. The riveted connection provided further load carrying capacity beyond failure of the adhesive connection. The force-displacement response of the riveted and hybrid configuration was comparable after adhesive failure. However, the failure mode of the SPR connection was different. In the hybrid configuration the rivet was pulled out from the bottom sheet (Figure 4b). This might indicate a slight reduction of the mechanical interlock due to the adhesive layer. The impact on the maximum force and failure displacement of the SPR connection was negligible. Failure of the adhesive layer was dominated by cohesive failure for both the adhesive and the hybrid configuration. However, small areas can be identified where no adhesive is present due to inclusions of air (Figure 4c). These air inclusions were introduced during the riveting process.

Figure 5 shows the force-displacement response under shear loading.

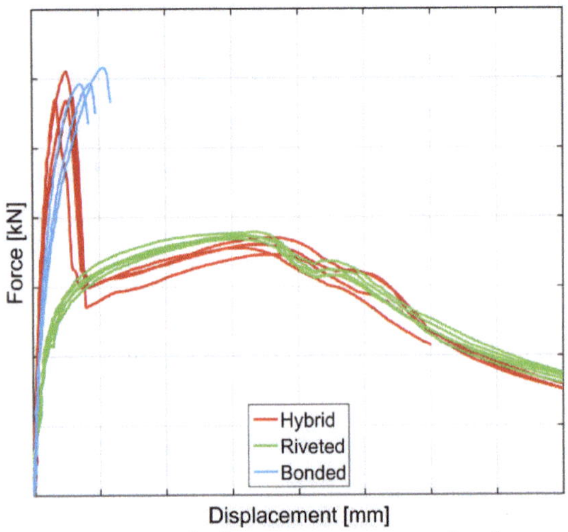

Figure 5: Force-displacement response under shear loading

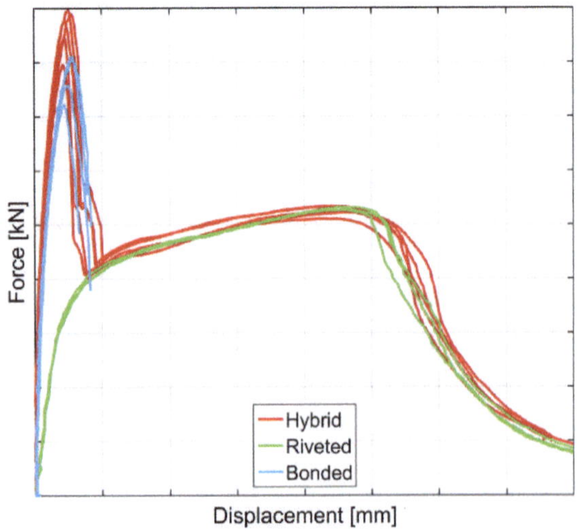

Figure 6: Force-displacement response under mixed-mode loading

In the riveted configuration the force increased almost linear in the beginning. The rivet started to rotate with increasing load. Pull-out of the rivet from the bottom sheet caused connection failure. The maximum force under shear loading was significantly higher than under tension loading. No sudden force drop can be observed even after the rivet was pulled out from the bottom sheet as the rivet was clamped in between both sheets. The initial stiffness of the bonded and hybrid configuration was higher compared to the riveted specimens. The shear loading resulted in a more uniform stress distribution inside the adhesive layer. The maximum force was approximately four times higher than under tension loading. Failure was characterized by a sudden crack inside the adhesive layer. As expected, the behaviour of the hybrid specimen beyond adhesive failure was comparable to the riveted specimen. Failure was also defined by rotation and pull-out of the rivet from the bottom sheet.

The force-displacement response under mixed-mode loading is shown in Figure 6. The deformation and failure behavior of all joint configurations was comparable to the shear load case. The joint strength under mixed-mode loading was in between the shear and tension load case.

3 COMPONENT TEST

3.1 Experimental setup

A new component test for hybrid connections was developed. Here, two HP sections are joined using adhesive bonding combined with SPR and tested in a 3-point bending setup (Figure 7).

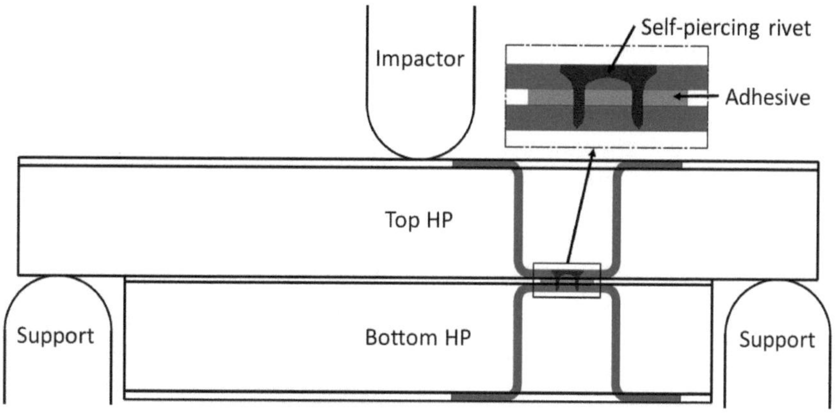

Figure 7: Principle of the new component test setup for hybrid connections

The SPR process requires access to the joining location from both sides. This is enabled by the open geometry of the HP sections. The well-defined boundary conditions of the 3-point bending setup ensure a simple correlation with numerical models. Ten SPR connections are distributed evenly over the length of the lower HP section. The adhesive layer has a nominal width of 15 mm and a nominal thickness of 0.3 mm. The geometry of the HP sections largely determines the loading introduced into the connections. It is of foremost importance to subject the connections to a wide range of different loading modes to enable a broad validation of numerical models. However, this can hardly be achieved with only one specimen configuration. Therefore, two configurations were developed, one for normal and one for shear dominated loading. In the normal configuration (Figure 8a), the top and bottom HP section have the same height.

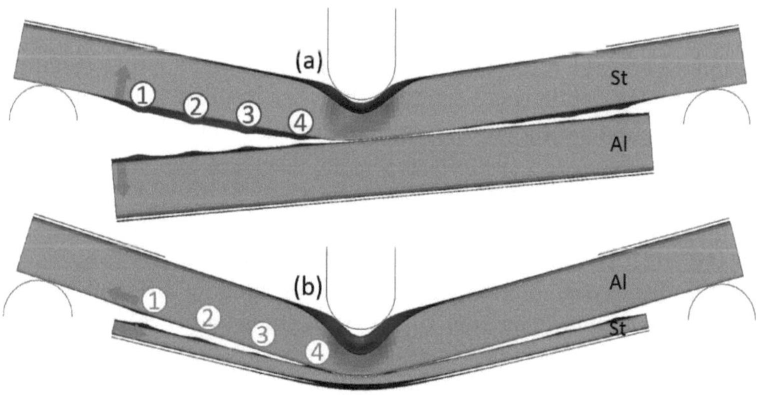

Figure 8: Bending behaviour of the normal (a) and shear (b) configuration

The top HP section is bent for increasing impactor displacement, while the bottom HP section remains straight. This results in a relative displacement normal to the connected surfaces. Hence, the connections are loaded in tension. In the shear configuration (Figure 8b), the height of the bottom HP section is significantly reduced. Furthermore, the top HP section is now made from the thicker aluminium sheet and the bottom HP section is made from the thinner steel sheet. Both changes reduce the bending stiffness of the lower HP section significantly. The relative displacement is now tangential to the connected surface, resulting in a shear dominated loading. Quasi-static experiments were performed for both specimen configurations.

3.2 Experimental results

Figure 10 shows the impactor force-displacement curves for both specimen configurations.

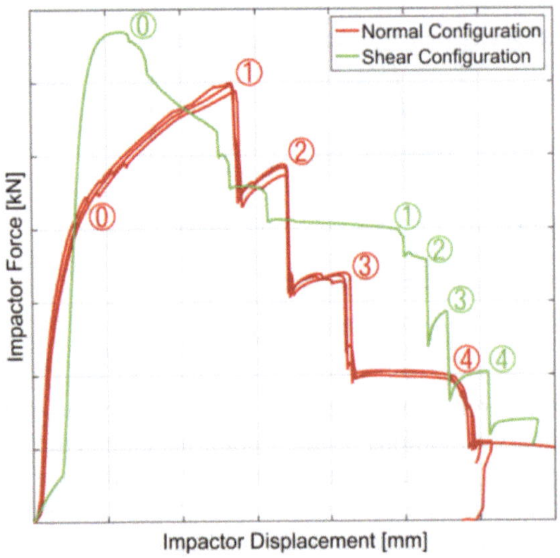

Figure 9: Experimental force-displacement curves of the normal and shear configuration

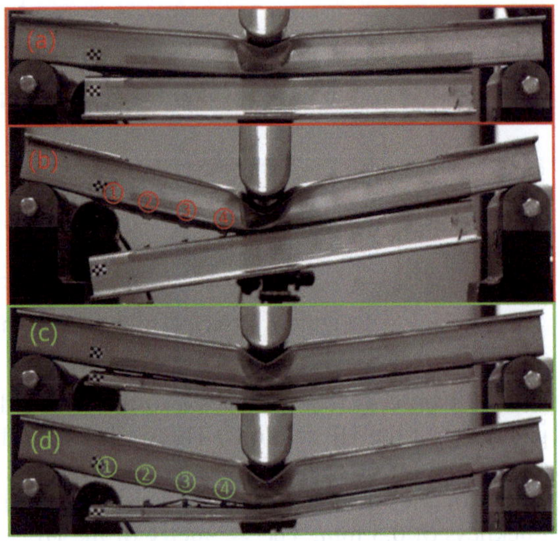

Figure 10: Failure sequence of the normal (a),(b) and shear (c),(d) configuration

In the normal configuration, the top HP section deformed plastically underneath the impactor for increasing force. At position 0, a crack was initiated inside the adhesive layer. The maximum force was reached right before failure of the first SPR connection at position 1 (Figure 10a). The crack inside the adhesive layer propagated towards the center of the specimen for increasing displacements accompanied by successive failure of three more SPR connections at positions 2 to 4 (Figure 10b). The maximum force in the shear configuration was slightly higher than in the normal configuration. Here, forces increased up to the point where the top HP section started to buckle underneath the impactor. At position 0, a crack was initiated inside the adhesive layer. The crack propagated towards the center of the specimen for increasing impactor displacement. Figure 10c shows the shear specimen right before failure of the first SPR connection. In total, four SPR connections failed in short succession marked by positions 1 to 4 (Figure 10d). The bottom HP section remained straight in the normal configuration but bent significantly in the shear configuration, successfully introducing different loading modes into the connections. Consequently, failure behaviour of the SPR connections differed significantly between both specimen configurations.

4 CONCLUSION

A new test setup was presented for characterizing connections under tension, shear and mixed mode loading. Quasi-static tests were performed on riveted, bonded and hybrid cross specimens. Hybrid specimens experienced an almost sequential loading of the adhesive and SPR connection. Here, the adhesive connection showed a stiffer response and a lower failure displacement. The strength of the adhesive connection under shear loading was significantly higher compared to the SPR connection. The SPR connection provided further load carrying capacity after failure of the adhesive. The presence of an adhesive layer had no significant impact on the behaviour of the SPR connection. The cross test results provided valuable information on the behaviour of hybrid connections and will further be used to calibrate numerical models.

A new component test to validate numerical models of steel-aluminium connections was presented. Test specimens consisted of two HP sections joined by adhesive bonding in combination with SPR. The specimens were tested in a 3-point bending setup up to failure of the adhesive and multiple SPR connections. Two specimen configurations were proposed, one for normal dominated and one for shear dominated loading. Both specimen configurations yielded successive failure of multiple SPR connections with different failure modes. The developed component test setup is therefore well suited for validating numerical models of hybrid connections. The setup will further be tested under impact loadings for validation of numerical connection models applied in vehicle crash analysis.

It was shown, that the mechanical behaviour of a connection can be successfully established through experimental tests. However, a full characterization involves an extensive and costly test program. Furthermore, the joint behaviour depends strongly on the material combination and a large amount of different materials are applied in a modern car body. Therefore, alternative methods are required. A virtual test setup enables an efficient joint characterization without cost and time consuming experimental tests. This virtual characterization approach will be studied in future research work.

REFERENCES

[1] R. Porcaro et al. The behaviour of a self-piercing riveted connection under quasi-static loading conditions. *International Journal of Solids and Structures*, **43**, 2006, pp. 5110–5131.

[2] L. Han et al., Mechanical behaviour of self-piercing riveted multi-layer joints under different specimen configurations. *Materials and Design*, **28**, 2007, pp. 2024–2033.

[3] D. Li et al. Influence of edge distance on quality and static behaviour of self-piercing riveted aluminium joints. *Materials and Design*, **34**, 2012, pp. 22–31.

[4] N. Hoang et al., The effect of the riveting process and aging on the mechanical behaviour of an aluminium self-piercing riveted connection, *European Journal of Mechanics A/Solids*, **30**, 2011, pp. 619–630.

[5] G. Dean et al., Prediction of deformation and failure of rubber-toughened adhesive joints. *International Journal of Adhesion and Adhesives*, **24**, 2004, pp. 295–306.

[6] M. May et al., Rate dependent behavior of crash-optimized adhesives – Experimental characterization, model development, and simulation. *Engineering Fracture Mechanics*, **133**, 2015, pp. 112–137.

[7] A. B. de Morais et al., Strength of epoxy adhesive-bonded stainless-steel joints. *International Journal of Adhesion and Adhesives*, **27**, 2007, pp. 679–686.

[8] R. Porcaro et al., Joining of aluminium using self-piercing riveting: Testing, modelling and analysis. *International Journal of Crashworthiness*, **9**, 2004, pp. 141–154.

[9] N. H. Hoang et al., Structural behaviour of aluminium self-piercing riveted joints: An experimental and numerical investigation. *International Journal of Solids and Structures*, **49**, 2012, pp. 3211–3223.

[10] D. Morin et al., A new cohesive element for structural bonding modelling under dynamic loading. *International Journal of Impact Engineering*, **53**, 2013, pp. 94–105.

[11] M. May et al., Predictive modeling of damage and failure in adhesively bonded metallic joints using cohesive interface elements. *International Journal of Adhesion and Adhesives*, **49**, 2014 pp. 7–17.

[12] J. K. Sønstabø et al., Testing and modelling of flow-drill screw connections under quasi-static loadings. *Journal of Materials Processing Technology*, **255**, 2018 pp. 724–738.

APPLICATION OF INNOVATIVE MATERIAL CONCEPTS FOR SAFETY LIGHTWEIGHT INSIDE CARS USING ALTERNATIVE POWERTRAINS

Stefan Lindner[1]

[1] Oberschlesienstraße 16, 47807 Krefeld, Outokumpu Nirosta GmbH, Stefan.Lindner2@outokumpu.com, www.outokumpu.com

Keywords:

Material concepts, Lightweight, Electric mobility, Crash-performance, Composite structure

ABSTRACT

New megatrends in the automotive sector like alternative powertrains, autonomous driving or car sharing but also continuous improvements like increased safety regulations or CO_2-emission standards can be influenced directly but also indirectly by an application-orientated selection of the used material or material combination. Thereby a new generation of material concepts can help to fulfil the partly divergent requirements and conflict of objectives consisting of strength, stiffness, energy absorption or lightweight. Further, the materials must be suitable for volume production and easy to integrate into established manufacturing processes like cold or hot forming and assembling, especially joining. Of course, new materials should be cost-effective, recyclable and completely simulatable.

To reach those targets, material scientists have different approaches like developing a monolithic metal or a compound structure, varying by fundamental basics like alloying elements, microstructure, number of phases, homogeneity, anisotropy, cross-sectional profile but also layer set-up and order [1].

The present paper takes up the mentioned diversity and introduces into different further developed material concepts which can be differentiated into opportunities for creating tailored properties of austenitic cold-hardening stainless steels, surface structured thin steel sheets and steel-polymeric composite structures. For every development, the focus is targeted to the combination of strength, stiffness and lightweight with the question how to increase every single value of the combination by using one of the new material concepts.

The target application is thereby the field of alternative powertrains, especially the application area of electric mobility. Therefore, three different concept ideas are given for this strategic part of automotive development. One element is to use significantly cross-industry innovations to ensure a fast integration combined with reliable experience into this new application field.

1 INTRODUCTION

In 1886 Gottlieb Daimler with his invention of a motorised carriage and Karl Benz with his invention of a motor vehicle established the automobile as one new individual private transport technology for passengers [2]. Parallel to the development of the automobile with combustion engines, researchers also developed successfully on electric vehicles. As two substantial examples Werner von Siemens with his electrically powered carriage (1882) or the electric cars developed by Ludwig Lohner und Ferdinand Porsche for the world exhibition 1900 in Paris can be pointed out. Passenger cars with combustion engines dominate the 20th century because of their significant expanded range, availability and price of the fossil fuels as well as the quick refuel process. During the last years, the frame conditions like the increased price and limitedness of fossil fuels but also the social desirability and acceptance changed. Therefore, electric vehicles experience a renaissance [3].

Simultaneously staged megatrends like autonomous driving or car sharing reinforced this trend. Further, regions inside Asia and Africa show an increased need for transportation concepts. Especially the

K. Dröder et al (eds), *Technologies for economical and functional lightweight design*, Zukunftstechnologien für den multifunktionalen Leichtbau, https://doi.org/10.1007/978-3-662-58206-0_17

topic of "last-kilometre" transportation for goods can be pointed out. In turn, the changing kind of mobility also results into demands for application-specific material concepts. As the main requirements for the mentioned application field of transportation, the key topics of lightweight, strength, stiffness and safety (resistance against impact) can be highlighted. At the same time, other important aspects like joinability, especially for multi-material-design, but also formability, corrosion resistance, an aesthetic impression of the surface, global availability, recyclability and the material-specific CO_2-footprint (life cycle engineering) must be also keep in mind during material selection and vehicle development [4, 5].

As one outcome of material development inside Outokumpu to support the efforts of the automotive industry and to realize the future expectations for transport systems, Figure 1 summarizes different material-related solutions to fulfil the alleged contra-diction between the four key-topics lightweight, strength, stiffness and crash-safety.

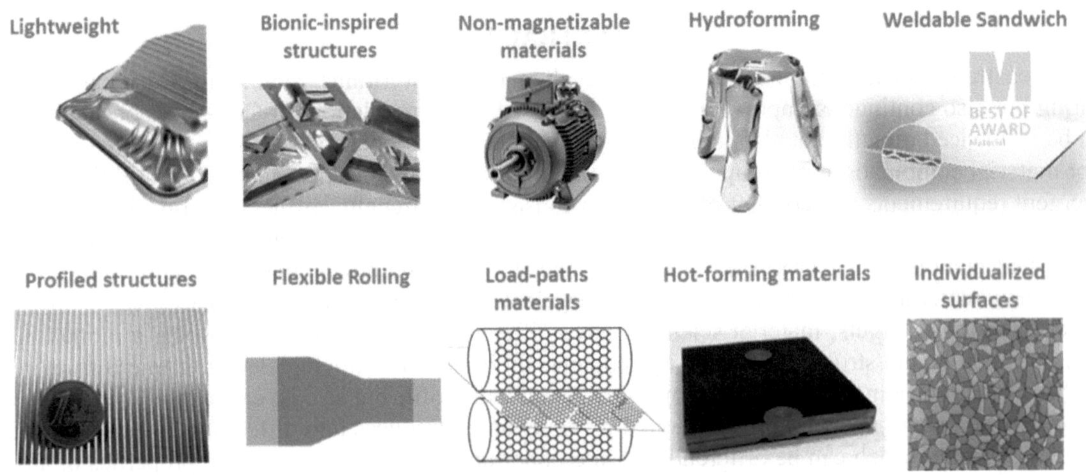

Figure 1: Overview of innovative material concepts by Outokumpu

2 AUSTENITIC COLD-HARDENING STAINLESS STEELS

Stainless steels with a stable one-phase, fully austenitic microstructure can be developed because of a special hardening mechanism called TWIP-effect (Twinning-Induced Plasticity) into the material category of ultra-high strength steels. Responsible for this characteristic profile are the coordinated alloying elements like chromium and manganese reaching in a specific stacking fault energy. As a result, the cold-formable material enables an intensive work-hardening during cold-rolling of the material but also during cold-forming of the component or during impact situation of the vehicle. At the same time an enormous energy absorption can be realized because of the outstanding relation of the mechanical-technological values. By taking advantage of the work-hardening effect, a complete material series could be created by cold-rolling with just one chemical analysis, view Figure 2. [6]

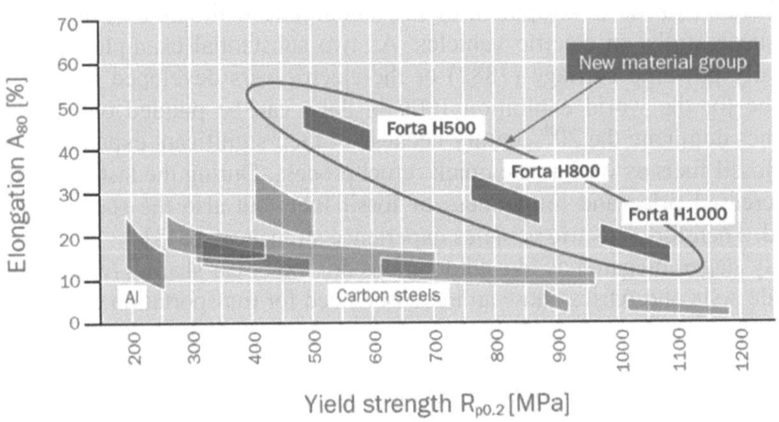

Figure 2: Mechanical-technological values of Forta H-series

The same effect supports the application during an impact or crash-situation. Furthermore, the component-manufacturer is able to adjust the local desired properties of the material in dependence of the cold-forming degree, view Figure 3. Design engineers can create higher strength or higher ductility areas inside one component where the usage conditions of the components require it.

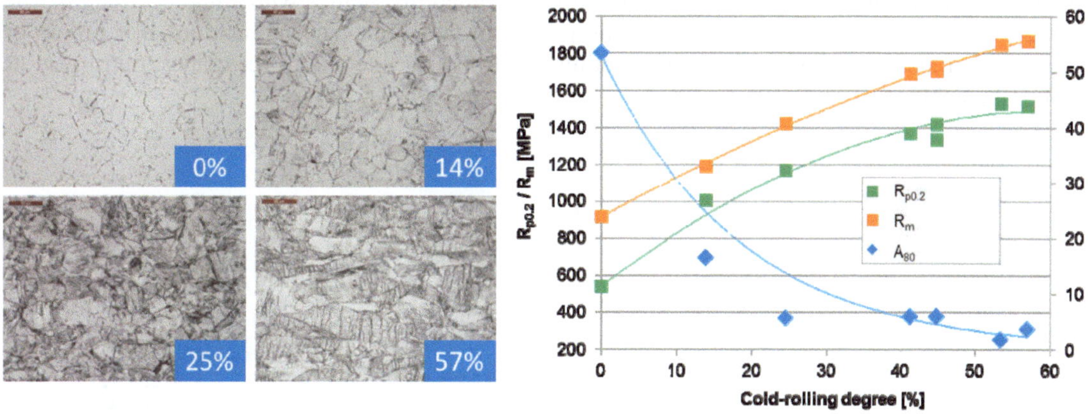

Figure 3: Resulting microstructure and mechanical values depending on the cold-forming degree.

In manufacturing of transport system, especially automotive car bodies, engineers use arrangements to have the right material at the right place and therefore to reach the relevant properties. Such approaches are called "multi-material design" or "using tailored products". [7]

A further developed approach in combination with the described material characteristic is therefore to initiate areas with a higher strength in combination with areas having a higher ductility. In relation to state-of-the-art material usage, this circumstance leads to an additional design criteria for materials beside a varying thickness with homogeneous mechanical values.

2.1 Integrated load-paths

The material concept of a TWIP-hardening and fully austenitic steel was now complemented by a process innovation to create ductility areas inside a high strength matrix having the same chemical composition and microstructure inside one material. Figure 4 points out one possible cold-rolling with a structured roll inspired by a honeycomb structure to integrate load-paths into the steel.

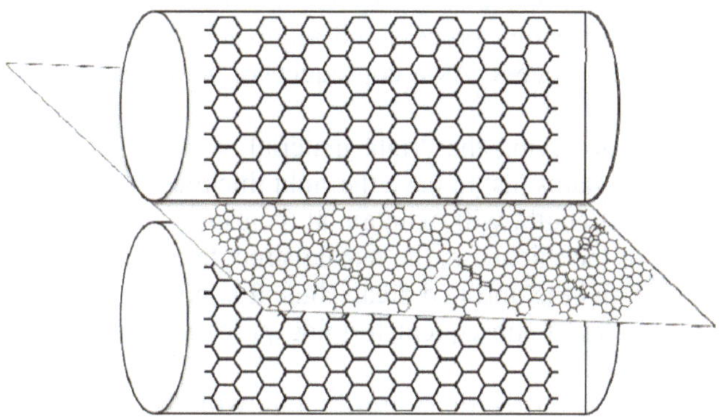

Figure 4: Cold-rolling with a structured roll to integrate load-paths into steels

Simple wave structure as 1st trials were carried out whereby the waves are displaced to each other to realize ductile areas called "combs" or "core" surrounded by higher strength indentations called "load paths" or "webs". The local-shaping is directly related to the above described cold-forming degree of the base material and results with a global view of sample properties in an increased yield strength but with nearly original ductility, view Figure 5. The deformed product with cold-formed indentations seems to have a better fatigue behaviour by creating a supporting effect of the ductile areas to the higher strengthened webs. Furthermore, the thesis of a particularly lower springback during forming operations when compared with state-of-the-art homogeneous materials can be formulated. Summarizing the product combines areas of higher ductility embedded in a matrix with a higher strength. The resulting product can be used because of its properties in crash-relevant applications like pillars or crash-boxes but also in fatigue and strength optimized structures like cantilever cranes, outriggers or agricultural machines.

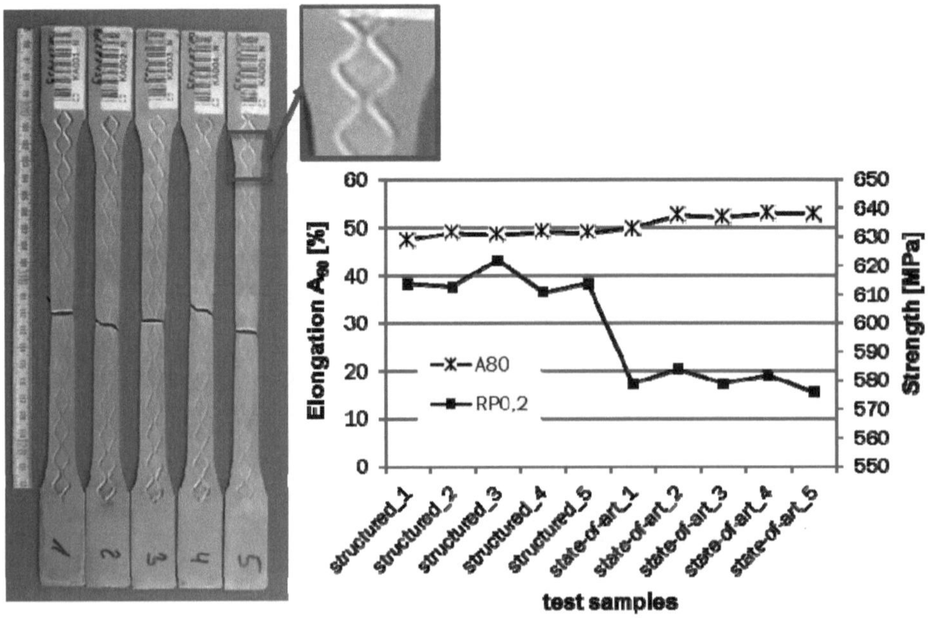

Figure 5: Effects on the mechanical properties because of inserting load paths

2.2 Flexible rolling

Another combination of a process and a material concept to create Tailored properties are flexible rolled blanks, which are metal products having different material thicknesses along its length. Usually, a subsequent recrystallization annealing process and a galvanizing step follow to the origin flexible cold rolling or eccentric rolling process. As a consequence, homogeneous mechanical-technological properties result.

Following the design rule of having the "right material at the right place", it means for flexible rolled blanks in state-of-the-art status just to have the right thickness at the right place. The mechanical properties, such as the tensile strength, will maintain at the same value as well as the ratio of the ultimate loads F as the product of the thickness, the tensile strength Rm and the width of the material between the flexible rolled area and the unrolled area. Thus, it is not possible to create areas with different strength and ductility, e.g. for a subsequent forming process or for fatigue depended component properties.

Using a stable one-phase, fully austenitic microstructure with the cold-hardening effect of TWIP in combination with the flexible rolling process, a combination of mechanical properties with thickness variation can be realized. As a result, the thickness reduction in the further cold deformed areas of the initial material is now combined with a specific and balanced local change in the mechanical properties of the material, such as yield strength, tensile strength and elongation, view Figure 6.

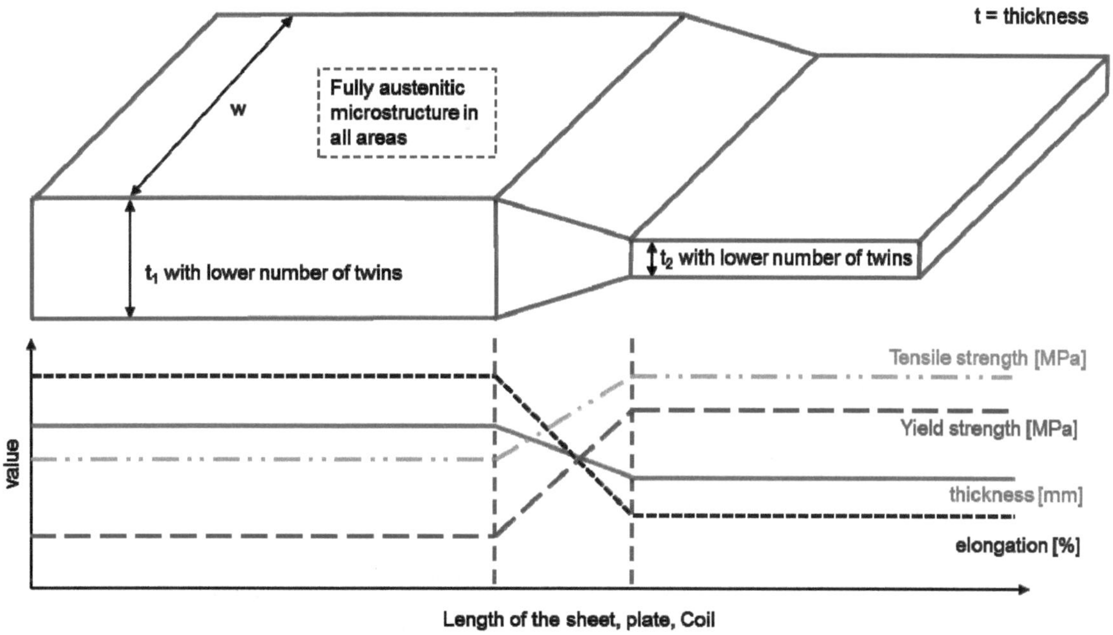

Figure 6: Relationship of thickness and technological values for enhanced flexible rolling

To point out the resulting benefits in mechanical properties, Figure 7 shows a simple way of calculation for a state-of-the-art flexible rolling process with annealing and galvanizing in relation to the enhanced process using a fully austenitic material with TWIP hardening and without a further annealing step.

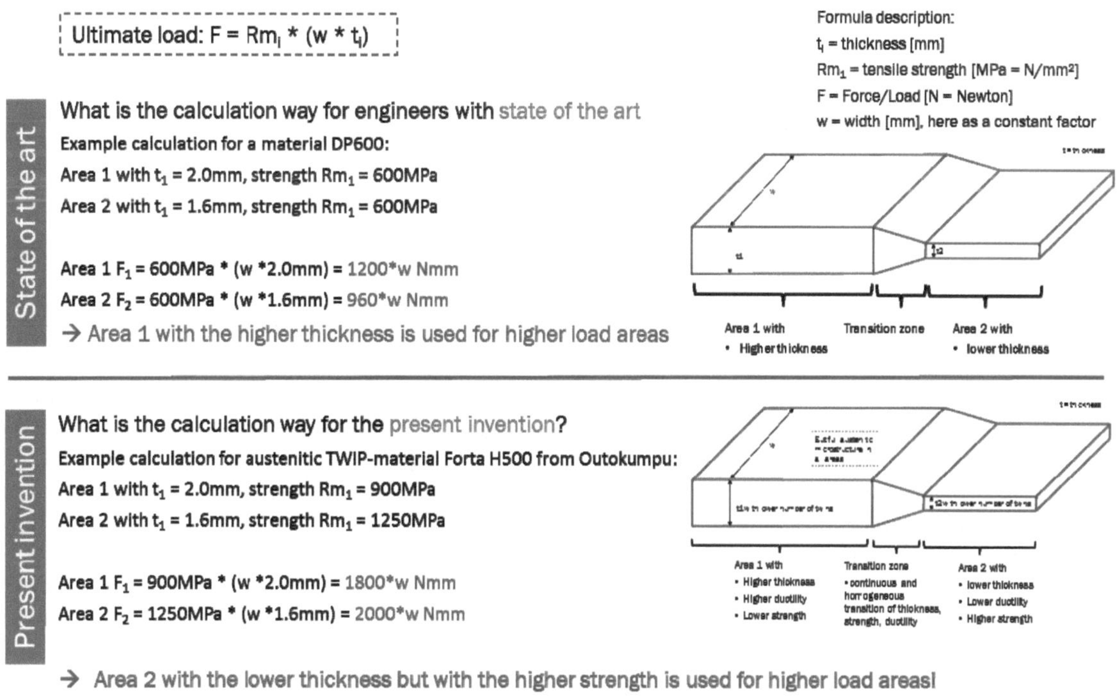

Figure 7: Relationship of thickness and technological values for enhanced flexible rolling

As a consequence, the method of construction changes: State of the art, the higher thick areas are used for higher load areas. With this kind of innovation, it is now possible that thinner areas can bear higher loads than other areas, depending on the applied material. One further benefit could be that the strength can be adapted to the following forming steps of the component manufacturer: Areas like flanges which will not be formed during component manufacturing can have a higher strength in a thin state. Other areas with a thicker initial thickness can be formed because of the higher ductility and therefore thin-out during forming operation.

3 PROFILED SHEETS

Beside the direct material related lightweight, there exist further lightweight designs like concept lightweight design, conditional lightweight design or form and structure lightweight design [8]. The last one was applied in a targeted manner to realize lightweight of thin steel sheets just because of a local change of the surface structure. One solution was a profiled sheet, in more detail an anisotropic corrugated sheet with the geometry represented in Figure 8. The profiled structure was manufactured with cold-rolling using double-sided sinusoidal structured rollers.

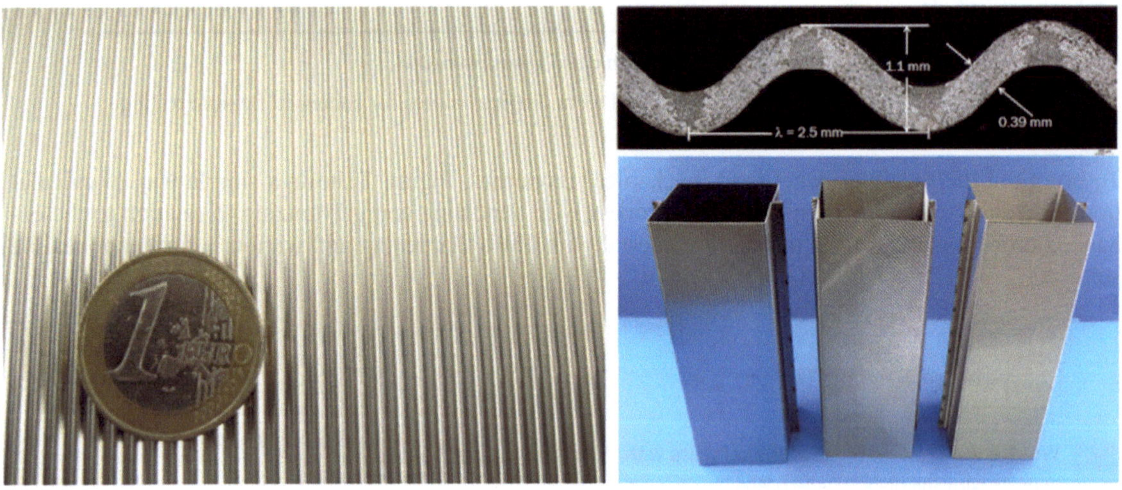

Figure 8: Profiled structured stainless steel sheet

As a result, a significant higher bending strength of +150 %, depending on the material direction, was able to create. At the same time, a lightweight of 27–29 % depending on the microstructure and material rigidity can be worked out. Further, an attractive appearance and a modified acoustic-behaviour can be added with the present structure. With a view to electric mobility and more specific to battery compartments, additional functionalities like sensor and measuring technologies as well as a defined thermal management can be integrated into the wave troughs. Setting-up the corrugated sheets as a package, functions of a heat exchanger or channels for cooling fluids can be realized. The profiled structure can be produced within a standard cold-rolling process for stainless steels. The solution can be adapted for all grades of stainless steels as coils or sheets with a width of 1250mm. Thereby annealed but also cold-hardened materials like pointed out in chapter 2 can be used. With the last aspect, it is further possible to combine different technologies and material concepts to reach at the end an increase of strength, stiffness and lightweight at the same time. Further profiles also with an isotropic rigidity are under development.

4 WELDABLE SANDWICH WITH 3D-PROFILED CORE

By using this profiled sheet as a 3D-profiled core material, a subsequent innovation for steel-polymeric composite structures, often called "sandwich structures", was possible to develop. The key of the material innovation is to combine the profiled metal with a polymeric material filled in from two sides. A dual-material core with an unidirectional supporting effect results, added by two thin stainless steel sheets as outer-layers, view Figure 9 and 10.

Former sandwich solutions had not been successful in spite of their enormous lightweight potential because of their insufficient weldability and a lower stiffness compared to monolithic metals.

Figure 9: Up-set of the different layers of Outokumpu's Weldable Sandwich

The resulting stainless sandwich structure called "Weldable Sandwich" is the first direct weldable steel-polymeric composite structure for automotive car body, transport and construction engineering. The developed sandwich structure is now able to combine the automotive challenges consisting of a reduced weight (–30 %) with an increased stiffness of +80 % compared to a monolithic material with the same thickness at the same time. Further, a better behaviour in point of acoustic, energy absorption and crash safety is enabled. Thereby no changes during manufacturing are necessary: the Weldable Sandwich can be handled like a conventional steel sheet during component manufacturing in point of joining, cutting or punching and limited in point of forming. The profiled core, pointed out in detail in Figure 10, enables at every time a metallic contact between booth thin stainless outer-layers with the result of having a constant electrical current flow during resistance welding. Thereby, no special clamping devices, welding machines or specific welding parameters like more-impulse welding or a previous displacing of the polymeric core material are necessary.

Figure 10: Final structure of the Weldable Sandwich with metallic contacts

Dissimilar welding combinations in a lap-joint configuration with monolithic metals are possible. Moreover, also similar sandwich-with-sandwich combinations are possible, view Figure 11. In such a case, the current flow can be enabled by the profile over in sum six metal layers. Using typical contact diameters for electrode caps of d_{EL} = 5.5 mm, the used profiled sheet core material enables at every location a minimum number of two waves as a contact for current flow because of the wave length λ = 2.5 mm.

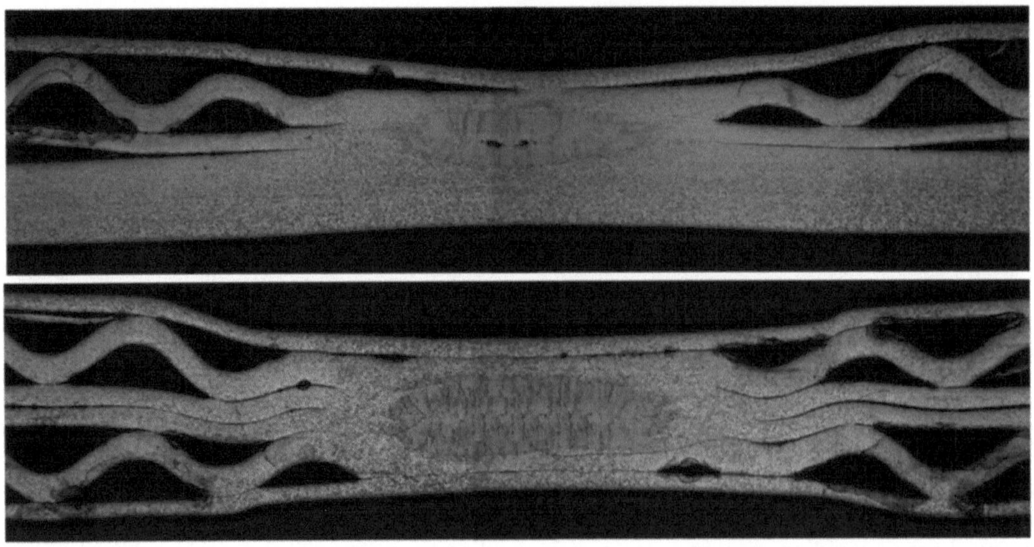

Figure 11: Dissimilar (above) and similar (below) resistance spot welding of the sandwich.

Furthermore, the sandwich solution offers also for mechanical joining processes like riveting advantages like a significant higher power transmission and compressive rigidity in comparison to other tested sandwich structures. The profiled core enables for these cases the possibility of a mechanical clamping and additional reinforcement between the core and the joining elements, view Figure 12.

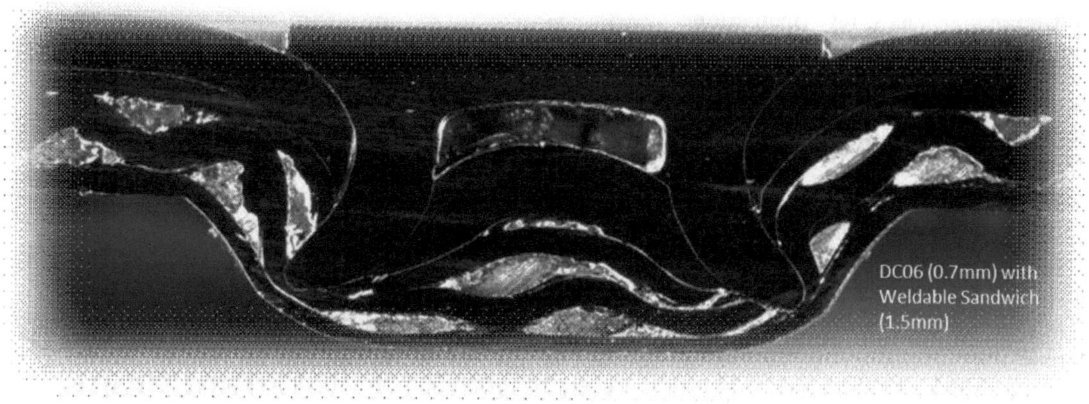

Figure 12: Structure of Weldable Sandwich after mechanical joining with a half-hollow rivet

Summarizing, the innovation of a profiled core solves the question how to fulfill the requirements with a geometry depending answer: Because of the continuous contact between the metallic outer-layers at every time stiffness and acoustic advantages combined with a direct resistance weldability can be enabled. The Weldable Sandwich with its described benefits was awarded with the Best-of-Award for the category "material" during the 15th Materialica Design- and Technology Award-Show for the most innovative material development in 2017 (Munich, 10/2017).

5 CONCEPT IDEAS FOR ELECTRIC MOBILITY

One effective method to create new products or solutions is to apply cross-industry innovations whereby the method deals with drawing analogies and transferring approaches between contexts. Target is to enlarge the border of the own business to other sectors and areas where the new innovation for the own application field is used since many years with a high level of experience and reliability. Therefore Vullings and Heleven [9] concluded that "the best way to develop ideas is to look at other places". This method is applied in the following subchapters to introduce new ideas into the market of electric mobility.

5.1 Individualized surfaces

The electric mobility in combination with other major trends like autonomous driving or car sharing changes the requirements of the interior of the future. Because of this, some sources are speaking about the "interior as number one factor for future's purchase decision" [10]. At the same time, the outer-skin remains a character-shaping identifying feature of a car.

Inside this environment, stainless steel with its versatile possibilities of a patterned surface will act one important role in human-to-material-contact. Thereby, the well-known advantages of stainless steel surfaces for design-applications remain valid and open the possibility for cross-industry innovations. In architectural applications like the Empire-State Building, New York, from the 1930ies or the Burj Khalifa as the highest building of the world, stainless steel surfaces are used because of their aesthetical and tasteful impression. Furthermore, stainless steel is used in medical and food technology because of its antibacterial properties and additionally in the application field of railway vehicles because of its easy cleanability and resistance against aggressive cleaning media. With a harmonious adjustment of optical properties like gloss level or reflection factor in combination with a specific patterned surface, the visual appearance of a scratch-resistant surface can be emphasized. Furthermore, the optical properties can be combined with safety benefits, e.g. used in series applications like the underride guard of the Mercedes Benz Actros where design and crash properties are successfully used since many years. Summarizing, further developed stainless steel surfaces offer unique features for future design concepts and beside that are highly customizable for passengers as well as for brands.

5.2 Spring steels as safety protection

Another cross-industry innovation for electric mobility can be divided with a view to the application of springs where ultra-high strength austenitic stainless steels are produced by cold-forming since many years. According to DIN EN 10151 [11], tensile strength levels with the EN-code C1900 reaches 1,900 up to 2,200MPa.

The properties of cold-hardened spring steels were used to design a crash management system for a battery compartment to protect the battery modules and cells inside. The system refers to the physical operating principle of a mechanical compression spring combined with a double floor system, view Figure 13.

The last aspect is an additional cross-industry innovation, inspired from maritime navigation, where a double-floor between the outer-skin and the inner bottom is statutory since the 19th century and serves the ship for an increased safety in the event of grounding or collision. Further, the whole ship stiffness increases because the double floor for ships is rigidly connected with inner beams. As a result of adapting this well-known concept of shipbuilding in combination with an inside compression spring effect now for Battery electric vehicles (BEV), the double-floor system will be yielded because of the spring effect and therefore absorb the crash energy without destroying of the inner bottom. Beside the fact that the vehicle battery will be directly protected, the forces and accelerations affected to the occupants will decrease significantly, too.

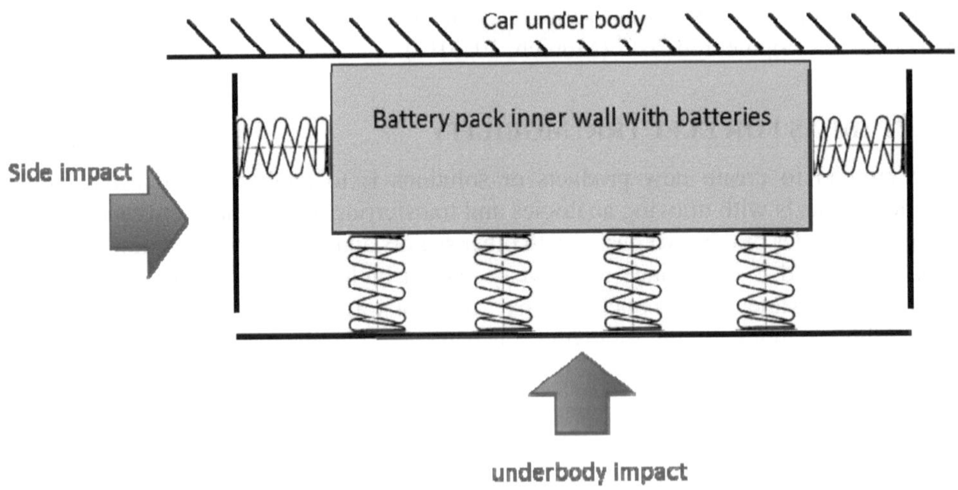

Figure 13: Schematically view of the spring system inside a double-floor arrangement

In the case of an underbody slitting, the sharp-pointed object will slip-off if the impact takes place with a slanted angle. The contact angle will decrease because of the spring effect. If the underbody impact of a sharp-pointed object takes place under a contact angle of 90° without the possibility of slipping, the outer-skin absorb the energy even if the sharp-pointed object will introduce into the space of the double-floor system but without damaging the battery system itself.

For every embodiment typical spring values can used to characterize the properties of the crash system. One parameter is the spring rate which defines the ratio of a force which contacts the spring and the therefore induced displacement of the spring. Depending on the desired crash behaviour, the ratio can be shown a linear or a non-linear characteristic. A non-linear progressive characteristic could be used if the OEM wants to absorb firstly the energy and at the end creating a high resistance against the impact. Conversely, a degressive characteristic can be used if initially a high resistance is desired and then the energy will be absorbed. Another spring characteristic is the block length which can be defined in general for a compression spring as the situation where all windings rest on top of each other. Adapted to the case of a vehicle crash, it is the situation where the energy absorption and the thereto related impact of a compression spring system is exhausted. Now the system is compressed and further energy must be dissipated over inner beams or cross members.

As one calculation example, the final way of the spring system inside the double-floor system reaching the block length can be calculated with 0.1 m in the case that a vehicle with 1.5 tonnes weight collides under side impact conditions with a speed before impacting of $v_1 = 60$ km/h.

The battery compartment using such a double-floor system can be executed as a repeatable non-destructed system which can be continued used after the crash without further effort by getting itself into its origin starting configuration.

One advantage of stainless spring steel systems is the benefit of corrosion resistance in the wet underbody area of an electric vehicle. Further, stainless steels have a higher acid-resistance in the case of a leakage with battery liquid and a higher heat and thermal resistance. The last point is important that the battery compartment can maintain the structure during a case of fire. Also this can be interpreted as a cross-industry innovation, in this case from exhaust systems.

5.3 Integrated safety cell

As an outlook, a further developed safety concept for a battery electric vehicle should be given. Having a view to passenger cars until the 1950ies, the disadvantage of having nearly the same material strength in longitudinal centre line of the vehicle affected the occupant safety significantly. Béla Barényi from Daimler AG defines in his DE patent application 854157C a passenger compartment whereby the strength decreases constant or in gradual phases in direction of the front and back end of the vehicle. The highest strength level of a material was reached inside the passenger compartment. Thereby, the security safety cell for passenger cars was invented to protect passengers during an accident or crash situation. Since then the front compartment and rear end of the car are known as deformable zones. Béla Barényi spoke about a centre cell for the higher strength passenger cell and outer cells for the softer front compartment and rear end. [2]

Today, the crash requirements change again by introducing a battery compartment into the vehicle floor which is critical against an impact and resulting deformation. In particular, 27 % of all accidents happened as side impacts whereby 2 % of all accidents showing a speed of $v \geq 64$ km/h. The number of side-impacts increases excessive for higher speed levels whereby 26 % of those side-impacts show a fatally effect and further 57 % with a severally injury of the occupants. [12]

The invention of Béla Barényi was adapted by using again the physical operation principle of a compression spring and extended also for side impact (y-orientation) and underbody impacts (z-orientation), view Figure 14. The battery compartment was integrated into the former passenger safety cell whereby the security cell works like a spring on the block. Further, the integrated security cell is surrounded by a lower strength energy absorbing area in cross direction as well as in height direction of the vehicle.

It is important to ensure that the physical protection according the UN R94 and R95 standards is given to protect against persons touching of the high-voltage components (IPXXB protection). Those components must be completely covered. For this, the integrated security safety cell can be designed into several inner rooms to separate the passengers from the engine. In this case the before described bulkheads known from ships but here in a horizontal orientation can separate the different spaces from each other.

During a rollover situation, the complete weight of the battery components works on the roof structure which must be stable without buckling to safe the passengers. This fact can be also integrated into the present concept. Also for the roof structure, the security cell can be defined as a non-deformable cell surrounded by an energy-absorbing area.

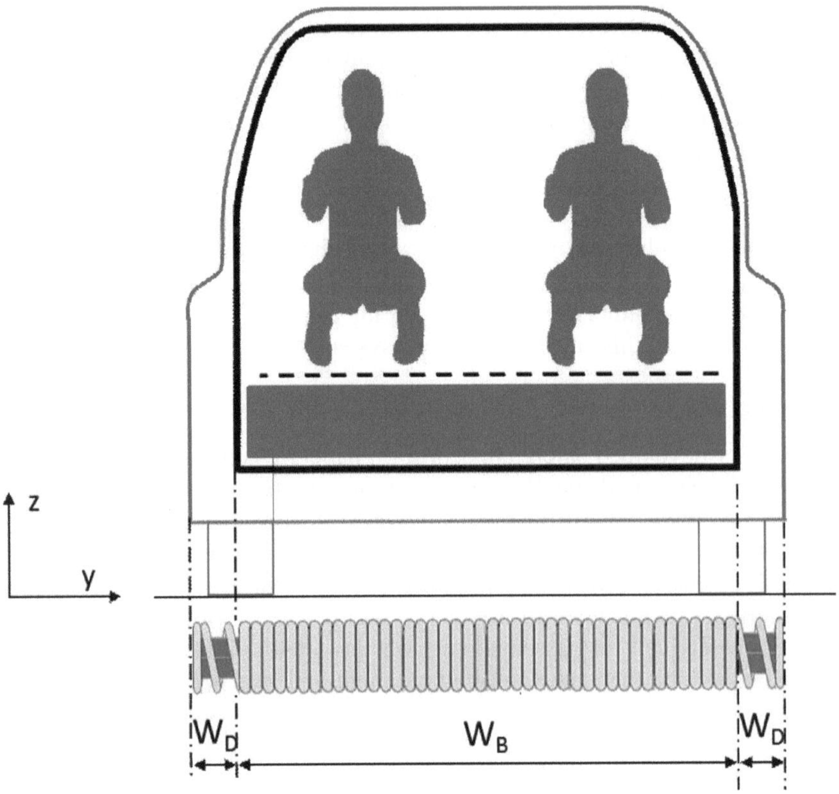

Figure 14: Schematically view of integrated safety cell for battery electric vehicles

Summarizing, this vehicle concept leads to a complete new design of vehicles, having a higher seating position for the passengers. In time of electric mobility in urban spaces, the traditional vehicle dimensions which are aerodynamically optimized and characterized by a longer length and lower height could be changed into dimensions which are more compact in length but enlarged in height. The necessity of consumption reduction by optimizing the aerodynamic seems to be not relevant for future battery electric vehicles in urban areas.

6 CONCLUSIONS

The present paper places further-developed material concepts within the context of changing customer demands and more challenging requirements in the automotive world. Innovative material concepts divided by cross-industry innovations could be one substantial enabler to solve these contrary goals by dealing with local adjustable properties. This characteristic unites different material concepts independent from their micro- or global structure, hardening mechanism or alloying concept. Further, all described material concepts combine general development objectives like lightweight, strength, stiffness and energy absorption. Summarizing, current developed material concepts must provide evidence about their ability for a worldwide and large scale production in the area of automotive over many years. But already nowadays, the outlook can be formulated that these material concepts result in new design possibilities, sometimes changing design rules and as demonstrated within the chapter about "concept ideas for electric mobility" in new ways of thinking automotive and transportation.

REFERENCES

[1] K. Stamm, H. Witte, Sandwich-Konstruktionen, Ingenieur-Bauten 3- Theorie und Praxis, Springer-Verlag, 1974.

[2] H. Niemann, Béla Barényi und seine Erben – Sicherheitstechnik made by Mercedes-Benz, Motorbuch-Verlag, 2016.

[3] A. Karle, Elektromobilität – Grundlagen und Praxis, Carl-Hanser-Verlag, 2015.

[4] Q. Zhou, Electric Mobility in China – Developments, Opportunities & Challenges, 30th International electric vehicle Symposium, Stuttgart, 10.09.2017.

[5] B. Müller, C. Zachäus, G. Meyer, European strategic processes towards competitive, sustainable and user-friendly electrified road transport, 30th International electric vehicle Symposium, Stuttgart, 10.09.2017.

[6] S. Lindner, Leichtbaupotenziale mit ultrahochfesten Mangan-Chrom-Stählen im Fahrzeugbau, Faszination Hybrider Leichtbau 2016, Wolfsburg, IST automotive nord e.V., 2016, pp. 356–375.

[7] J. Hillmann, The new Golf VII – Lightweight Design in Large Scale Production, Aachen Body Engineering Days, 2012.

[8] H. Friedrich, Leichtbau in der Fahrzeugtechnik, ATZ/MTZ-Fachbuch, Springer Vieweg, 2013.

[9] R. Vullings, M. Heleven, Not-invented here – cross-industry innovations, BIS Publishers, 1st edition, 2015.

[10] H. Hendriks, Automobil Produktion, volume 04/2017, p. 27.

[11] DIN EN 10151, Stainless steel strip for springs – Technical delivery conditions; German version EN 10151:2002.

[12] M. Büchsner, Integration of occupant safety systems into seating environment in the light of autonomous driving, presentation at 2nd Annual Seating Innovation Summit, Berlin, 6th April 2017.

BIO-BASED MATERIALS FOR EXTERIOR APPLICATIONS – PROJECT BIOHYBRIDCAR

Ole Hansen, Christoph Habermann, Hans-Josef Endres

[1] Heisterbergallee 10A, 30453 Hanover, Fraunhofer WKI, https://www.wki.fraunhofer.de/, hofzet@wki.fraunhofer.de

Keywords:

Hybrid material concepts, Sustainable mobility, Natural fibre reinforced plastics

ABSTRACT

The objective of this work is to evaluate the possible uses of bio-based materials (natural fibres, matrix systems) and hybrid material concepts for sustainable mobility. In the first stage, this has been elaborated for a door frame and door panel of a Porsche Cayman GT 4 Clubsport (981) as a test vehicle. On this demonstrator vehicle, the material concepts optimised in the laboratory for selected components are intensively tested in practice in the VLN racing and their suitability for serial production of various exterior components will be analysed. The holistic view is divided into three main points:

- Technical evaluation (mechanics, vibration, semi-finished products, design, calculation)
- Ecology (LCA)
- Economy (supply chain, availability, production process, processability)

This article is focused on the first results of the technical evaluation for the door frame and gives a first outlook on the ecology and economy aspects for the used materials. As part of the project, various exterior components are manufactured under different weighted aspects that allow the comparison of carbon, natural and, especially, hybrid fibre reinforced plastics, thereby highlighting the specific advantages of different material combinations, e.g. using a combination of flax and carbon fibres for reinforcement. In addition to the technical feasibility in terms of material properties and the processing methods as well as an ecological consideration, in particular the economic feasibility of these materials is considered. The following chapters present the fundamental strategy of the project which is to transfer the results to a sustainable mobility in the future.

1 INTRODUCTION

In the automotive industry "fibre reinforced plastics" (FRP) have prevailed in recent years. FRPs are materials that are made up of at least two different components. They consist of a homogeneous carrier component, the matrix, the plastic and, embedded therein continuously or discontinuously, the reinforcing component, the fibres. These composites are of particular importance in the lightweight construction sector and, due to their high-grade density-specific properties, are being used more and more frequently to replace materials such as steel and aluminium (Schürmann, 2007). When two or more types of reinforcing fibres are combined in a matrix in an FRP, the term hybrid fibre reinforced plastic (HFRP) is used (Neitzel et al., 2014). Hybridisation fuses the characteristics of two types of fibres in one component, and the proportion of their respective fibres allows the ratio between their advantages and disadvantages to be steered in a desired direction (John and Thomas, 2008, Amico et al., 2010).

In the automotive industry, conventional reinforcing fibres such as carbon fibres and glass fibres cover the majority of fibre composite technologies due to their good mechanical properties (Flemming

K. Dröder et al (eds), *Technologies for economical and functional lightweight design*, Zukunftstechnologien für den multifunktionalen Leichtbau, https://doi.org/10.1007/978-3-662-58206-0_18

and Roth 2003, Neitzel and Mitschang 2004, Schürmann 2007). The steadily growing public demand for sustainability, resource-saving and recyclable materials is moving more and more into focus, alongside mechanical properties such as the CO_2 balance or durability (Müller and Krobjilowski 2003). In particular, the hybridisation of conventional reinforcing fibres with natural fibres promises a good compromise between mechanical properties and environmental compatibility, as the oil-based fraction in an HFRP is substituted or reduced by renewable raw materials (Dhakal et al., 2013, Ramesh et al., 2013). High-performance fibres such as carbon or glass fibres used as reinforcement induce high strengths and stiffness, which decisively determine the mechanical properties of the fibre composite material. However, the disadvantageous, very cost and energy-intensive production with a correspondingly poor CO_2 balance and the exclusively conventional availability of petroleum-based carbon fibres is significantly compensated by the combination/substitution with natural fibres, because natural/organic fibres offer a number of advantages compared to the conventional (high-performance) reinforcing fibres (FNR 2007):

- low density
- better splinter behaviour under crash load
- advantageous acoustic damping properties
- low raw material costs
- lower production-related energy consumption and consequently a better carbon footprint
- physiologically easier handling
- renewable raw materials, which guarantees a recurring availability
- advantageous disposal properties

The aim of the overall project is to develop high-performance hybrid materials and components and to investigate their technical, economic and ecological potential for use in the context of sustainable mobility. In addition to classic material development, the focus is on the analysis of recyclability and economic considerations, taking into account raw materials and production costs, as well as the technology assessment (including ecological assessment) of hybrid materials from the point of view of the automotive industry.

The project aims to produce a partially bio-based composite material that meets the component specification of the chosen highly stressed demonstrator parts in order to obtain the information required for a feasible serial production. In addition, the engineering parameters for component design for the newly developed bio-hybrid materials are determined. If, for example, a complete material substitution of carbon by bio-based fibres for the fulfilment of the individual points of the specifications for every part is not feasible, synergies through intelligent combinations are used to provide better or at least functionally equivalent materials/components with an improved ecological performance. Bottlenecks of the bio-based fibres in the area of the mechanical characteristics are compensated by functional integration and stress-optimised hybridised structures. Appropriate concepts should also focus on the highest possible utilisation of production residues with regard to the conservation of resources and the improvement of the CO_2 balance. Politics, industry and science are provided with a basis of data for evaluation in order to favour ecologically advantageous hybrid constructions with materials based on renewable raw materials.

The project objective of the holistic evaluation is divided into three key points: the technical, the ecological and the economic valuation, which are systematically processed and build on each other, as shown in Figure 1. The technical assessment has the largest share in generating the necessary database, which is both component-specific and universal. During the project chosen components are mounted at the test vehicle and characterised due the racing season. The mechanical characterisation according to the corresponding standards of the automotive sector, as well as the recording of the price range depending on the purchase quantity supplements the market research with real values, and the aimed procedures and methods contain the technical part. As a result, there is a large selection of different materials that can easily be compared, supplemented or exchanged, simultaneously thereby providing information on cost effects, etc. for other parts in serial production.

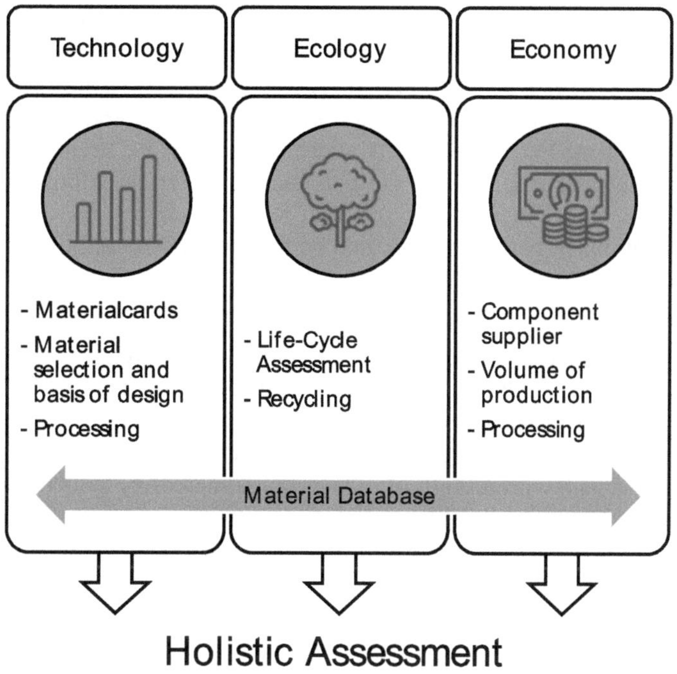

Figure 1: Core points of holistic assessment of natural fibre reinforced plastics (NFRP)

The technical, ecological and economic evaluations add up to the overall evaluation basis. Each of the three core topics is subdivided into further subtopics, which may look at the same aspect from a different point of view. For the ecological and economic parts, it is additionally necessary to figure out the data for the whole life cycle of a component part, especially the process data. This is content of further investigation: summarising the overall results in an evaluation basis that can individually set priorities in its assessment with regard to, for example, ecological advantage, price etc. and thus representing a decision-making aid for the manufacturer.

This article presents the results for the theoretical consideration of a selected door for the Porsche Cayman GT 4 Clubsport (981) with a technical view. In the context of the consideration, the effects of the individually set priorities are to be compared against each other.

2 STATE OF THE ART

In particular, the hybridization of synthetic fibres with natural fibres promises a good compromise between mechanical properties and environmental compatibility by replacing the petroleum-based fraction in a hybrid fibre reinforced plastic (HFRP) with renewable raw materials (Dhakal et al., 2013). In the following, selected mechanical properties of an HFRP with natural fibre content will be presented. When unidirectional scrims are combined into a biaxial scrim or tissue is used, the fibres in the different layers are loaded parallel and perpendicular to the fibre direction; a first failure always occurs in the transverse stress loaded layers (Hull and Clyne 1996). The fundamental, mechanical processes in a hybrid fibre reinforced plastic can be transferred from established FRP and are also subject to the conditions defined by Puck (Ehrenstein 2010):

- The breaking strength of the reinforcing fibre must be greater than the breaking strength of the matrix $\sigma_{fB} > \sigma_{mB}$
- The modulus of elasticity of the reinforcement fibre must be greater than the modulus of elasticity of the matrix $E_f > E_m$
- The breaking elongation of the matrix should be greater than the breaking elongation of the fibre $\varepsilon_{mB} \geq \varepsilon_{fB}$

The resulting mechanical properties depend, amongst other things, on the fibre volume content, the individual fibre length, their orientation and the fibre-matrix interaction (Sreekala et al., 2002, Amico et al., 2010, Fiore et al., 2012, Dhakal et al. Angrizani et al., 2014). For example, Davoodi and co-workers (2010) produced a bumper beam from a kenaf fibreglass hybrid and were able to demonstrate a significantly higher modulus of elasticity and increased tensile strength compared to the conventional, pure glass structure, underscoring the potential for using structural components in hybrid construction. Providing a high magnitude of research results, publications of NFRP and especially HFRP only a few applications of natural fibres could be found in the automotive sector, e.g. BMW i3.

Tensile properties

The tensile strength of an HFRP is significantly dependent on the strengths of the reinforcing fibres (Mallik P.K. 2007), (Drzal and Madhukar 1993). Various studies show that the influence on the tensile strength of an HFRP by the layer sequence is slight (Amico et al., 2010), (Pavithran et al., 1991). Various authors investigated the tensile strength of HFRPs with carbon and flax fibres in an epoxy matrix and were able to show that as the proportion of carbon fibres increases, the tensile strength, as well as the modulus of elasticity, increases, whilst the elongation at break decreases. The reason for this is the higher stiffness of the carbon fibre compared to the flax fibres (Dhakal et al., 2013), (Fiore et al., 2012).

Flexural strength

The flexural strength part in general is determined by the tensile and compressive properties of the components as well as by the structure of the HFRP. Material failure occurs either by compressive and/or tensile loading or shear failure around the neutral fibre (Figure 8) (Kim and May 1998), (Adams et al., 1990), (Drzal and Madhukar, 1993), (McKenna, 1975). If the stresses in the outer layers are higher at a bending load, the logical consequence is to process high-strength fibres into the outer layers in order to achieve higher flexural strengths (Angrizani et al., 2014) (Pavithran et al., 1991). (Dhakal et al., 2013) investigated the influence of the hybridisation of flax fabrics with carbon fibres on flexural strength and showed that flexural strength increases compared to flax fibre reinforced epoxy due to the higher strength of carbon fibres, but reduces elongation at break. In addition, (Fiore et al., 2012) a hybridisation has been shown to increase flexural strength. The influence of different fibre volume fractions of carbon fibres on a sisal/carbon polyester HFRP was investigated (Noorunnisa Khanam et al., 2010) and showed an increase in flexural strength with increasing carbon fibre content.

3 MATERIAL AND METHODS

3.1 Material

The natural fibres used in this study were flax fibres supplied by Bcomp, Lineo, Microtex and Composite Evolution. Using different surface weights in the range of 150–600 g/m^2, various textures (twill, sateen, plain weave and biaxial non-crimp fabric) were used as well. The carbon fabrics used in this study were twill weave Hexel 43200 UE and a non-crimp X-C-408 g/m^2 supplied by Seartex. The matrix was epoxy resin supplied by Sika as well as a bio-based epoxy resin supplied by Sicomin. The mould release agent used was Marbocote 625x eco.

3.2 Component and transaltion to 2D test specime

The focus in this paper lies on the door. Beside an iterate testing of the door within the racing season in the car for lay up and quality comparison three testing points were defined. To simplify and accelerate the comparison between the different lay ups a translation from 3D structure to 2D laminates has been carried out. Further the translation enable the use of international standards e.g. DIN. Given that the door is manufactured with both an inner and an outer shell, three testing points were localised, as depicted in Figure 2.

1. Door frame external shell
2. Door panel external shell
3. Door frame internal shell

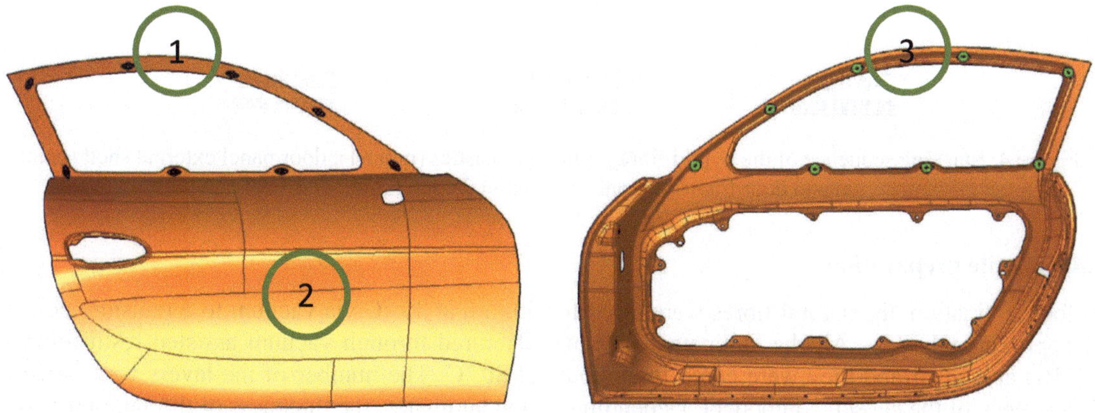

Figure 2: Points for material characterisation of inner and outer door shell © Porsche AG

The comparison therefore neglected load paths and 3D structures in the first step. The framework for the comparison of the NFRP and HFRP lay-up was defined by minimum requirements on the flexural and tensile properties imposed by Porsche AG that is represented due the 100 % mark in the results.

3.3 Stacking sequences for tested samples

As shown in Figures 3 and 4, with the exception of the door frame external shell of the HFRP, all laminates were manufactured by stacking four layers of different fabrics via vacuum assisted resin infusion. For confidential reasons, only the orientation and type (carbon fibre = black, natural fibre = brown) of fibre is published. All laminates contain fabrics of all orientations 0/90° as well as +/–45° to create a quasi-isotropic laminate. A directional design takes place later in the selected component structures. In Figure 3, the stacking sequence of the natural fibre reinforced plastics is shown. While using various fabrics for the door frames, a core (orange layer) made from renewable material was additionally used for the door panel.

Figure 3: Stacking sequence of the natural fibre reinforced plastics for door frame external shell (point 1), door panel external shell (point 2) and door frame internal shell (point 3), see Figure 2 as well

In Figure 4, the stacking sequence of the hybrid fibre reinforced plastics is shown. While using various fabrics for the door frames, a core made from renewable material was additionally used for the door panel, similar to the NFRP doors.

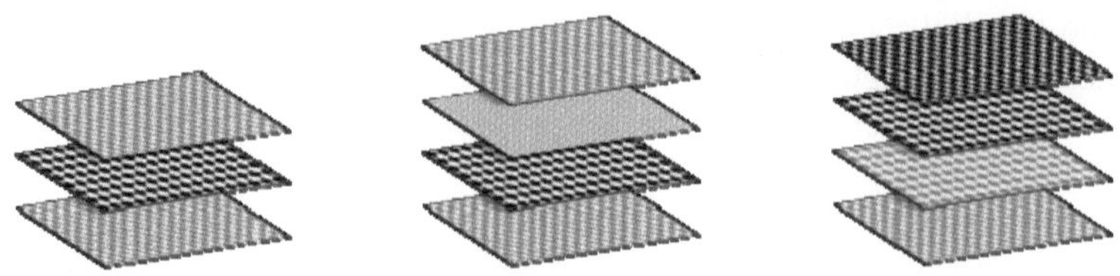

Figure 4: Stacking sequence of the hybrid fibre reinforced plastics (point 1), door panel external shell (point 2) and door frame internal shell (point 3), see Figure 2 as well

3.4 Sample preparation

Before lamination, the natural fibres were dried for 60 min at 30 °C via vacuum to a moisture content of approximately 3 %. All the laminates were manufactured through vacuum assisted resin infusion (VARI) at 4 mbar and they were cured for 8 hours at 80 °C. The number of the layers was varied in dependence of the chosen component. Depending on the laminates, the thickness of the samples varied. For mechanical testing, the samples were cut with a precision saw DIADISC 5200.

3.5 Experimental characterisation

Five samples per laminate were tested for each lay-up through four-point bending tests according to an automotive standardisation based on DIN EN ISO 14125 by using a 100 kN Universal Testing Machine, model Z100 (Zwick/Roell – Germany) and setting the crosshead speed to 1 mm/min. Depending on material thickness, samples were sized 20 mm × 80 mm (thickness ≥ 3.5 mm, span length 60 mm) and 20 mm × 100 mm (thickness ≤ 3.5 mm, span length 80 mm). Tensile properties were performed according to an automotive standardisation based on DIN EN ISO 527-4 by using the same machine and testing parameters as were used for the bending tests. Five samples per laminate (25 mm × 250 mm) for each lay-up were tested. All samples were conditioned for 24 h in standardised conditions (23 °C / 65 % humidity) according to DIN EN ISO 291.

3.6 Standardisation of fibre content

To compare the different lay-ups, the fibre mass content was standardised to 50 %. For determination of fibre mass content, the different lay-ups were weighted after drying as well as the consolidated board after vacuum assisted resin infusion. Fibre mass content defined by ψ according to the equation:

$$\Psi_z = w_{fibre}/(w_{fibre} + w_{matrix})$$

where w_{fibre} is the weight of the dry fabric and w_{matrix} is the weight of used matrix after manufacturing.

Based on these results, all laminates were standardised to a fibre mass content defined by ψ_{50} according to the equation:

$$\Psi_{50} = (x_\psi / \Psi_z) * 0{,}5$$

where x_ψ is the measured value of a selected property with determined fibre mass content ψ_z.

3.7 Relativisation of mechanical properties

The different mechanical properties were relativised to allow a verification of the specification of each area of the door. Within these requirements, various cases, for example no buckling of the outer shell or bending of the door frame, were analysed. All measured values of bending and tensile properties with the exception of elongation of break were relativised to the specification of each area of the door.

The different properties were defined by x_{rel} according to the equation:

$$x_{rel.} = (100/x_{req}) * x_{meas}$$

where x_{req} is the given value of a selected property and x_{meas} is the measured value of this selected property.

3.8 Comparison of prices

Prices of the used dry fabrics were provided by the different suppliers. The matrix and the exact pattern size of the door or the required fabric amount were initially neglected at this point. This means all presented prices correspond to 1 m^2 of dry fabric combination when ordering an amount of at least 2,000 m^2.

4 RESULTS AND DISCUSSION

The presentation of the results follows the preceding objective of comparing an NFRP with an HFRP. In the first part, the requirement is to achieve a minimum of specified characteristic values, while the second part compares the effect of different structures on the selected properties (tensile and flexural properties, price and amount of renewable material). In addition to selected mechanical properties, the proportion of renewable raw materials in the form of the natural fibres used is compared, as well as the price for the dry laminate construction with an order quantity of 2,000 m^2 and the change in weight using different materials and stacking sequences. It should be noted that both the proportion of renewable raw materials and the prices refer to 1 m^2 of dry semi-finished products. The matrix system used and the component-specific area of semi-finished products are not considered in this illustration. Furthermore, the share of renewable raw materials makes no statements about the biogenic fraction or its influence on the carbon footprint.

4.1 Directional flexural properties of fibre reinforced plastics

Figure 5 compares the flexural properties of carbon composites with those of flax composites as a function of direction respectively fibre orientation. It has to be noted that these values are not adjusted to material thicknesses but correlate with each constructed door frame. The fibre mass content and thickness of the samples are summarised in Table 1. The load deformation curve for the carbon fibre 0/90 sample is linear with a small non-linear region at the end of the test and a sudden drop after a peak load ending with a breaking at approximately 1.5 % deformation. Changing the direction within sample CFRP +/–45, the load deformation shows a linear beginning ending with expanded peak without a sudden drop of load. Compared to CFRP 0/90, the maximum load is reached on average at one quarter of load, while deformation increases by one quart. For both flax samples, a non-linear behaviour is observed. Notable differences are observed in the near-to-failure zone of the load deformation curves.

The sample NFRP +/–45 showed a similar behaviour to that of CFRP +/–45 with an expanded peak of maximum load and no breaking into two parts. However, the max load of NFRP +/–45 is three times as much and the deformation more than double that of CFRP +/–45. While CFRP 0/90 showed a sudden drop ending in a break, NFRP 0/90 reached its peak of maximum load at the same level, observing a gradual failure before breaking. On the other hand, the deformation of NFRP 0/90 is more than triple that of CFRP 0/90. In comparison, the NFRP showed only little change of load depending on the fibre direction, while the CFRP lost a third of its load by changing the fibre orientation by 45 °. These observed effects support the idea of a directional usage of high-performance fibres in combination with natural fibres.

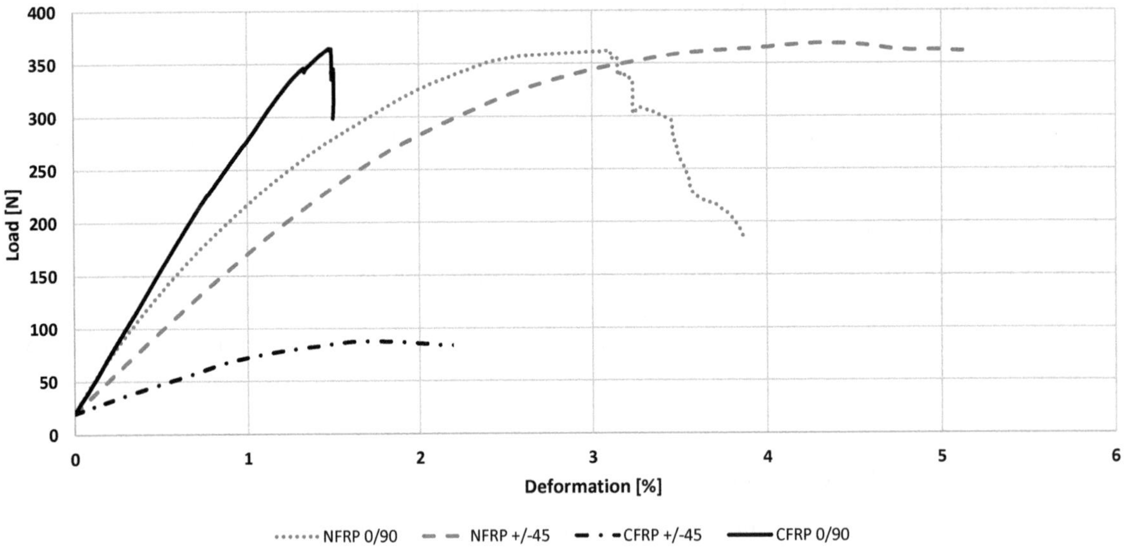

Figure 5: Load deformation curves for different fibre reinforced plastics

Table 1: Properties of different specimens

Sample	Average max load [N]	Thickness [mm]	Fibre mass fraction [%]
NFRP 0/90	397.03	2.083	47
NFRP +/–45	352.07	2.312	47
CFRP 0/90	326.75	1.08	57.6
CFRP +/–45	80.77	0.998	56

4.2 Flexural properties of fibre reinforced plastics

The results of the flexural strength and the flexural modulus of the three selected characterisation points are summarised in Figure 6. The minimum requirements imposed by Porsche AG on the flexural properties is represented due the 100 % mark. In principle, the values of flexural modulus and the values of flexural strength were able to reach the specifications of the 100 % mark. In particular, the door frames showed an increase in properties due to hybridisation with carbon by up to 100 %. The increase in flexural properties is as expected due to the higher properties of the carbon. Due to the prescribed test rigs, a test until breakage was not always possible depending on the test specimen strength, as otherwise a risk of trapping the specimen between the bearing and the pressure fin was risked.

Contrary to the results of other researchers (Dhakal et al., 2013), in almost all cases the hybrid fibre-reinforced plastics used in the door frame have a higher deformation respectively outer fibre strain than the natural fibre-reinforced plastics. Caused by a termination of the testing through a defined traverse path of 15 mm the samples showed no failure. Calculating the outer fibre strain according to DIN EN ISO 14125, the result of outer fibre strain ε depends on the thickness of the sample according to the equation:

$$\varepsilon = 4{,}7sh/L^2$$

where s is the flexure in mm, h the thickness of the sample in mm and L is the span in mm.

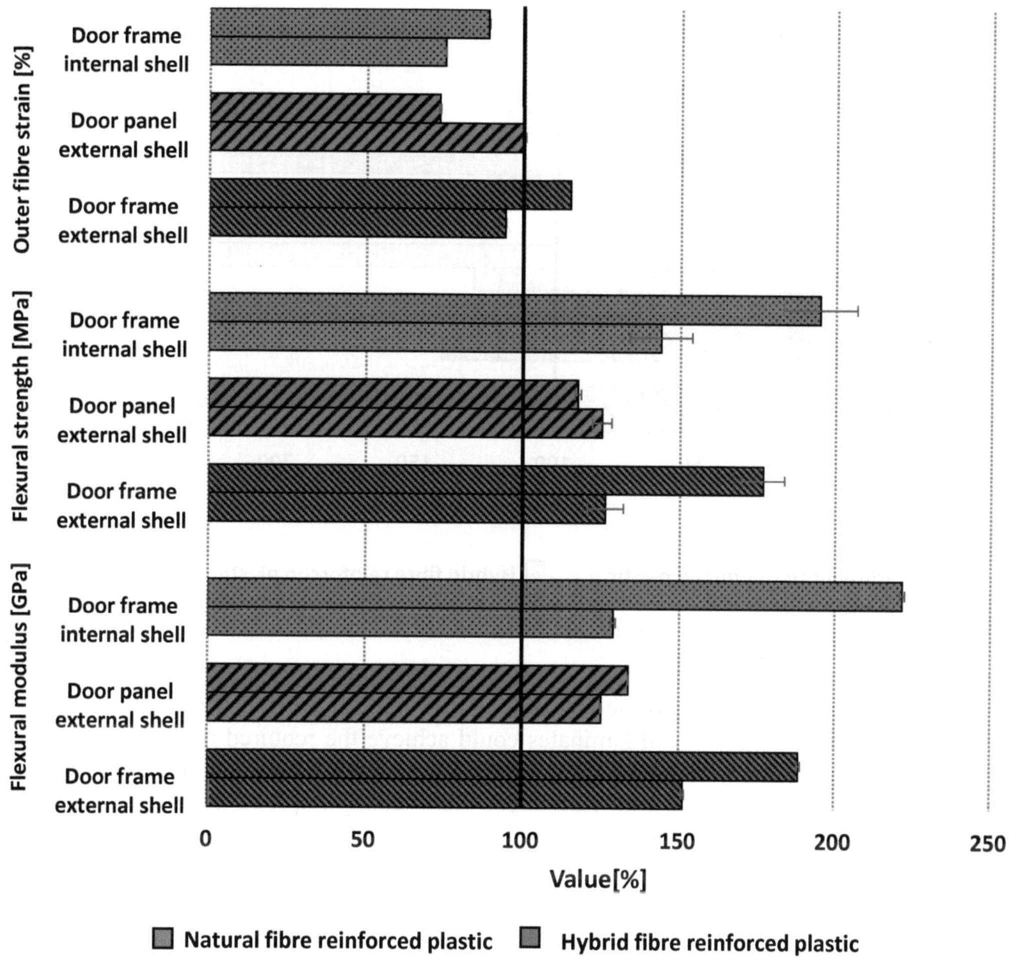

Figure 6: Comparison of flexural properties of NFRP and HFRP located in different parts of a door

However, the NFRP of the door frame internal shell and the door frame external shell samples showed an averagely greater sample thickness than HFRP, causing a minor outer fibre strain of NFRP when interrupting the testing before failure. Both samples of the door panel external shell showed a failure through the flexural testing. The flexural modulus and flexural strength of the door shell are only minimally different between HFRP and NFRP. In contrast to the door frames, a core material was used to exploit the sandwich effect. In return, the samples of the door shell had a significantly higher material thickness. During the test, it was observed that a failure of the sandwich structure is caused by interlaminar detachment of the cover layers from the core layer. As a result of hybridisation of the NFRP with one layer of carbon, the flexural modulus is little higher, but the outer fibre strain is lower.

4.3 Tensile properties of fibre reinforced plastics

The components examined in the project are exposed to various stresses during their use phase. Since this is a racing car, these are not to be compared with the approval criteria of a vehicle for road traffic. Nevertheless, the stresses during a race are the same, if not higher, since the car is often driven at its load limit. Also identical are the weather conditions that affect the material. Therefore, the first requirement for the comparison between NFRP and HFRP was defined so that specified values for tensile and flexural properties given by Porsche had to be met. In the following comparisons, these specified properties are indicated by the 100 % mark.

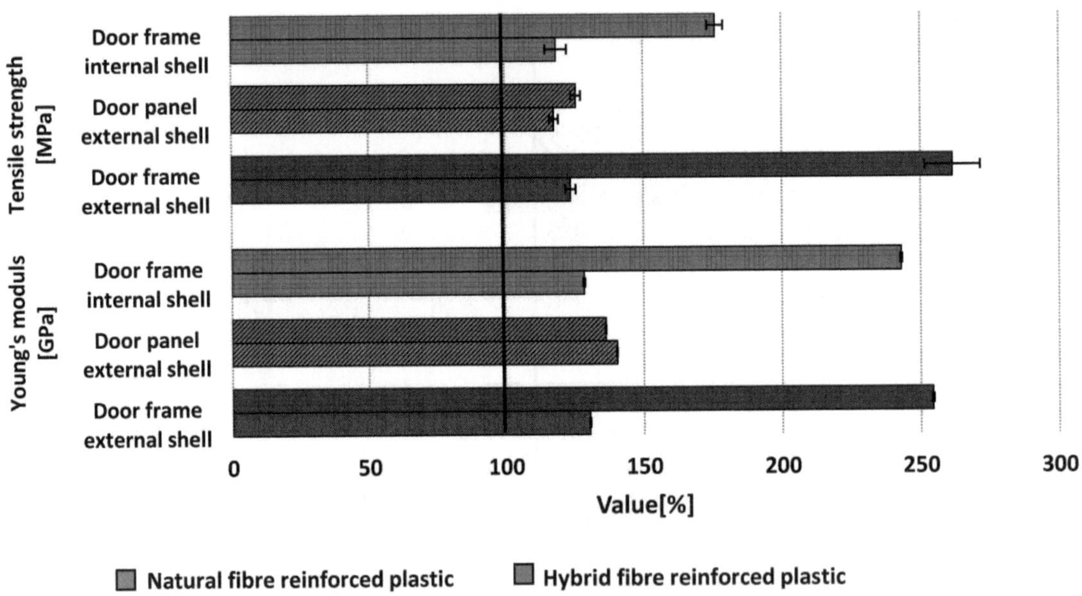

Figure 7: Comparison of tensile properties of NFRP and HFRP located in different parts of a door

The results of the tensile strength and the modulus of elasticity of the four selected characterisation points are summarised in Figure 3. All laminates could achieve the required minimum requirements. Hybridisation with carbon provides much stiffer, stronger properties over natural fibre constructions, demonstrating that flax hybridisation with carbon can significantly improve tensile properties. In particular, the hybrid structures of the door frame, both the inner and outer shell, show an increase in the modulus of elasticity of more than 100 %. The same could be seen for the tensile strength, where improvements can be detected of up to 100 %.

Because the fibres take up a majority of the load in a fibre composite plastic, in particular under tensile load parallel to the fibre direction, it is not surprising that through hybridisation with stiffer and stronger carbon fibres, these properties increase accordingly. Because of used fabric and scrim structures, tensile stresses are also applied transversely or at an angle greater than 0 ° to the fibres. With tensile stresses across the fibres, the fibres act as a filler rather than absorbing forces. Additional consideration of the effect of the expansion magnification and different number of reinforcing layers or basis weights results in a different range of improvements. In particular, the tensile properties of the door shell in the outer shell show no or only a very small increase in tensile properties. A core material was used for the door shell and, regardless of the top and core layers, the tensile test showed interlaminar failure between the layers. The use of a shear-resistant core material to create a sandwich prevents full exploitation of the possible elongation at break of the fibres, leading to premature failure.

5 CONCLUSIONS

The aim of this work was the investigation of a hybridisation of NFRP with carbon fibres. This was performed by way of example on a car door for a Porsche Cayman GT 4 Clubsport (981). Three areas were chosen, which require different material performance. While isotropic material behaviour is advantageous for use in the door shell, anisotropic behaviour is advantageous in the door frame. The investigations show similar mechanical behaviour for the NFRP and HFRP door panels. Figure 8 shows, in addition to the mechanical parameters, the price for 1 m^2 of dry laminate, the proportion of renewable raw materials and the change in component weight with NFRP as 100 %. The graph shows a higher proportion of renewable raw materials, as well as a lower commodity price, with almost identical parameters. By contrast, the component weight is minimally increased. These preliminary results support the use of an NFRP in the outer door panel area.

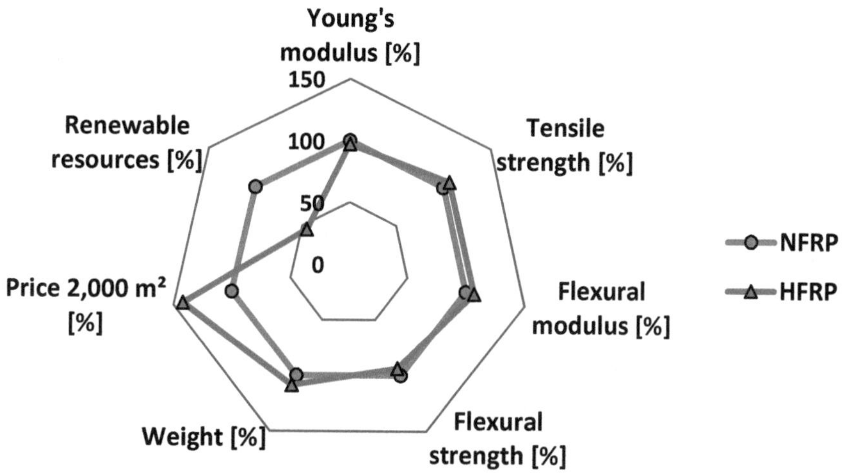

Figure 8: Comparison of selected properties of the external door panel manufactured from NRFP and HFRP

Figures 9 and 10 show comparatively different properties of the inner and outer door frame. Both graphs show a significant increase in mechanical properties due to hybridisation. Since these are converted two-dimensional test plates and not the real component structures, the anisotropic design of the test plate is theoretical. Compared to the NFRP, the price level of the HFRP remains at a similar level, as does the component weight. Only the share of renewable raw materials in the form of natural fibres is, of course, reduced, similar to the situation in the figure before. However, for both superstructures, the share is 40 % above the target share of 30 %. It should also be noted that so far, the matrix has not been included. By using a bio-based matrix system, the proportion of renewable raw materials can be further increased. Against the background of distributing the force under load of the door frame as possible in the door leaf, this offers great potential for a hybrid construction. In the next step, the hybridisation takes place on the basis of the real component structures in order to design them according to the load path.

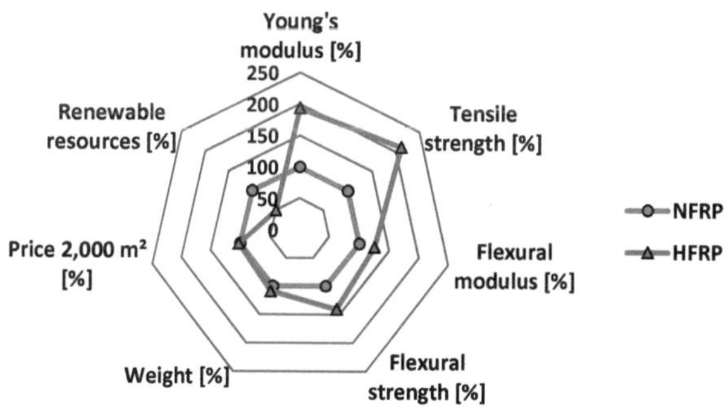

Figure 9: Comparison of selected properties of the external door frame manufactured from NRFP (100 %) and HFRP

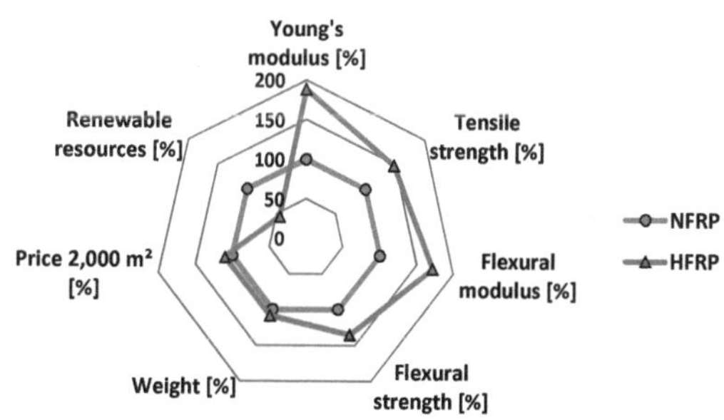

Figure 10: Comparison of selected properties of the internal door frame manufactured from NRFP (100 %) and HFRP

ACKNOWLEDGEMENTS

The authors would like to thank the German Federal Ministry of Food, Agriculture and Consumer Protection and the Agency for Renewable Resources (German acronym: FNR) for the financial support.

REFERENCES

Adams, Donald F.; King, Todd R.; Blackketter, Donald M. (1990): Evaluation of the transverse flexure test method for composite materials. In: Composites Science and Technology 39 (4), S. 341–353. DOI: 10.1016/0266-3538(90)90080-O.

Amico, S.C.; Angrizani, C.C.; Drummond, M.L. (2010): Influence of the Stacking Sequence on the Mechanical Properties of Glass/Sisal Hybrid Composites. In: Journal of Reinforced Plastics and Composites 29 (2), S. 179–189. DOI: 10.1177/0731684408096430.

Angrizani, C. C.; Cioffi, M. O.; Zattera, A. J.; Amico, S. C. (2014): Analysis of curaua/glass hybrid interlayer laminates. In: Journal of Reinforced Plastics and Composites 33 (5), S. 472–478. DOI: 10.1177/0731684413517519.

Dhakal, H.N.; Zhang, Z.Y.; Guthrie, R.; MacMullen, J.; Bennett, N. (2013): Development of flax/carbon fibre hybrid composites for enhanced properties. In: Carbohydrate Polymers 96 (1), S. 1–8. DOI: 10.1016/j.carbpol.2013.03.074.

Drzal, L. T.; Madhukar, M. (1993): Fibre-matrix adhesion and its relationship to composite mechanical properties. In: J Mater Sci 28 (3), S. 569–610. DOI: 10.1007/BF01151234.

Ehrenstein G. W. (2006): Faserverbund-Kunststoffe Werkstoffe – Verarbeitung – Eigenschaften. 2., vollständig überarbeitete Auflage: Carl Hanser Verlag GmbH & Co. KG.

Fiore, V.; Valenza, A.; Di Bella, G. (2012): Mechanical behavior of carbon/flax hybrid composites for structural applications. In: Journal of Composite Materials 46 (17), S. 2089–2096. DOI: 10.1177/0021998311429884.

Flemming, Manfred; Ziegmann, Gerhard; Roth, Siegfried (1996): Faserverbundbauweisen. Berlin, Heidelberg: Springer Berlin Heidelberg.

FNR (2007): Marktanalyse Nachwachsende Rohstoffe Teil II. Hg. v. Fachagentur für nachwchsende Rohstoffe (FNR). Online verfügbar unter https://mediathek.fnr.de/marktanalyse-nachwachsenderohstoffe-teil-ii.html, zuletzt geprüft am 21.04.2018.

John, M.; Thomas, S. (2008): Biofibres and biocomposites. In: Carbohydrate Polymers 71 (3), S. 343–364. DOI: 10.1016/j.carbpol.2007.05.040.

Kim, Jang-Kyo; Mai, Yiu-Wing (1998): Measurements of interface/interlaminar properties. In: Engineered Interfaces in Fiber Reinforced Composites: Elsevier, S. 43–92.

Mallik P. K. (2007): Fiber Reinforced Composites: Materials, Manufacturing and Design. 3. Aufl. Boca Raton: CRC Press. Online verfügbar unter https://books.google.com/books?hl=de&lr=&id=eGLLBQAAQBAJ&oi=fnd&pg=PP1&ots=oNTR0ZXhDf&sig=0roFapdzarDzk9ACKK4qgvpS5qQ, zuletzt geprüft am 21.04.2018.

McKenna, Gregory B. (1975): Interlaminar Effects in Fiber-Reinforced Plastics-A Review. In: Polymer-Plastics Technology and Engineering 5 (1), S. 23–53. DOI: 10.1080/03602557508063092.

Mueller, Dieter H. and Krobjilowski, Andreas (2003): New Discovery in the Properties of Composites Reinforced with Natural Fibers. In: Journal of Industrial Textiles, Vol 33, Issue 2, pp. 111–130. DOI: 10.1177/152808303039248.

Neitzel, Manfred; Mitschang, Peter; Breuer, Ulf (2014): Handbuch Verbundwerkstoffe. München: Carl Hanser Verlag GmbH & Co. KG.

Noorunnisa Khanam, P.; Abdul Khalil,H. P. S.; Jawaid, M.; Ramachandra Reddy, G.; Surya Narayana, C.; Venkata Naidu, S. (2010): Sisal/Carbon Fibre Reinforced Hybrid Composites: Tensile, Flexural and Chemical Resistance Properties. In: J Polym Environ 18 (4), S. 727–733. DOI: 10.1007/s10924-010-0210-3.

Pavithran, C.; Mukherjee, P. S.; Brahmakumar, M.; Damodaran, A. D. (1991): Impact properties of sisal-glass hybrid laminates. In: J Mater Sci 26 (2), S. 455–459. DOI: 10.1007/BF00576542.

Schürmann, H. (2007): Konstruieren mit Faser-Kunststoff-Verbunden. Berlin, Heidelberg: Springer Berlin Heidelberg.

Sreekala, M.S.; George, Jayamol; Kumaran, M.G.; Thomas, Sabu (2002): The mechanical performance of hybrid phenol-formaldehyde-based composites reinforced with glass and oil palm fibres. In: Composites Science and Technology 62 (3), S. 339–353. DOI: 10.1016/S0266-3538(01)00219-6.

IMPROVING THE DURABILITY OF BIO HYBRID FIBER REINFORCED PLASTICS BY PLASMA TREATMENT

Florian Bittner[1], Martin Bellmann[2], Madina Shamsuyeva[1], Hans-Josef Endres[1,3], Wolfgang Viöl[4]

[1] Application Center for Wood Fiber Research HOFZET®, Fraunhofer Institute for Wood Research, Wilhelm-Klauditz-Institute WKI, Heisterbergallee 10A, 30453 Hannover, Germany, florian.bittner@wki.fraunhofer.de, www.wki.fraunhofer.de
[2] Fraunhofer Institute for Surface Engineering and Thin Films IST, Von-Ossietzky-Straße 100, 37085 Göttingen, Germany, martin.bellmann@ist.fraunhofer.de, www.ist.fraunhofer.de
[3] Institute for Bioplastics and Biocomposites, University of Applied Science and Arts Hannover, Heisterbergallee 10A, 30453 Hannover, Germany, hans-josef.endres@hs-hannover.de, www.ifbb-hannover.de
[4] HAWK University of Applied Sciences and Arts, Faculty of Natural Sciences and Technology Göttingen, Von-Ossietzky-Straße 99, 37085 Göttingen, Germany, wolfgang.vioel@hawk.de, www.hawk.de

Keywords:

Bio hybrid fiber reinforced plastics, Atmospheric-pressure plasma, Interface quality, Long-term stability

ABSTRACT

Natural fiber-reinforced polymer composites have already been successfully established in various lightweight applications subjected to moderate mechanical stress, e.g. automotive interior. Recently, the development of biobased hybrid composites containing both, natural and high-performance fibers, gained the attention of research institutions and industry. These new composites offer optimized density, mechanical performance, have reasonable cost and low environmental impact. The bio hybrid composites enable utilization of the advantages of cellulose-based and high-performance fibers in the same composite and minimize the limitations of the individual composites. Although this approach seems to be a promising solution for several drawbacks, the deficiency of the approaches to improve the durability of these composites in outdoor applications limits their market penetration.

In this project, a specially developed plasma treatment approach, cascade-atmospheric pressure plasma is being adapted in order to improve the mechanical performance of the biobased (hybrid) composites for the application-oriented use under various environmental conditions, e.g. high humidity absorption and temperature variance. This study presents selected results of this project including the assessment of the plasma treatment of multi-ply textiles and the corresponding influence on the mechanical performance of the composites. UD flax tapes and balanced, more isotropic flax fabrics were treated under various plasma parameters and integrated into a partially biobased epoxy resin via vacuum infusion. The analysis of the textile surface and composite mechanical properties were used for the assessment. It has become evident that the selection of the plasma parameters has a significant effect on the packing density and wettability of the treated textiles. Controversial effects have been observed with regard to the mechanical properties of the composites. An improvement of weathering durability is still remaining.

1 INTRODUCTION

Nowadays, the amount of industrially produced natural fiber-reinforced polymer composites (NFPC) and their advanced derivatives, e.g. biobased hybrid fiber-reinforced polymer composites (bio-HFPC)

K. Dröder et al (eds), *Technologies for economical and functional lightweight design*, Zukunftstechnologien für den multifunktionalen Leichtbau, https://doi.org/10.1007/978-3-662-58206-0_19

is dynamically rising [1]. In this context, the leading application field for the textile composites containing natural fibers (NF) is automotive and sport industries. Low density, good mechanical properties with regard to the density, cost-efficiency, high-energy absorption and ecological aspects make NFPCs attractive for various industry applications. Similar to conventional composites made from synthetic fibers, the type of used fibers, textile configuration as well as the matrix type define the properties of the product. Exemplary, a unidirectional flax fabric integrated into a thermoset resin would provide low undulation and thus a higher reinforcing effect of the fibers than a weaved fabric. At the same time, these textiles offer low isotropy, thus offering this higher mechanical performance in a certain direction. On the other hand, a thermoset resin reinforced with a weaved fabric offers, depending on the binding and used yarns, a certain isotropy of the mechanical properties. However, weaved fabrics suffer from undulation and may cause a higher areal weight. Consequently, the selection of the textile depends significantly on the manufactured composite and desired product properties.

Although NFPCs are already being used efficiently in automotive interior, their overall mechanical performance, which is not comparable with that of composites containing synthetic high-performance fibers, e.g. carbon or aramid, limits their application. On the other hand, e.g. carbon fiber-reinforced polymer composites (CFPC) are expensive and suffer from splintering. In this context, the synergetic integration of both natural and carbon fibers in one composite appears to be a promising approach to combine effectively the advantages of both types of reinforcing fibers in one composite and minimize the individual limitations. This approach led to the development of the bio-HFPCs, where fibers of diverse origin can be combined synergistically in different ways [2]. The simplest approach to manufacture bio-HFPC is stacking, where textiles made of different fibers are placed on top of each other in a certain sequence and integrated into a polymer matrix, resulting in a laminate. This approach offers a homogeneously distributed mechanical performance within the whole construction part and is suitable for the fabrication of flat components, like tailgates, snowboards, skateboards, etc.

A further drawback for the application of composites containing natural fibers, especially in outdoor applications, is there poor resistance to environmental influences, most notably moisture. The hydrophilic nature of the fibers leads to fiber swelling and delamination under moisture exposure. Besides water diffusion inside the matrix and matrix defects such as pores, the fiber-matrix interface has a significant influence on moisture-induced degradation [3].

In order to broaden the use of the NFPCs and bio-HFPCs in outdoor applications, the mechanical durability of these materials in various environments and especially under extreme weather conditions, e.g. high temperature variance, stress cycles, exposure to water and coldness should be optimized.

Plasma surface functionalisation on organic substrates tends to homogenise the surface properties. Concerning the heterogeneous composition of natural materials, plasma seems to be an interesting tool for the optimisation of bio hybrid fiber reinforced plastics. By homogenizing the fiber surface and influencing the surface polarity, an improved fiber-matrix interaction and a reduction of weak points for degradation are targeted. Low process costs and easy incorporation into continuous processes make plasma treatments attractive for industrial applications [4].

This study represents selected results of the project aiming to develop a new plasma treatment method in order to optimize the durability of NFPC and bio-HFPC.

2 EXPERIMENTAL

2.1 Materials

Initially two flax textiles were selected for plasma treatment; an unidirectional flax fabric (LINEO FLAXPLY UD150, 150 $g{\cdot}m^{-2}$) with sizing and a balanced flax fabric (Biotex Flax 200 $g{\cdot}m^{-2}$ 2×2 Twill) without sizing. The fabrics were cut to plies of 300 × 300 mm^2.

A biobased resin system (epoxy resin Sicomin Greenpoxy 56 + epoxy hardener Sicomin SD8822) was applied for composite manufacturing by vacuum infusion.

2.2 Plasma treatment

A plasma source as sketched in Figure 1 was used [5]. Based on a jet induced dielectric barrier surface discharge (dbd), a nearly homogeneous plasma cloud is formed. By generating a dbd jet between the high voltage electrode and the dielectrically insulated upper ground electrode, the flow of the process gas blows the remote plasma out of the nozzle. The remote plasma generates a so-called floating potential on the nonconductive substrate surface, which is taken over by the dielectrically insulated ground electrode in the disc shaped bottom of the plasma source. The so formed closed circuit enables the generation of reactive species directly on the substrate surface. This discharge circuit consists of surface discharges, which enable effective activation up to a diameter of 100 mm. Furthermore, the surface discharges allow diffusion into the smallest crevices. This feature leads to the suggestion, that this plasma principle could be an effective solution for a homogeneous fabric activation.

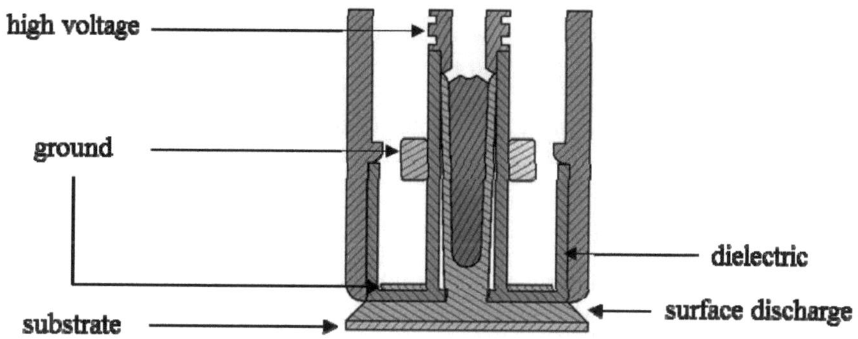

Figure 1: Discharge principle [5]

Figure 2 shows the plasma cloud at a treatment distance of 2 mm at an exposure time of 1/100 s. To determine the discharge characteristics, high-speed video recordings were performed at an exposure time of 1/6.000 s. The high-speed recording shows isolated surface filaments, which slide out of the nozzle (1) and ignite back to the disc shaped bottom of the plasma source (2). The whole plasma cloud consists of these filaments and surrounds every cavernous substrate, for example fabric plies.

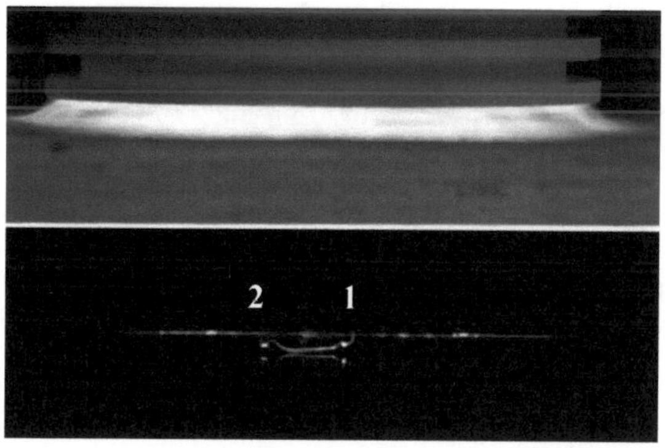

Figure 2: Plasma cloud at a 2 mm discharge gap (above), frame of a high-speed video recording of the same discharge (below)

Flax fabrics were plasma treated with either argon or a mixture of argon and hydrogen (10 % H_2) as process gas. With argon as process gas (gas flow 20 L·min^{-1}), either single fabric plies or 8 stacked fabric plies were treated. The distance of plasma source to fabric surface was set to 3 mm, the plasma power was adjusted to 150 W. The dbd jet moved with a speed of 4 m·min^{-1}. The treatment process is illustrated in Figure 3.

With Ar/H_2 (gas flow 30 L·min^{-1}) as process gas, only single fabric plies were treated. The distance of plasma source to fabric surface was set to 6 mm, all other parameters remained unchanged.

Figure 3: Simultaneous plasma treatment of 8 fabric plies (Biotex Flax 200g/m^2 2×2 Twill) with Ar plasma

2.3 Surface energy measurements

To determine the plasma penetration throughout the fabrics and the possible amount of simultaneously treated plies, nine polypropylene substrates (used as pads) with dimension of 50 × 200 mm^2 were plasma treated. The first PP pad was left without a fabrics ply and from the second to the last pad, the amount of fabric plies (Biotex Flax 200g/m^2 2×2 Twill) was increased up to eight. After the plasma treatment, the fabric plies were removed from the PP pads and the surface energy of each PP pad was verified. In Figure 4 the experimental setup is shown.

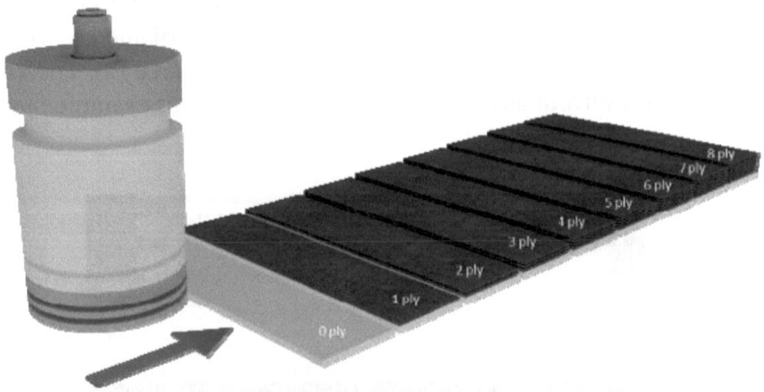

Figure 4: Treatment of an increasing amount of fabric plies applied on PP surfaces to identify the plasma penetration by measuring the surface energy of the covered PP surface

To verify the surface energy of plastic surfaces, a mobile surface analyzer by company KRÜSS based on the OWRK method was used.

The measurement of the dynamic advancing contact angle according to the modified Washburn's method is described in details by Pucci et al. [6]. For the measurement a rolled piece of textile is packed tightly into a cylindrical holder and can be described as a bundle of capillary tubes. A tensiometer is used to track the rise of the weight of the liquid penetrating the textile. The measurement contains two parts: in the first part, the geometric porous medium factor, C, is defined using a fully wetting (zero contact angle) liquid and in the second part, the dynamic wetting angle, θ, is calculated using the defined factor and the measured weight rise with regard to a test liquid. The factor C is an empirical material constant depending on the particle size and degree of packing and determines the "wicking rate", if the textile is modelled as a powder. The following equation (1) is used for the calculation [7]:

$$m^2 \cdot t^{-1} = [(c \cdot \bar{r}) \cdot \varepsilon^2 \cdot (\pi R^2)] \cdot \rho^2 \cdot \gamma_L \cdot \cos\theta \cdot (2\eta)^{-1} \tag{1}$$

where m is the weight of the penetrating liquid, ρ the density of measuring liquid, ε the relative porosity and R the inner radius of the measuring tube. The term in square brackets is a constant materials factor defined as a packing constant, C. The derivation of the equation is described in literature [5,6].

In this study, the factor C is defined using n-hexane and the dynamic angle is calculated with 2-propanol for the UD flax tape and ethanol for the balanced flax fabric. The size of the analysed textile samples is 2×30 cm^2 (h x l). The weight of the balanced flax fabric and of the UD flax tape is 1.3 g and 1.0 g, respectively. At least two measurements were performed for each sample.

2.4 Composite fabrication

To preserve the plasma effect the composites were fabricated on the day of plasma treatment. The composites were composed of 8 fabric plies each. The unidirectional flax fabric was aligned in 0° direction, the balanced flax fabric was aligned in 0°/90°.

Vacuum infusion with the resin (ratio resin:hardener 100:31) was followed by curing at room temperature for 24 h and tempering at 80 °C for 2 h.

2.5 Environmental testing

Accelerated environmental testing was conducted with an environmental stress chamber (Espec ARS-1100) on non-treated and plasma-treated balanced flax fabric composites according to the testing standard DIN EN 60068-2-38:2009 [8] with application of cold subcycles. During the testing procedure the samples are exposed to a number of temperature changes in the range of –10 to +65 °C under a relative humidity of 80 to 93 %. One cycle consisting of 10 subcycles (24 h each) was applied to the samples.

2.6 Mechanical testing

Mechanical performance of the composites was characterized by tensile and 3-point bending tests. Tensile testing was conducted based on the testing standard DIN EN ISO 527-4. Due to fabrication issues, the sample dimensions of the UD flax tape composites were restricted to 150 mm × 25 mm. The samples of the balanced flax fabric composite conform to the testing standard with dimensions of 250 mm × 25 mm.

3-point bending tests were conducted according to the testing standard DIN EN ISO 14125 with sample dimensions of 100 mm × 15 mm.

2.7 CT measurements

CT measurement of composite samples were performed on a Procon X-Ray CT-AlphaDuo device. The samples were scanned at an x-ray voltage of 60 kV and a current of 200 µA with a voxel resolution of 5.2 µm. For visualization and analysis of the volume data VGSTUDIO MAX 3.1 was used.

3 RESULTS AND DISCUSSION

3.1 Plasma penetration depth

In Figure 5 the surface energies of PP pads covered with up to 8 balanced flax fabric plies are shown. The results show that all treated polypropylene pads reach the same surface energy (about 50 mN·m^{-1}) despite of the increasing amount of fabric plies. For comparison, the untreated polypropylene pads only achieve surface energies of roughly 30 mN·m^{-1}. Only at an amount greater than 7 plies, the surface energy is slightly decreasing. This leads to the assumption that an individual treatment of each ply is not necessary.

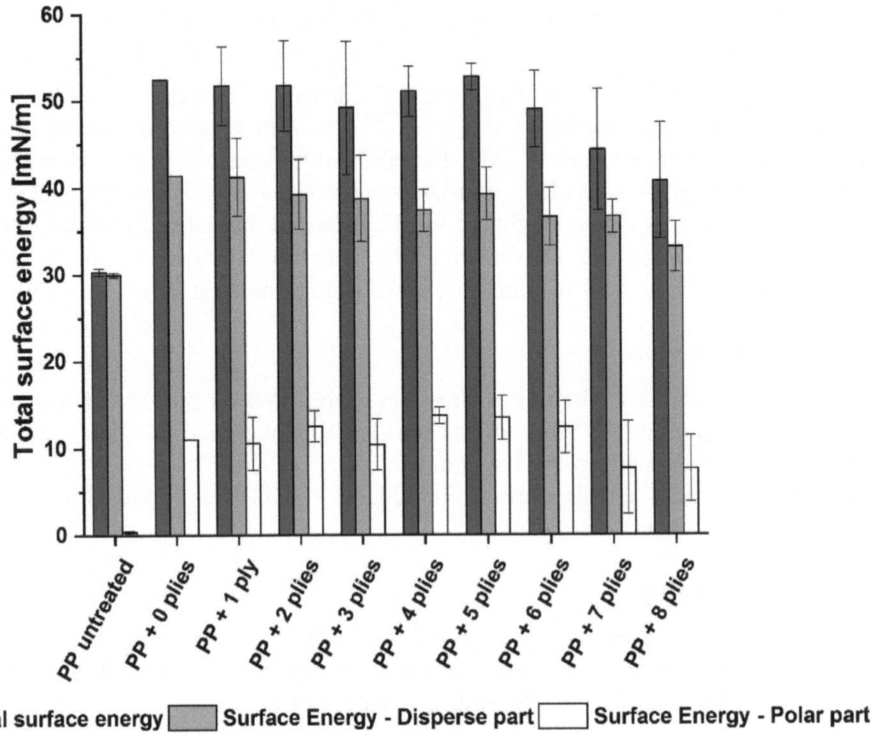

Figure 5: Dependence of the PP surface energy investigation on the number of fabric plies between PP surface and dbd jet

3.2 Effect of plasma treatment on textile surface properties

The analysis of the untreated UD flax textiles via tensiometry showed that the porous medium constant depends significantly on the orientation of the measured sample. The mean values are presented in Table 1. A constant *C* nearly twice as high as with parallel direction is achieved when the samples are measured with perpendicular direction to the reinforcing fibers. When the test liquid rises along the fiber the porosity of the sample is higher or the binding between fibers is lower. At the same time, this sample shows a significantly lower dynamic contact angle than the untreated samples, which means a better wettability. The UD flax sample treated with Ar plasma showed that factor C is significantly reduced after the treatment, meaning that the binding between the fibers is higher, i.e. the packing density has been increased. The decreased dynamic contact angle after plasma treatment indicates a better wettability and thus a higher polarity of the fiber surface.

Table 1: Mean geometric porous medium constant and advancing contact angle of the UD flax samples

	C [cm^5]	$\Theta_{propanol}$ [°]
untreated (parallel)	$3.4 \cdot 10^{-4}$	73.1 ± 1.5
untreated (perpendicular)	$7.1 \cdot 10^{-4}$	56.0 ± 1.7
treated with Ar plasma (perpendicular)	$2.5 \cdot 10^{-4}$	20.4 ± 4.7

The mean values of the balanced flax fabric measurements are shown in Table 2. For the balanced flax fabric it has been shown that in contrast to the UD flax, the porous medium constant rises after the plasma treatment.

Further important objective of this measurement was to determine the effectiveness of the plasma with regard to the homogeneous treatment within different layers of the textiles. All of the fabric plies show very similar values of the factor C and the dynamic contact angle, independent of the location in the stack. Thus, it has been demonstrated that the plasma achieves a homogeneous treatment of the textiles and a good penetration. All of the three analysed samples possess decreased binding between the fibers and increased dynamic contact angle compared with the untreated sample, meaning a lower polarity. Physical and chemical changes occurring on the fiber surface after plasma treatment have possibly led to this results. The increase of the microroughness and elimination or rather decrease of the oxygen amount on the fiber surface is expected to promote these results. The sample treated with an Ar/H_2 plasma showed the same packing density as in the case of the samples treated with Ar plasma. However, the dynamic contact angle increased showing a lower wettability.

Table 2: Mean geometric porous medium constant and advancing contact angle of the balanced flax fabric samples

	C [cm^5]	$\Theta_{ethanol}$ [°]
untreated	$1.0 \cdot 10^{-4}$	45.0 ± 6.4
treated with Ar plasma ply 1	$1.8 \cdot 10^{-4}$	49.4 ± 8.6
treated with Ar plasma ply 4	$1.7 \cdot 10^{-4}$	51.5 ± 5.0
treated with Ar plasma ply 8	$1.9 \cdot 10^{-4}$	51.9 ± 4.6
Treated with Ar/H_2 plasma	$1.8 \cdot 10^{-4}$	60.7 ± 7.4

3.3 Composite morphology

CT measurements reveal the inner morphology of the composite samples. The UD flax tape composites (Figure 6) show a high infiltration of the tapes with only small porosity. There is no obvious difference in morphology between the untreated and simultaneously plasma-treated samples.

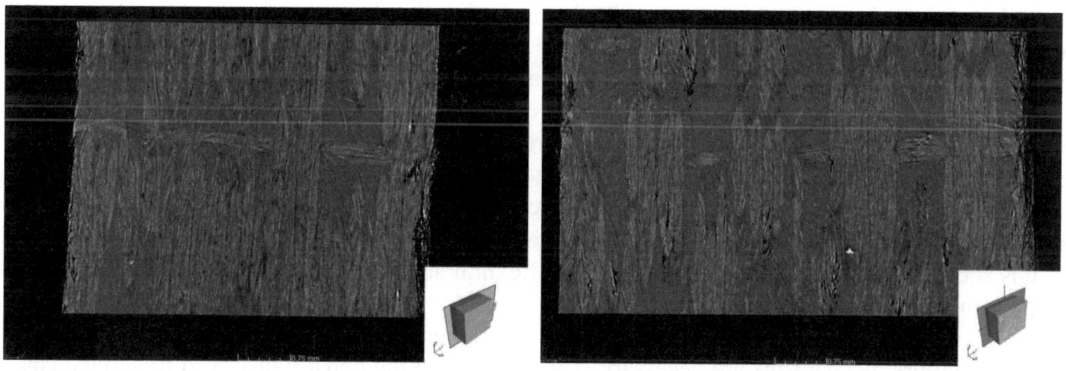

Figure 6: CT images of the untreated (left) and Ar plasma-treated (right) UD flax tape composites

In contrast, the balanced flax fabric composites (Figure 7) show a high porosity, that reflects the lower impregnation properties of the unsized fabrics. The impregnation inside the yarn differs between the untreated and the plasma-treated sample. While the untreated sample exhibits cavities inside the yarn, the plasma-treated sample shows no such cavities. This indicates an improved micro infiltration of the treated yarns. Additionally the matrix porosity is distributed homogeneously inside the untreated sample, while it is concentrated at the upper side of the plasma treated sample. With 2.88 vol-% compared to 3.36 vol-% the overall porosity is slightly lower for the treated sample.

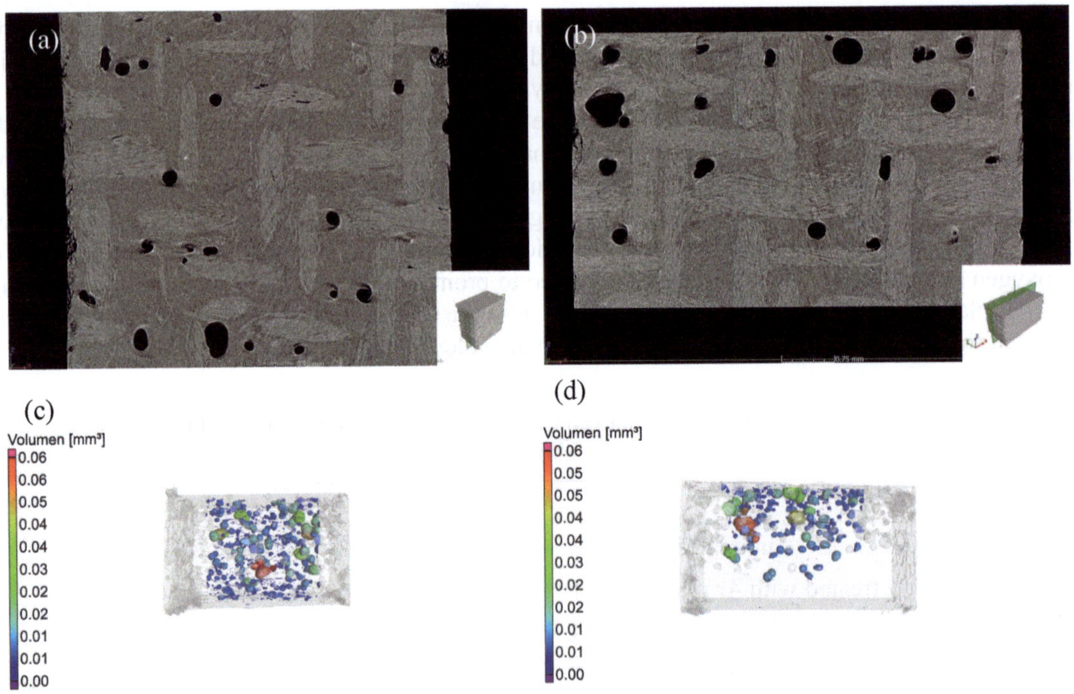

Figure 7: CT cross sections (parallel to fabric) and porosity distributions (top view) of the untreated (a & c) and Ar plasma-treated (b & d) balanced flax fabric composites

3.4 Influence of plasma treatment on mechanical properties

The argon plasma treatment of the unidirectional flax fabric with sizing results in an improved tensile strength of the composite (Figure 8). This is in agreement with the observed improved wettability of the textile (cf. 3.2). Interestingly this effect is obtained after simultaneous plasma treatment of all flax plies. The treatment of each ply individually would be expected to ensure a homogeneous exposure of the textile to the plasma and consequently to show a plasma treatment effect at least as intense as after the simultaneous treatment. Instead, the composite containing the separately treated flax plies exhibits mechanical properties roughly the same as the reference material. The reason for this behaviour is still under investigation. One assumption is that by increasing the amount of flax plies, the amount of plasma filaments could be forced to divide into several more filaments by making contact to each roving of the fabrics. This could lead to a better penetration of the whole fabrics bundle. While the tensile modulus shows no effect of plasma treatment, the flexural strength and modulus are lowered slightly after the simultaneous flax fiber treatment.

To understand this contrary behaviour in tensile and flexural testing further investigations about the dominating failure mechanisms are necessary. The higher packing density observed in surface characterization of the textiles might result in a more distinct unidirectional reinforcement of the composite leading to improved tensile and deteriorated flexural properties.

After Ar/H_2 plasma treatment, this behaviour is even more distinct, with a significantly increased tensile modulus of the composite while the tensile strength remains stable. In contrast, the flexural properties decrease significantly.

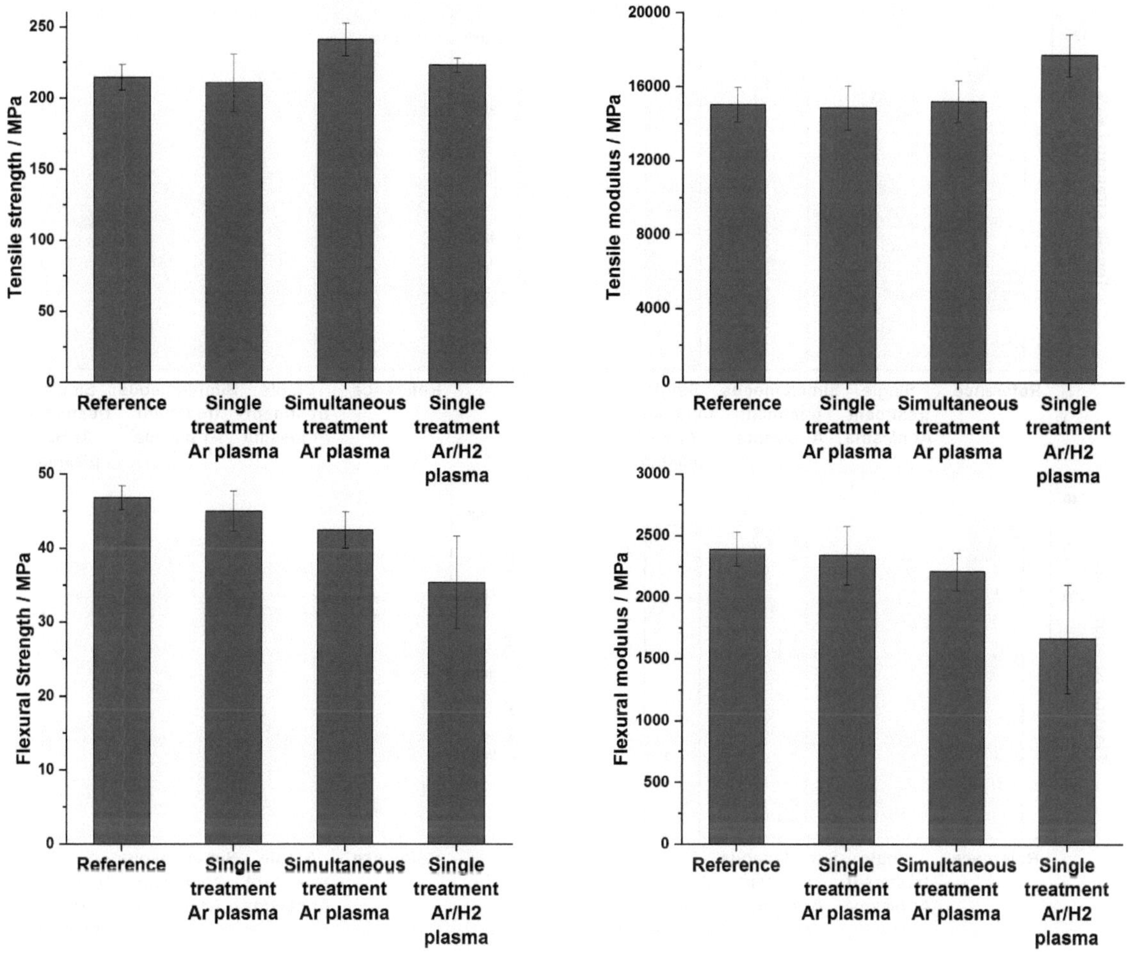

Figure 8: Influence of argon plasma treatment on mechanical properties of the UD flax tape composite

The Ar and Ar/H_2 plasma treatment of the unsized balanced flax fabric shows no improvement in tensile and flexural properties of the composite (Figure 9). Similar to the treatment of the UD flax tapes the separately treated plies of the balanced fabric exhibit lower values than the simultaneously treated plies.

The slight influence of the plasma treatment on the properties of the unsized textiles leads to the assumption that the positive plasma effect is coupled with a modification of the sizing present in the UD flay fabric. This is supported by the results of the contact angle measurements, where the unsized balanced fabric did not show a distinct plasma effect. A modification of the plasma treatment parameters might be necessary to achieve a distinct plasma effect.

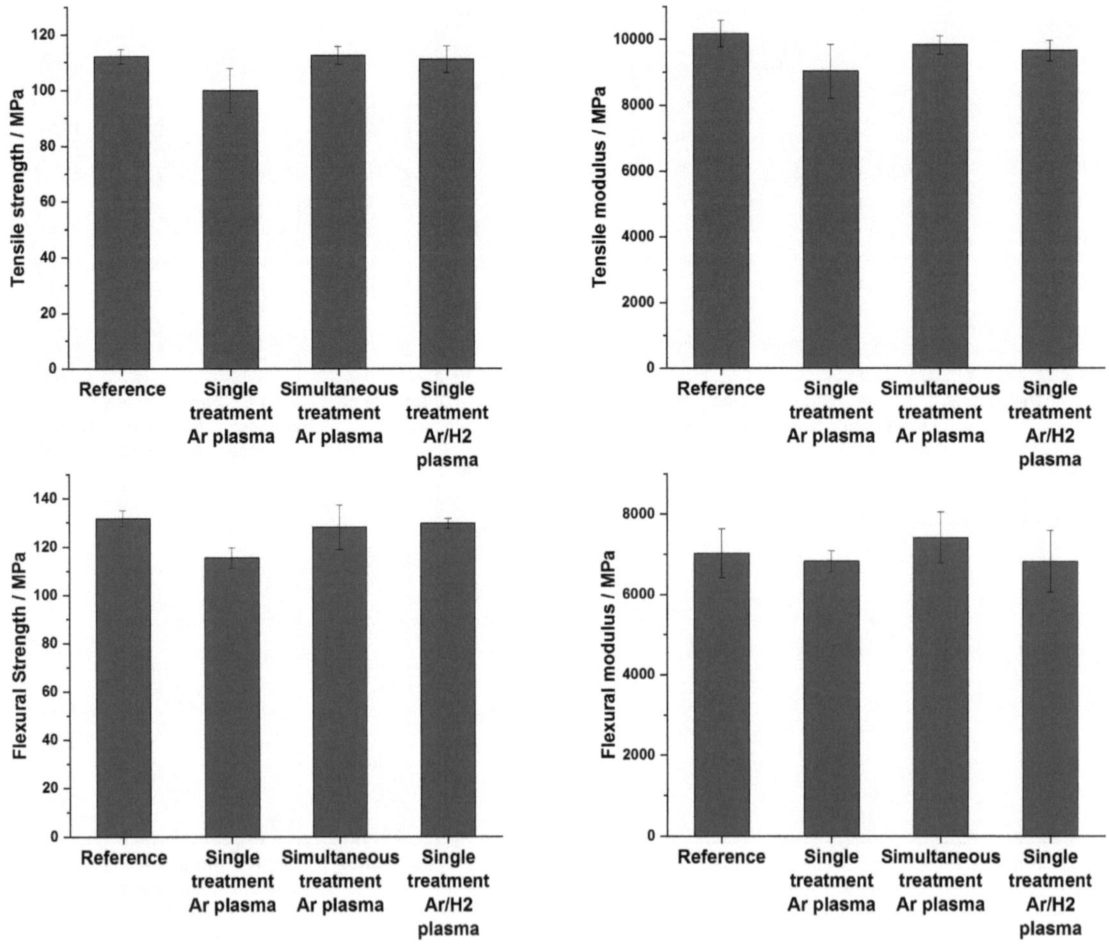

Figure 9: Influence of argon plasma treatment on mechanical properties of the balanced flax fabric composite

3.5 Composite durability under accelerated environmental testing

Until now, accelerated environmental testing as part of this study was possible only for the balanced flax fabric composite and Ar plasma treatment. Results of mechanical testing after the temperature/moisture exposure are depicted in Figure 10.

The untreated reference material as well as the plasma-treated sample show a significant decrease of tensile and flexural properties. The decrease in tensile modulus of about 50 % is especially significant. No obvious effect of plasma treatment is visible. The high porosity of the samples of about 3 vol-% might determine the low weathering resistance by aiding water uptake into the samples. Morphological studies on the ageing and failure mechanisms will follow. The present results indicate the importance of preventing high porosity especially in natural fiber bases composites during manufacturing.

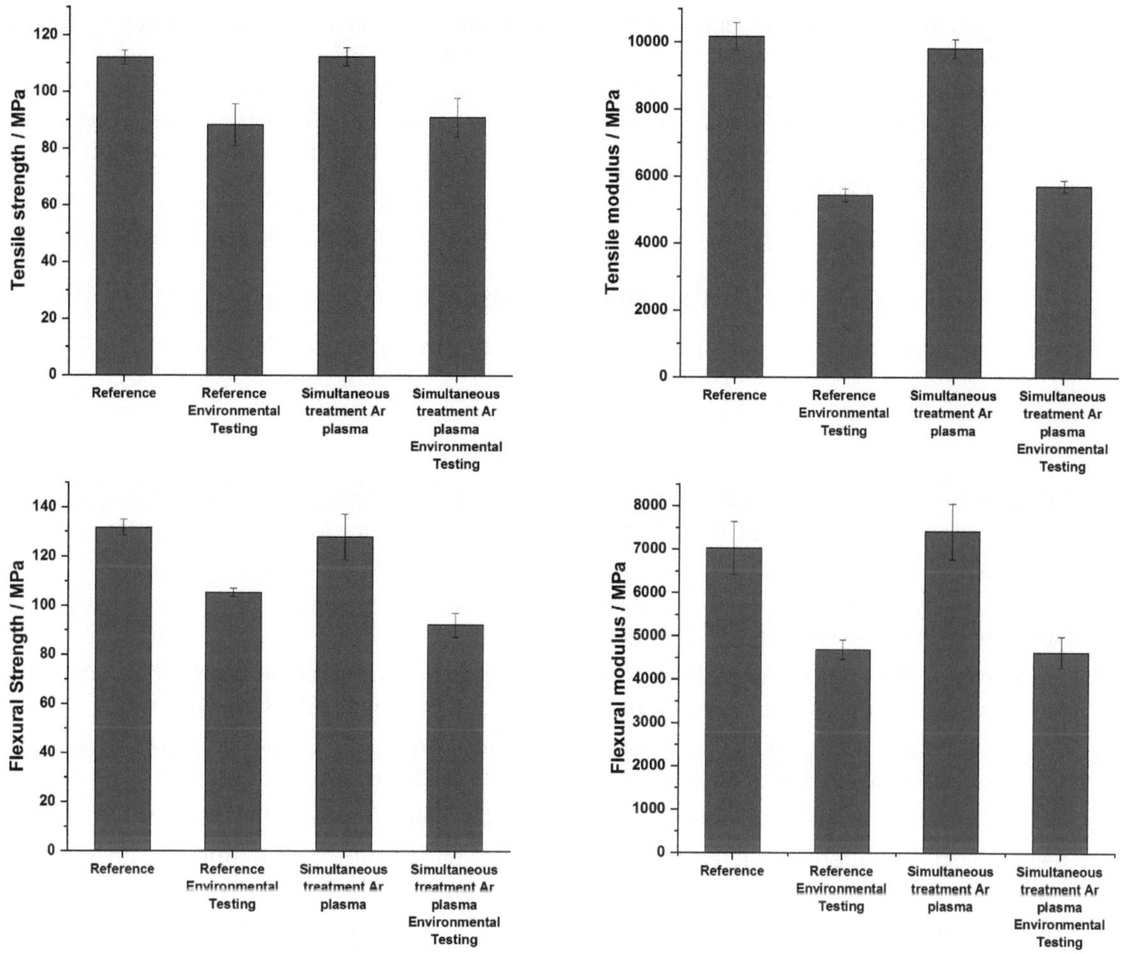

Figure 10: Effect of accelerated environmental testing on mechanical properties of the balanced flax fabric composite

Considering the more obvious plasma effect and lower porosity a more satisfying weathering durability is anticipated for the plasma-treated UD flax tape composites.

4 CONCLUSIONS

The presented results demonstrate the suitability of a plasma process based on a jet induced dielectric barrier surface discharge to influence the surface properties of natural fibers. The plasma process is able to treat several plies of fabric in one step, highlighting its potential for use as inline process.

Surface characterization tests show that the plasma effect is more pronounced for textiles with sizing than without sizing. An optimization of plasma parameters for unsized textiles might be necessary.

Tensile properties of the composites based on the sized UD flax tapes can be improved by plasma treatment while the flexural properties decrease.

In agreement with the surface characterization the plasma treatment of the unsized balanced flax fabric has no positive effect on the mechanical properties of the composite.

Accelerated environmental testing of the balanced flax fabric composites resulted in a significant decrease of mechanical properties, independent of a plasma treatment. An increased porosity of the samples is expected to determine the low durability of the samples.

The observed plasma effect on the sized UD flax tapes provides the basis for the anticipation of a considerable influence on the durability of the composites, which is the main objective of the study.

These tests are currently underway and results will determine the direction of further modification of plasma treatment. For example results could lead to the evaluation of different process gases.

When a positive effect on the NFPC durability by plasma treatment is achieved a transfer of the method to hybrid configurations containing natural and carbon fibers will succeed. The influence of the electrical conductivity and dense structure of carbon textiles on the plasma process will provide a challenge heading towards durable bio-HFPCs for outdoor applications.

ACKNOWLEDGEMENTS

This work was supported by the Fraunhofer Internal Programmes under Grant No. MEF 835 501.

REFERENCES

[1] D. Ray State-of-the-Art Applications of Natural Fiber Composites in the Industry. In: Natural Fiber Composites. R.D.S.G. Campilho (editor), CRC Press Taylor & Francis Group, Boca Raton, 2016, pp. 319–340.

[2] M. Shamsuyeva, H.-J. Endres ,Various Sectors Benefit from Bio-Hybrid Fiber Composite, bio-fibre Magazine **12/2016**, 2016, pp. 218–19.

[3] Z.N. Azwa, B.F. Yousif, A.C. Manalo, W. Karunasena, A review on the degradability of polymeric composites based on natural fibres, Materials and Design **47**, 2013, pp. 424–442.

[4] A. Meiners, G. Ohms, M. Leck, U. Vetter, B. Abel, Modifying Glass Fiber Size by Plasma Treatment, Journal of Adhesion Science and Technology **26**, 2012, pp. 1611–1627.

[5] M. Bellmann; C. Ochs, M. Harms, W. Viöl, Plasmadüse, DE 102016209097 A1, 2017.

[6] M.F. Pucci, P.-J. Liotier, S. Drapier, Capillary effects on flax fibers – Modification and characterization of the wetting dynamics, Composites: Part A **77**, 2015, pp. 257–265.

[7] DataPhysics, Determination of surface free energy of powders with DataPhysics DCAT 11, Application Note 10, Available online: http://www.dataphysics.de/2/start/application-centre/application-notes/

[8] DIN EN 60068-2-38, Umgebungseinflüsse – Teil 2-38: Prüfverfahren – Prüfung Z/AD: Zusammengesetzte Prüfung, Temperatur/Feuchte, zyklisch, Beuth Verlag Berlin, 2010.

THE BEST PROPERTIES FROM THERMOPLASTIC AND THERMOSETTING RESINS COMBINED IN FIBER REINFORCED PLASTICS FOR IMPROVED PRODUCTIVITY AND PROPERTIES

Andreas Niepel[1], Woo Jin Choi[2], Thomas Kowalik[1], Andreas Hartwig[1]

[1] Fraunhofer-Institut für Fertigungstechnik und Angewandte Materialforschung IFAM
Wiener Straße 12, D-28359 Bremen, correspondence to: andreas.hartwig@ifam.fraunhofer.de
website: www.ifam.fraunhofer.de
[2] Unitech Deutschland GmbH, Mündelheimer Weg 51a–53, 40472 Düsseldorf
website: unitech99.co.kr

Keywords:

Prepreg, Thermoset, Thermoplastic resin, FRP

ABSTRACT

Structural parts made of fiber reinforced plastics do not play a predominant role in automobile mass production due to the costs associated with storage condition, low level of automation, long processing times and safety and health regulations. The idea was to develop a holistic approach which can solve the problems with understanding the whole process chain from molecule to the finished product and integration in common technical infrastructure. A promising strategy was to create a matrix resin which combines easy processing of thermoplastics with superior mechanical properties of thermosets. The new matrix can pre-applied on all type of fibers. The resulting tack-free prepreg has a long shelf life at room temperature in consequence of phase separation of matrix raw materials which can be eliminated by raising the temperature. The resulting material has high reactivity and can be cured to a thermoset within a few minutes. Individual specification of the material like glass temperature, cure rate and toughness can be adjusted.

1 INTRODUCTION

Fiber reinforced plastics (FRP) are produced with either reactive resins or thermoplastics as a matrix polymer. The reactive resins form thermosets with good mechanical properties but require more complex processes to produce FRPs compared to the use of thermoplastics. The goal of this development was to combine the advantageous properties of both thermosets and thermoplastics.

These new fiber-based semi-finished products, such as prepreg, are characterized by matrices that can be processed thermoplastically and crosslinked to thermosets. The properties of the resins can be formulated in such a way that they can be used in a wide range of applications. Solvents are not needed when manufacturing prepreg using water-based systems. High temperatures, as in the case of cost-intensive melt impregnation with conventional thermoplastics, can be avoided, making it possible to use heat-sensitive fibers such as polyethylene or cellulose. The innovative resins are dried at only moderate temperatures.

Manufacturing the semi-finished products does not require any new infrastructure. The fiber coating can be applied roll-to-roll by immersion, knife coating, or spray. After drying, the fibers are protected against slipping and thus facilitate insertion in the tool that is suitable for the load path. Below the softening temperature of approximately 60 °C, the material is not sticky, rather it remains flexible and drapable. In contrast to conventional prepregs, it can be stored stably at ambient conditions for many months. At temperatures below 60 °C, the shape is stable, and the semi-finished product is still workable between the softening temperature and the hardening temperature (above 120 °C). Curing to

K. Dröder et al (eds), *Technologies for economical and functional lightweight design*, Zukunftstechnologien für den multifunktionalen Leichtbau, https://doi.org/10.1007/978-3-662-58206-0_20

transfer the prepreg into a thermoset component is carried out either during or after forming, as a subsequent step. To do this, the dry layered structure can be formed in a heating press and cured. Another advantage is that the component can be removed from the press while still hot, but the shape is already stable (Figure 1). We name this new kind of method for FRP production TvD wich means „thermoplastically processable thermoset" or in German „thermoplastisch **v**erarbeitbare **D**uromere".

Figure 1: TvD process steps: Prepregs without tack (left), thermoplastically processed semifinished part (mid) and thermoset structural part (right) (last two steps preferably carried out in one step).

During component manufacture, as with other procedures for FRP production, mold release agents must be used. For this the FlexPlas® release films that have been developed by Fraunhofer IFAM have proven particularly useful. These films are shaped with the fiber sheets and have permanent release layers with different release properties on both sides. The highly effective release layer leads to easy removal of the part from the mold, whereby the side of the film attached to the part does not release as quickly and the film remains attached to the component until further use. After removing the film, the component can be processed further without the usual surface treatment.

2 EXPERIMENTAL

2.1 Materials

Polyamid 6 homopolymer Ultramid® B27 (BASF SE, Ludwigshafen, Germany) was used as thermoplastic matrix material **PA6**. Granulate was cryomilled with liquid nitrogen to a fine powder of < 200 µm. Powder dried at 100 °C, 10^-3 bar 24 hours and stored under dry Argon atmosphere. The conventional thermoset matrix **EP** comprised 100 parts by weigth (pbw) of epoxy resin L and 25 pbw of curing agent EPH 161 (R&G Faserverbundwerkstoffe GmbH, Waldenbuch, Germany). Both components were mixed directly before use with a planetary mixer at 3500 rpm for 1 minute to the appearance of a clear, slightly viscous liquid. Thermoplastic processable thermoset matrix **TvD-0, -1, -10** and **-25** (Fraunhofer IFAM, Bremen, Germany) was an opaque waterborne dispersion with a solid content of 50 % and particle size < 1 µm. The different versions contain 0, 1, 10 and 25 pbw of a toughener, respectively. Biaxially oriented carbon fiber fabrics (±45°, 300 g/m^2, 12,000 filaments per fiber) purchased at C. Cramer, Weberei, GmbH u. Co. KG used as reinforcement material without further treatment.

2.2 Preparation of Carbon Fiber Reinforced Plastics (CFRP)

In general a single carbon fiber fabric layer with the dimensions of 0.3 m × 0.3 m (0.09 m^2) and a weight of 48 g was combined with 48 g of the different matrix materials. CFRP's for mechanical benchmarking were made of 6 layers. The PA-, EP- and TvD sheets were made in a heated press in a closed mold with different curing / hardening conditions.

Carbon fiber reinforced PA6 was made by powder coating of each fabric in the mold at room temperature. In the next step the mold was closed with a force of 1.111 kN/m^2 and heated to 250 °C with a heating rate of 5 °C/min. After 30 minutes at 250 °C the composite material was cooled down to 30 °C with a cooling rate of 3 °C/min.

Carbon fiber reinforced EP was made by liquid coating of each fabric in the mold at room temperature. In the next step the mold was closed with a force of 1.111 kN/m^2 and heated to 130 °C with a heating rate of 5 °C/min. After 10 hours at 130 °C the composite material was cooled down to 30 °C with a cooling rate of 3 °C/min.

Carbon fiber reinforced TvD was made by preparing a dry prepreg first. Fabric was preimpregnated with waterborne TvD-dispersion. Prepreg was dried 12 hours at room temperature to remove residual water. In the next step dry prepregs were placed in mold at room temperature followed by closing the tool with a force of 1.111 kN/m^2. In the next phase the temperature was rised to 200 °C with a heating rate of 5 °C/min. After 20 minutes at 200 °C the composite material was cooled down to 30 °C with a cooling rate of 3 °C/min.

2.3 Characterization of Thermal and Mechanical Behavior

Surface topography was investigated with Scanning Electron Microscopy. 1st sample of TvD-10 matrix has been dried for 30 minutes at room temperature, and lowered pressure of 10^-3 bar on a sample plate. 2nd sample has been dried by the same procedure but additionally cured for another 30 minutes at 150 °C. Topographic imaging was generated by using a FE-SEM Leo 1530 Gemini Microscope (Zeiss, Jana, Germany) with a accelerating voltage of 7.00 kV and a working distance of 6.1 – 7.3 mm.

Thermal tests were conducted by differential scanning calorimetry and dynamic mechanical analysis. Calorimetric analysis of (3 – 6 mg) uncured samples was performed using differential scanning calorimeter Discovery DSC (TA Instruments – Waters LLC, New Castle, USA) in a temperature range of –90–250 °C with a heating rate of 10 °C/min. Melting and curing behavior was determined by integration of the endothermic melting and exothermic reaction peak.

To study the accelerating effect of higher temperature on the cure behavior single layers of uncured CFR with TvD-10-matrix was cured in a pre-heated oven at 150 and 200 °C within 0, 10, 30, 60, 120, 300 and 600 seconds. Remaining reaction enthalpy was measured by DSC.

Dynamic mechanical analysis was performed with cured multilayer CFRP samples with dimensions of 30 mm × 10 mm × 3 mm (±0.5 mm) by using DMA-Analyzer Q800 (TA Instruments – Waters LLC, New Castle, USA) using single cantilever mode with a stress frequency of 1 Hz, in a temperature range of –150–250 °C with a heating rate of 3 °C/min. Glass transition temperatures were determined by local maxima of the loss modulus. Glass transition and storage modulus of a carbon fiber reinforced thermoplast, thermoset and thermoplastically processable thermoset was investigated with PA6, EP and TvD-10. To study the plasticizing effect of water on the matrix materials DMA-specimens of carbon fiber reinforced PA6 and TvD-10 were aged 7 days in water and tested by DMA.

Mechanical tests were carried out with five specimens of each CFRP sheet with a Universal testing machine 100kN (Zwick GmbH & Co. KG, Ulm Germany) at 22 ± 2 °C. Tensile tests according to standard DIN EN ISO 527-5 [1], sample type 3 with as test speed of 2 mm/min with a force sensing unit of maximum 100 kN. Bending tests conducted in agreement with DIN EN ISO 14125 [2], sample dimensions 100 mm × 15 mm × 3.0 mm (±0.5 mm) and a maximum gage load of 50 kN.

Fracture mechanics of a carbon fiber reinforced thermoplast, thermoset and thermoplastically processable thermoset was carried out with PA6, EP and TvD-10 as matrix polymers. To study the effect of toughening on fracture mechanics a formulation without toughening (TvD-0) was compared to matrices (TvD-1, -5, -10 and -25) with increasing amount of toughening agent (1, 5, 10 and 25 pbw).

3 RESULTS AND DISCUSSION

Dried TvD-10-matrix at room temperature has a topology of separated phases dominated by spheres of 1–5 µm surround with globes of < 0.5 µm. After curing the separation has been terminated totally (Figure 2).

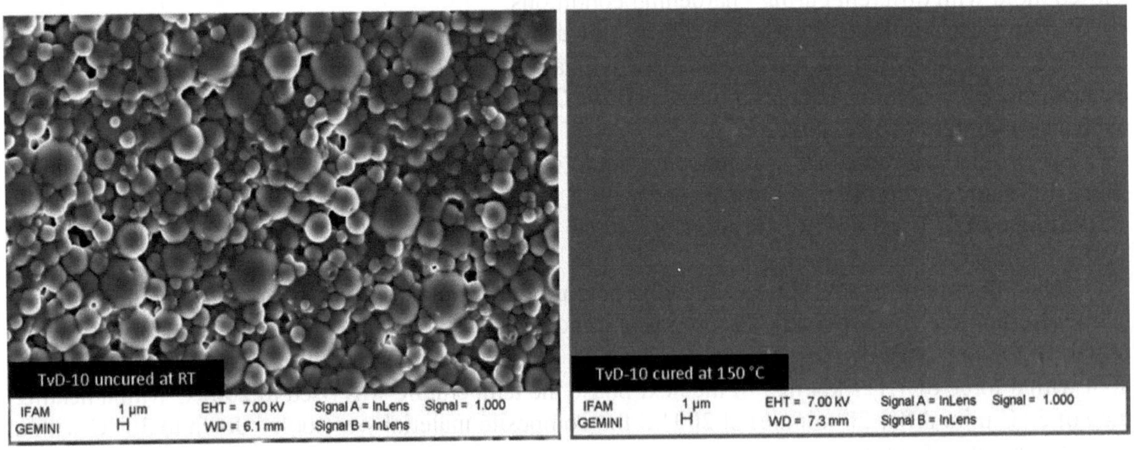

Figure 2: SEM images of the uncured, phase separated, TvD-10 layer at RT (left) and the at 150 °C cured counter piece with total coalescence (right)

Thermogram of TvD-10 (Figure 3) shows that conglomeration takes place at 55–70 °C followed by a sharp exothermic reactionpeak with an onset of ca. 125 °C. The difference in temperature of 55–70 °C should achieve a sufficent wetting of fiber surface before curing takes place.

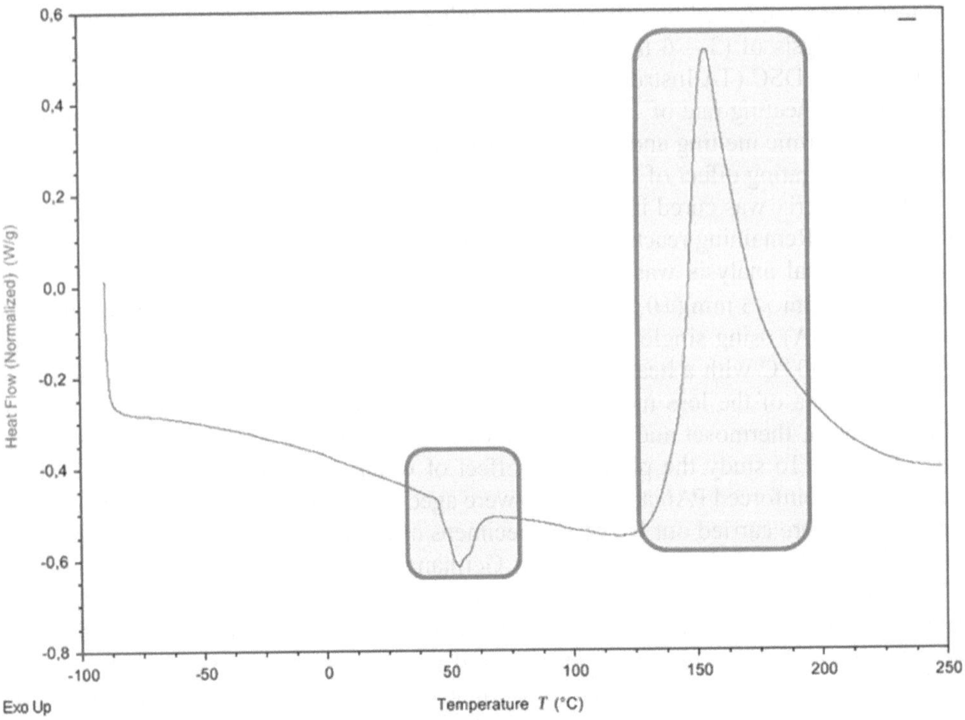

Figure 3: DSC-thermogram of TvD-matrix – film forming temperature: 55 – 70 °C and curing reaction, onset T ~ 125 °C

Raise the curing temperature from 150 to 200 °C leads to an unexpected strong conversion increase of CFR-TvD-10 Prepreg (Table 1). A dimension stable CFRP was obtained after only 10 seconds at 200 °C. Nearly full conversion was achieved within 60 seconds. Within 60 seconds at 150 °C there is almost no curing reaction. With increasing thickness of the laminate time at which the laminates core is hot enough increases. Therefore, heat transport is limiting curing reaction and thick laminates need a longer time.

Table 1: Reaction conversion of TvD-10 at two different curing temperatures

Time [s]	T_{cure} 150 °C Conversion [%]	T_{cure} 200 °C Conversion [%]
0	0	0
10	2	27
30	2	78
60	5	93
120	15	98
300	73	98
600	80	99

PA6 and TvD-10 provide similar thermal behavior in the temperature range of –150–220 °C. Above 220 °C TvD-10 shows still high storage-modulus due to high crosslinking. In contrast to the other CFRP with highly crosslinked common epoxide matrix (EP), TvD does not show a pronounced T_g due to the crosslinking density (Figure 4).

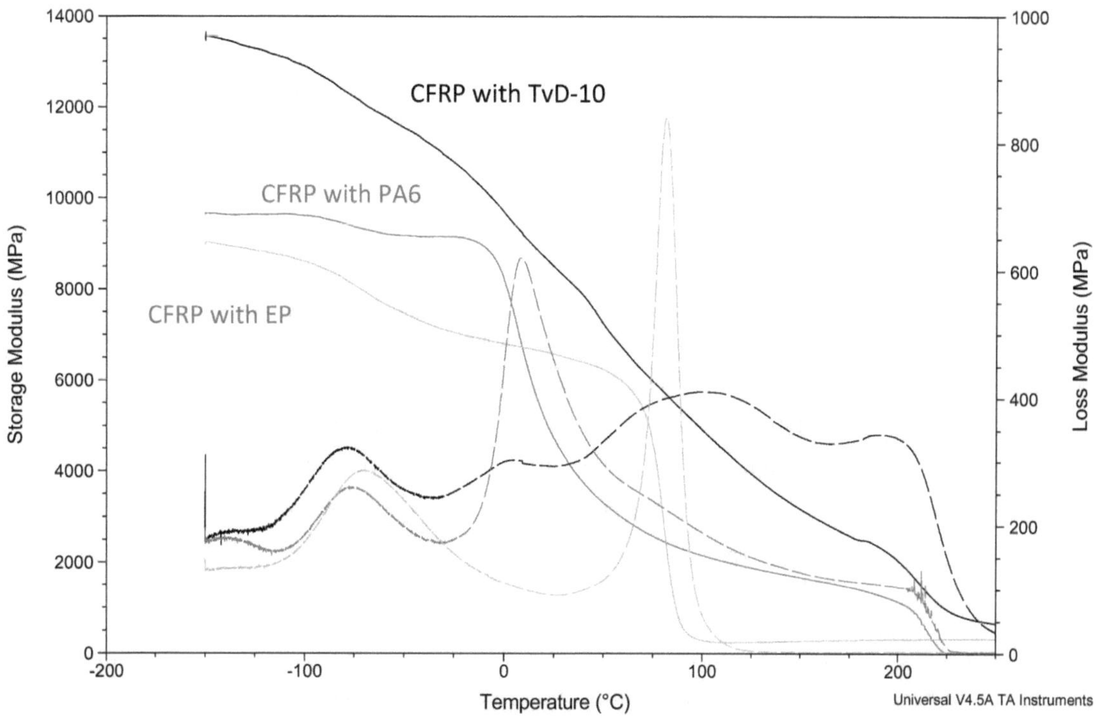

Figure 4: DMA-thermograms (3 °C/min) showing thermal behavior three different CFRP's made from a thermoplastic matrix (PA6), a conventional thermoset matrix (EP) and athermoplastically processable thermoset (TvD-10)

The investigation of water absorption on CFRP's showed only small changes in storage modulus of materials made with TVD-10 matrix. In contrast materials prepared from PA6 showed a plasticization by water indicated by a decreasing glass transition temperature of more than 20 °C (Figure 5). This effect can be explained by crosslinkage of TvD-10 and the high content of amide groups in PA 6.

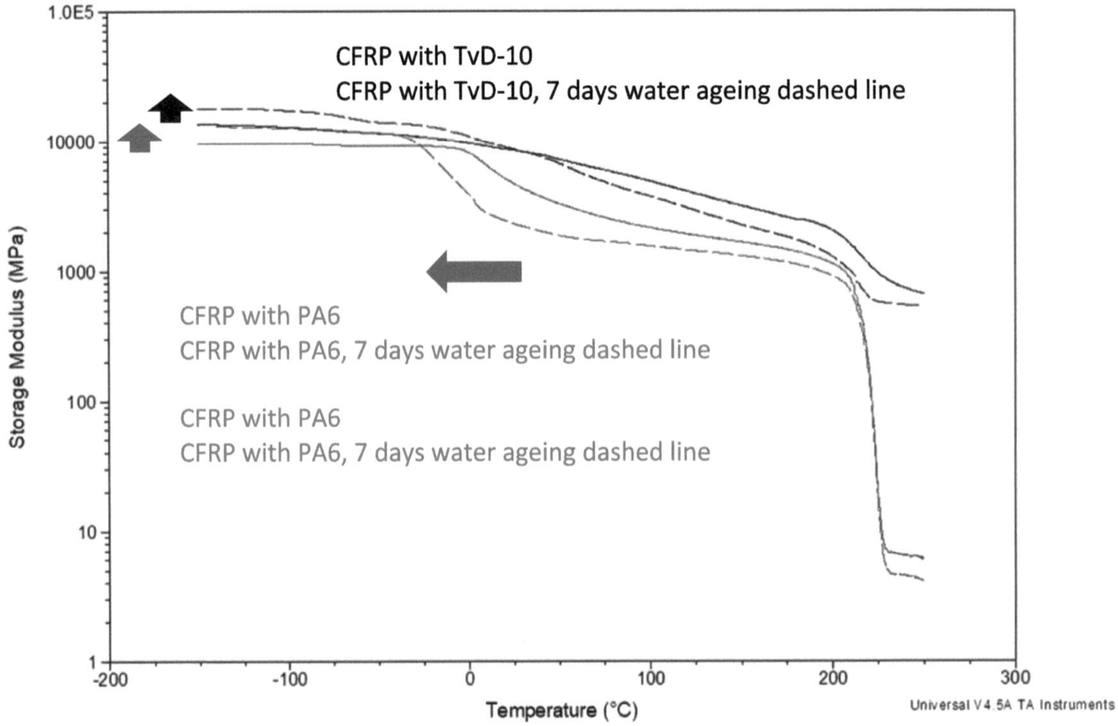

Figure 5: DMA-thermograms (3 °C/min) shows changes of the storage modulus for two different CFRP's before and after water storage

Mechanical properties of TvD-10 are similar or better compared to common resins (PA6 an EP). This is shown for the bending as well as tensile strength in Table 2.

Table 2: Tensile and bending strength determined for the CFRP's with different matrix resins at ambient temperature

Matrix	tensile strength [MPa]	bending strength [MPa]
PA6	53,2 ± 1,2	3215 ± 159
EP	107,3 ± 3,8	7259 ± 1673
TvD-10	110,4 ± 1,6	11376 ± 555

Increasing the amount of toughening agent up to 25 pbw in the TvD-matrix leads to a slight decrease of tensile and bending strength. There is almost no effect in mechanical behavior of TvD with 1 pbw compared to material without toughening agent (Table 3).

Table 3: Rising amount of toughening agent leads to an only low loss of tensile and bending strength

Matrix	tensile strength [MPa]	bending strength [MPa]
TvD-25	83 ± 2,2	7892 ± 656
TvD-10	110,4 ± 1,6	11376 ± 555
TvD-1	109 ± 1,1	13638 ± 1184
TvD-0	119,6 ± 1,4	13198 ± 635

4 CONCLUSIONS

TvD is a smart concept for automatized production of FRP components. Advantages of thermoplastic and thermoset properties are combined in one material. Individual specification of the resins (T_g, cure rate, toughness etc.) could be adopted through formulation variation. Special resins for different fibers and production processes can be formulated. Fixation of prepreg layers relative to each other is possible by melting the resin locally with short heat treatment. Uncured material cuttings can be recycled and used as raw material in other processes. Additional advantages of TvD which are not shown here is its high potential for class A surfaces and applicability with pure white woven fiber appearance.

REFERENCES

[1] Plastics – Determination of tensile properties – Part 5: Test conditions for unidirectional fibre-reinforced plastic composites (ISO 527-5:2009).

[2] Fibre-reinforced plastic composites – Determination of flexural properties (ISO 14125:1998 + Cor.1:2001 + Amd.1:2011).

FAUST: MATERIAL CHARACTERIZATION OF LOW-COST FOAM MATERIALS UNDER REAL BOUNDARY PROCESS CONDITIONS FOR RTM LARGE-SCALE PRODUCTION

Mark Opitz[1*], Dominic Bertling[1], Nico Liebers[1]

[1] German Aerospace Center, Institute of Composite Structures and Adaptive Systems, Lilienthalplatz 7, 38108 Braunschweig, Germany, www.dlr.de/fa
*Corresponding author (Mark.Opitz@dlr.de).

Keywords:

Foam core material, Experimental material characterization, Resin transfer molding, Measuring system, Ultrasound sensor, Material-adapted process strategies

ABSTRACT

The Resin Transfer Molding (RTM) process is the first choice for large-scale production of continuous fiber reinforced composite structures due to its capabilities of industrialization and automation at low price. However, the process is currently limited to monolithic structures. Low-cost and yet powerful foam materials do not seem to be compatible with the manufacturing conditions of the RTM process. Available measuring methods do not sufficiently analyze the foam behavior during processing, so that expensive preliminary manufacturing tests are necessary. The use of high-performance foam material, as known in aerospace applications, is not an alternative due to their high price.

In order to enable the use of low-cost foam materials, it is important to match material and process. For this reason, a simple but highly efficient method based on ultrasonic sensors has been developed and patented by the Institute of Composite Structures and Adaptive Systems at DLR. The Foam Analysis Ultrasound System (FAUSt) enables a quantified property description of foam materials under realistic manufacturing conditions for the first time. Without contact to the sample the time-dependent deformation of foam materials depending on temperature and pressure can be determined. In addition to the material characterization itself, the measurement results benefit primarily the development of efficient, material-adapted impregnation strategies. Also process parameter identification for ideal processing and quality assurance is supported. Furthermore, the data can be used for numerical simulation methods in the early development process.

1 INTRODUCTION

One of the main goals for tomorrow's mobility is to reduce absolute energy consumption of vehicles and avoiding harmful emissions, such as CO_2 and noise. To overcome these challenges research is done on future lightweight vehicle concepts for use in urban and interurban environments in the DLR research project Next Generation Car (NGC) (Figure 1). The Interurban Vehicle (IUV) is designed as a holistic vehicle concept oriented towards the road vehicle market in 2025. The main features are a fuel cell and a CFRP body structure for increased comfort, reduced energy consumption for maximum range, energy efficiency and safety. One focus of the work is the realization of particularly light and high-strength composite sandwich structure, which can be produced in large quantities [1]–[4].

K. Dröder et al (eds), *Technologies for economical and functional lightweight design*, Zukunftstechnologien für den multifunktionalen Leichtbau, https://doi.org/10.1007/978-3-662-58206-0_21

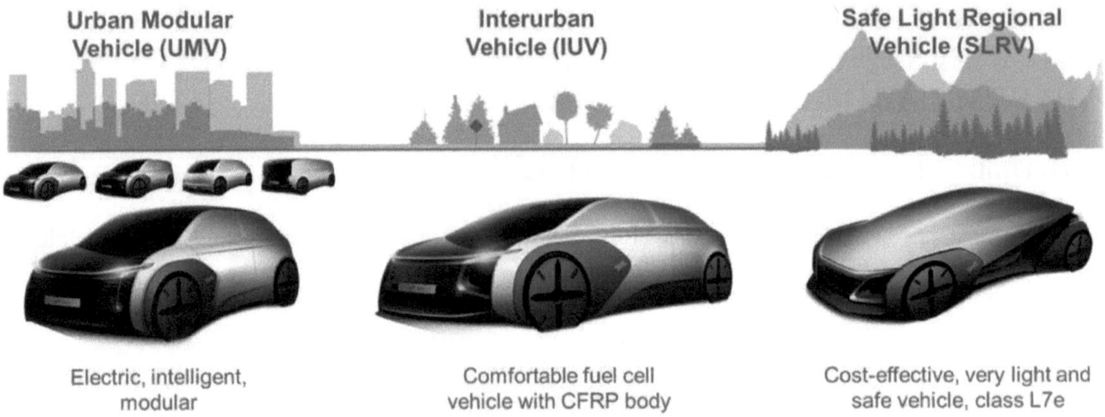

Figure 1: Next Generation Car (NGC) – Lightweight vehicle concepts for use in urban and interurban environments

2 RESIN TRANSFER MOLDING

Resin Transfer Molding (RTM) is a closed-mold vacuum-assisted process, where dry fiber material is placed and impregnated with reaction resin in a fixed cavity [5]. RTM is primarily used to mold components with large surface areas, complex shapes and smooth finishes. The advantages over other methods are the achievable component properties as well as the processability. Especially the use of reaction resin systems that crosslink by heat supply, allows longer filling times at lower injection pressures. With a variety of different and highly automated technologies the process is robust and well-suited for a mass production up to 100.000 parts/year [6]–[8]. By using semi-finished products process costs can additionally be reduced. The finished component typically has a constant thickness, is near net shape with high dimensional stability and surface quality.

Typical cycle times can be less than 10 minutes per part with 30 s for impregnation and 1.5 minutes for curing. The impregnation takes place at up to 15 bar, while the post-injection pressure can be up to 120 bar. The exothermic reaction of the resin produces both a sufficient reaction rate for short cycles and a high glass transition temperature at low process temperature [9].

The overall process stability is largely dependent on the performance of the used materials. Since pressure and temperature are applied during RTM-Process for injection and curing, when using a foam core it must absorb the occurring loads. Therefore the foam core limits the process window. As a result, the foam core is responsible for the final component quality as well as the process costs [10].

3 SANDWICH STRUCTURES

The use of sandwich structures enables a high bending stiffness with overall low density by attaching two thin, but stiff face sheets to a thick, but lightweight core. The face sheets carry tension and compression loads while the core carries shear loads and supports the face sheets to be in position. The performance of a sandwich structure depends primarily upon the surface skin and the distance between them. A large distance of the face sheets leads to a higher geometric moment of inertia, which results in a higher bending stiffness. Due to this arrangement layout with relatively small stress levels in the core, and the low density of the core material itself a high stiffness-to-weight ratio can be achieved [11]–[13].

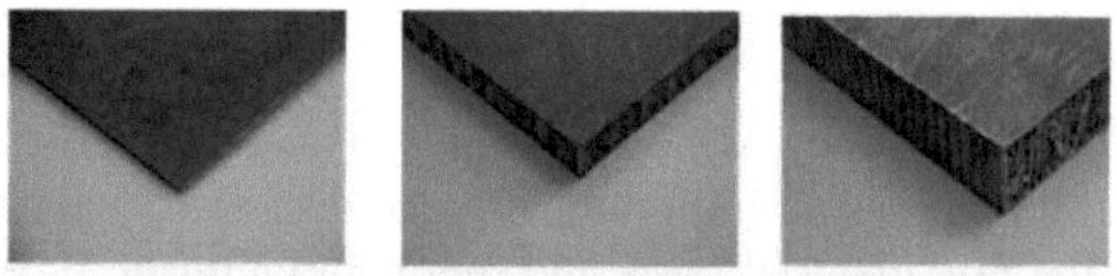

	Monolitic (t)	Thin (2t)	Thick (4t)
Bending Stiffnes (E*J)	100%	bis 700%	bis 3700%
Weight	100%	103 – 108%	106 – 120%

Figure 2: Principle of sandwich structures: Increase of the specific bending stiffness [14]

4 CORE MATERIALS

Widely used core materials are rigid foam core, balsa, and honeycomb. Since the dry fibers are impregnated with liquid resin during the impregnation process, only closed-cell foam materials are used in RTM processes. When using e.g. Honeycomb as a core material, the core would be completely filled with resin during the impregnation process [15]. Thus, no significant weight or cost reduction can be achieved.

4.1 Foam core material

There are many foam cores available on the market, which are of different plastics and produced by different processing techniques. Most common are rigid foams, or integral foams with densities from 30 kg/m^3 to 150 kg/m^3. The choice of raw material determines the properties and therefor the performance of the final foam and its suitability as a structural material. In addition to the mechanical properties, other relevant criteria are the maximum operating temperature, necessary post-processing steps, reusability and fire, smoke and toxicity properties. For example foams made of polyurethane (PUR) and polyvinyl chloride (PVC) yield good mechanical performance at very low weight in low temperature applications, while polyethylene terephthalate (PET) possesses also excellent thermal stability. Polyethersulfone (PES), and polymethacrylimide (PMI), offer a unique high service temperature, very low water absorption, superior damage and impact performance and improved dielectric properties [11], [12], [16].

4.2 Foam core manufacturing

The foam core is made from a variety of melted or granulated plastic in a continuous or discontinuous process. The foaming operation is performed by physical, chemical or mechanical methods whereby air is inserted into the melted plastic. The air is a result of the blowing agent mixed together with the raw material or expanding gas during injection molding. Together with the manufacturing conditions (vacuum, pressure, temperature, time, mold filling level) and various process steps, like preexpansion, dwell time and post processing, the foam characteristic can be customized. For this purpose, the structure of the foam is changed by the number and size as well as distribution and structure of the cells. For the homogeneity of the foam a good mixing is crucial as well as constant filling and processing. Since process sensitivity increases with decreasing density, meaning larger cells, deviation in the dimensions and mechanical properties of the foam core increase as well [17]. For large scale production foam cores are often produced by Reaction Injection Molding (RIM), whereby the inhomogeneity of the so called integral skin foam is expressly intended. The liquid plastic is injected into a colder mold and the temperature gradient decreases the cell size on the surface. This results in a part with a low-density core and cell-free edges.

For small quantities, the final foam core geometry is made from sheets by bonding, milling, thermal forming or pressing. During thermoforming or pressing, the foam material is shaped under pressure and temperature. Large or complex degrees of deformation can damage the cell structure. The consequences are reduced mechanical properties and penetration by the resin system during injection.

4.3 Low cost foam core

Beside the material efficiency, which is usually defined as a mechanical performance per weight, the performance per cost plays an important role for cost sensitive applications. The cost of foam materials depends on the raw material costs and the manufacturing process. This also takes into account the geometry and quantity. According to *Pflug* [18] the PET foam with a price between 2 €/kg and 15 €/kg belongs to the low cost core materials.

The core used in this investigation is AIREX T92.100 PET from 3A Composites with a thickness of 12 mm and 20 mm. The thermoplastic structural rigid foam with closed cells is highly suitable for all types of resins and processes. In addition to its very good mechanical properties the foam has an outstanding price-performance ratio. With an average density of 105 kg/m³ the compressive strength is about 1,4 N/mm². It is also characterized by high temperature stability during processing (up to 150 °C) and long-term applications (up to 100 °C) without post-expansion. [19]

5 MATERIAL CHARACTERISATION

Rigid foams can be analyzed with common test methods. Properties of interest are the long term thermal stability, the heat resistance, the compressive strength and the roughness [20]–[22]. These properties are usually tested by tactile measuring methods based on small, standardized specimens. However, the operating conditions or manufacturing constraints deviate from these specified values. Furthermore as uniaxial testing is the most common mechanical test used, the results are insufficient to characterize foams for three dimensional loads. Although consistent properties, high temperature stability and low resin uptake are promised, research shows the opposite. As noted by [23]–[28], the material properties in terms of density, strength, thermal expansion and porosity show a different behavior.

5.1 Preliminary tests

To proof the mechanical properties of an AIREX T92.100 test specimen, the foam was subjected to a simple preliminary test. For this purpose, a foam plate was placed between press plates pre-heated at 115 °C and applied with 1 MPa compression for approximately 5 minutes. Examination of the panels showed that the initial thickness was reduced from 12 mm to 9.62 mm (19.8 % plastic deformation).

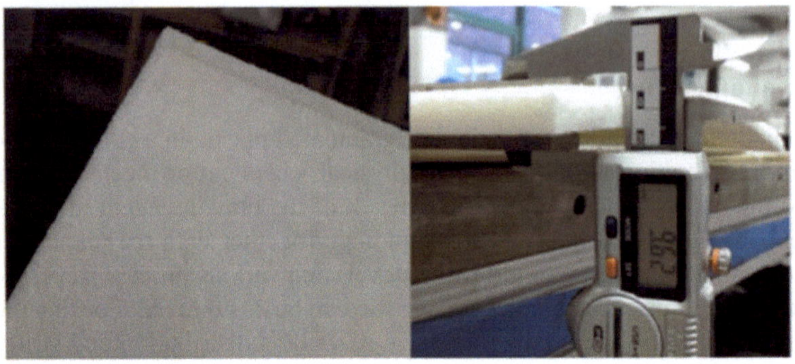

Figure 3: Preliminary test Compressive strength by means of a press

The determination of the linear coefficient of thermal expansion was performed according to ISO 11359-2 from 10 °C to 130 °C with a rate of 3 K/min and reverse. The samples were taken directly from supplied plates (original) or additionally thermoformed by DLR. Each sample was tested twice. All tests were performed using a thermomechanical analysis TMA/SFTA841^e provided by METTLER TOLEDO. The sample characteristics and test results are shown in Table 1. For different samples a non-linear, directional and production-dependent behavior of the coefficient of thermal expansion (CTE) was identified. Furthermore, the CTE values varied up to a factor of 5 for the same sample between Test Run 1 and Test Run 2. Consequently, the supplied data is only valid to a limited extent, since the specified values are determined under standard conditions.

Table 1: Sample characteristics and test results: CTE

Sample	Sample Dimensions w × l × t [mm]	CTE Test Run 1 [10^{-6} K^{-1}]	CTE Test Run 2 [10^{-6} K^{-1}]	AVG CTE [10^{-6} K^{-1}]	Standard Deviation [10^{-6} K^{-1}]	CTE Data Sheet [10^{-6} K^{-1}]
Airex T92.100-original A	4 × 4 × 12	68,98	139,92	104,45	35,47	76
Airex T92.100-original B	4 × 4 × 20	66,15	210,11	138,13	71,98	76
Airex T92.100-thermoforming A	4 × 4 × 2,9	200,07	1025,97	613,02	412,95	-
Airex T92.100-thermoforming B	4 × 4 × 2,9	199,04	1069,93	634,49	435,445	-

5.2 Measurement methods

Using an optical technique based on speckle interferometry, *Ferreira* [23] investigated the behavior under thermal loads. For PVC foams, a density variation over the cross section of up to 25 % was identified. Furthermore, a symmetrical density gradient of the examined samples was obtained, which is due to the manufacturing process. Also, there is no linear coefficient of thermal expansion across the thickness of the sample. *Scherble* [24] analyzed PMI foams and found a dependency between surface porosity and material density, pore shape, manufacturing process and finishing. The reduction in porosity results in lower resin uptake and reduced peel strength. *Viot* [25] developed a measuring chamber for the deformation analysis of PP foams under hydrostatic compression. The deformation of the sample is accompanied by high speed image techniques. The foam response under hydrostatic com pression shows a non-linear elastic stage, followed by a plastic plateau and densification. *Deng* [29] used a dilatometer for thermal expansion coefficient measurement at low temperatures.

These studies show that single properties of foams can be analyzed under certain conditions. The ratio of sample size to density does not allow valid statements regarding the CTE gathered from TMA measurements, since it is not a characteristic cross-section of the real part. However, no measurement or analysis method for determination of the foam characteristics under realistic process boundaries, both pressure and temperature was found.

Also process influences and additional boundary conditions of fiber, resin (e.g. exothermic heat release) or component geometry are not considered by the manufacturer. There is no general incompatibility between low cost foam cores and RTM processes. For the processing of foams, however, the associated material properties are required. Therefore the process window, regarding pressure and temperature, should be determined at first.

6 REQUIREMENTS FOR THE MEASURING SYSTEM

The measuring system has to fulfil the following requirements:

- robust and simple measuring method
- simultaneous and all-sided occurrence of boundary conditions: pressure and temperature
- controlled pressure application up to 20bar
- temperature control in a range of 20–180 °C
- continuous and contactless measurement of specimen size
- simultaneous measurement of dimension in at least three main axes
- free movement and elongation of the test sample

- adequate ratio of sample size to pore size of the foam
- variable sample height
- measuring fluid with matrix-compatible properties in terms of viscosity and wettability

7 MEASURING SYSTEM: FAUST

7.1 Measurement principle

The measuring method is based on ultrasound technology without contact to the sample. Position and deformation are calculated from distance measured by travel time of sound waves between sensor and sample, as seen in Figure 4.

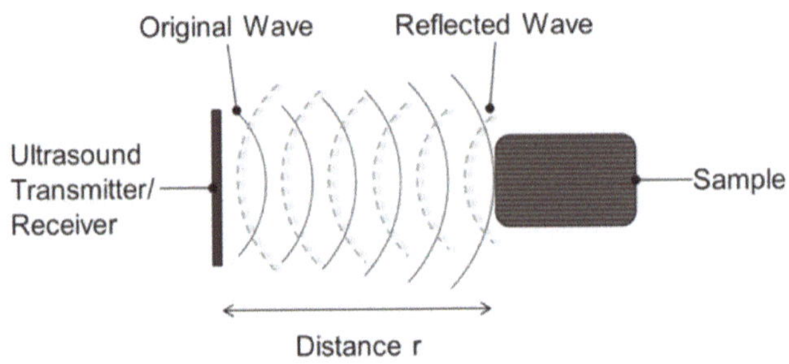

Figure 4: Principle of distance measurement with ultrasound [30]

By measuring the difference in time Δt between the pulse being transmitted and the echo being received, it is possible to determine the distance r for a given velocity of sound c.

$$r = \frac{c * \Delta t}{2} \tag{1}$$

The velocity of sound c of a fluid is depending on the current temperature and pressure, parameters which are dynamically varied during the experiments. Therefore this effect has to be compensated for precise deformation measurement.

For the signal generation bare piezoelectric ceramics by PI Ceramic (Figure 5) are used, which are used at the Institute FA for the monitoring of fiber composite processes [31]–[33]. Piezoelectric ceramics are cheap ultrasonic sensors and easy to integrate.

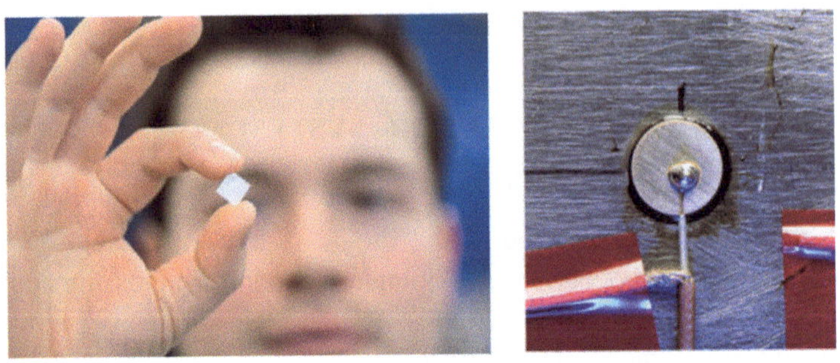

Figure 5: Bare piezoelectric ceramic (left); Sensor mounted and contacted (right)

7.2 Measuring system setup

The measuring system (Patent No. DE102015114492) consists of four components: measuring chamber, temperature control unit, hydraulic unit and control unit [34], [35]. Figure 6 shows the schematic structure and simplified operation.

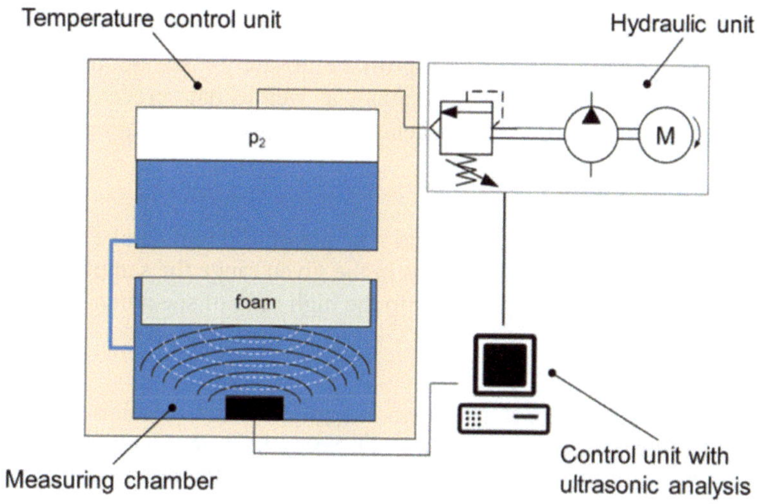

Figure 6: Setup of the measuring system

The measuring chamber contains the test cell in which the sample is analyzed (Figure 7). 24 ultrasound sensors are mounted on the inner chamber wall for "three-dimensional" measurement of the specimen size. To increase the accuracy and reliability, the sensors are arranged in pairs opposite. Due to the fact that ultrasound velocity depends on temperature and pressure an internal sound velocity measurement for compensation is integrated in the lower cover. The punctual fixing of foam sample by spike leads to a free deformation in all directions. For easy handling of the measurement, maintenance and additional functions, the measuring chamber is designed as a modular system.

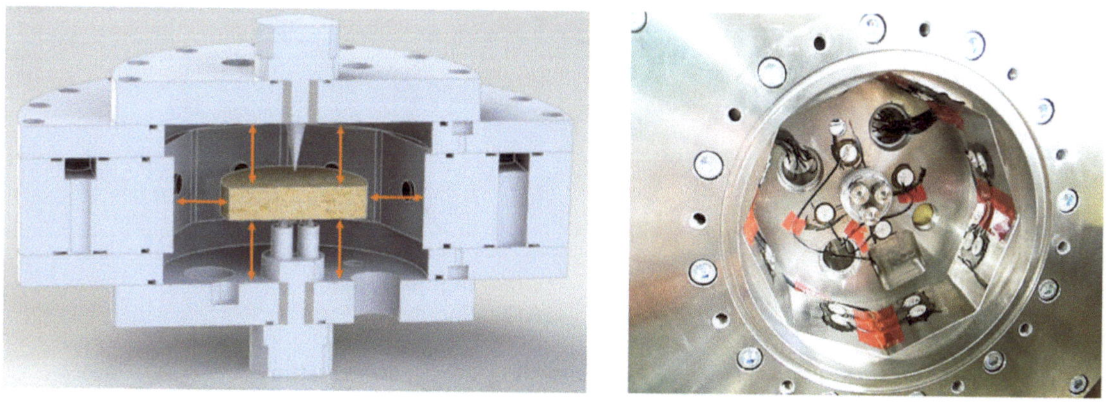

Figure 7: Sectional view of measuring chamber with test sample (left) and ultrasound sensors (right)

The process conditions of the RTM process are supplied by the temperature control unit and hydraulic unit. The thermal behavior of the chamber is controlled by a water-operated temperature control STW 200/1-36-60-H2.1 from SINGLE Temperiertechnik up to 200 °C. The integrated thermocouple close to the sample enables a precise temperature control. Pressure is applied in the chamber and thus on the sample via the hydraulic unit up to 20 bar designed by DELTA-Fluid Industrietechnik. A resin equivalent fluid is pumped from a storage container into the test cell. The pressure is controlled via the pressure sensor KISTLER Type 4001A integrated in the upper chamber cover. Thus, both continuous deformation and sudden failure of the foam can be considered for various high pressure changes and rates of change. Two independently scalable pressure measurement channels with different ranges ensure high resolution even with small pressure changes. The integrated temperature sensor provides information on the temperature conditions inside the chamber and allows also automatic temperature compensation of the pressure signal during the process.

The control unit includes the stored program control (SPC) for the test sequence including automatic program step, temperature and pressure profile and safety installations. The ultrasonic analysis performs the data evaluation of the ultrasonic signals as well as data processing for deformation visualization. In parallel, the central storage of all generated data of the measuring system takes place.

7.3 Test specimen

Specimens with a diameter of up to 120 mm and a height of 100 mm can be inserted. No special sample preparation is required. With the dimensions in the given range the sample represents the final part size. It also represents the cellular structure due to the high ratio of specimen-to-pore size.

7.4 Description of the test sequence

The measuring process includes four steps shown in Figure 8. First, a rotationally symmetric specimen with a digital cutting system is cut out of a plate (1). Second, the measuring chamber is filled with the sample and closed (2). Third, the test cell is filled with fluid and compression and thermal load is applied (3). Fourth, the ultrasound measurement and the deformation calculation of the specimen takes place simultaneously (4).

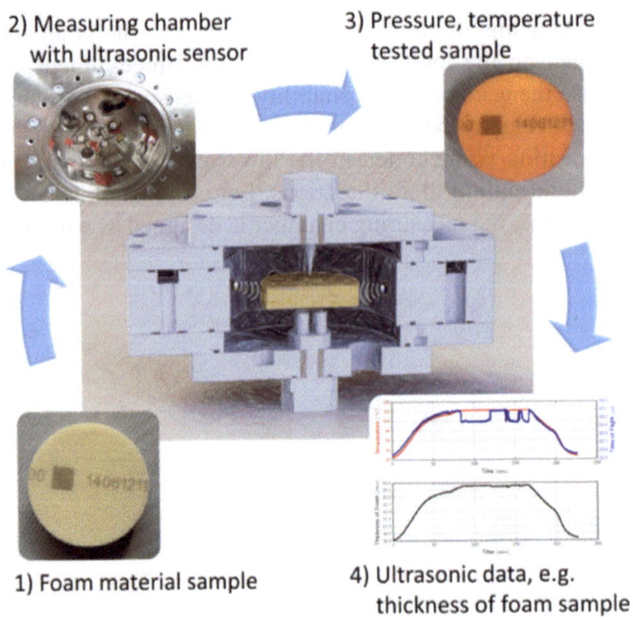

Figure 8: Test sequence for foam analysis by measurement system FAUSt

8 FOAM DEFORMATION MEASUREMENT BASED ON SOUND TRAVEL TIME

8.1 Calibration measurement

The size of the foam sample under the coupled load of pressure and temperature is measured by ultrasound waves. The ultrasonic sensors, mounted on the inner wall of the test chamber, send sound impulses through the fluid towards the foam sample. The waves are then detected by the same sensor after reflecting from the sample. The time of flight depends on the distance between sensor and test specimen and the fluid's sound velocity. The latter is also dependent of the current temperature and pressure, parameters which are dynamically varied during the experiments. Therefore this effect has to be compensated, which is done by a dedicated calibration setup.

In Figure 9 the result of a temperature calibration measurement is shown. The fluid temperature was varied from ambient temperature to 120 °C and simultaneously the sound velocity was acquired. The correlation shows a strong, but linear dependency between the sound velocity and the temperature.

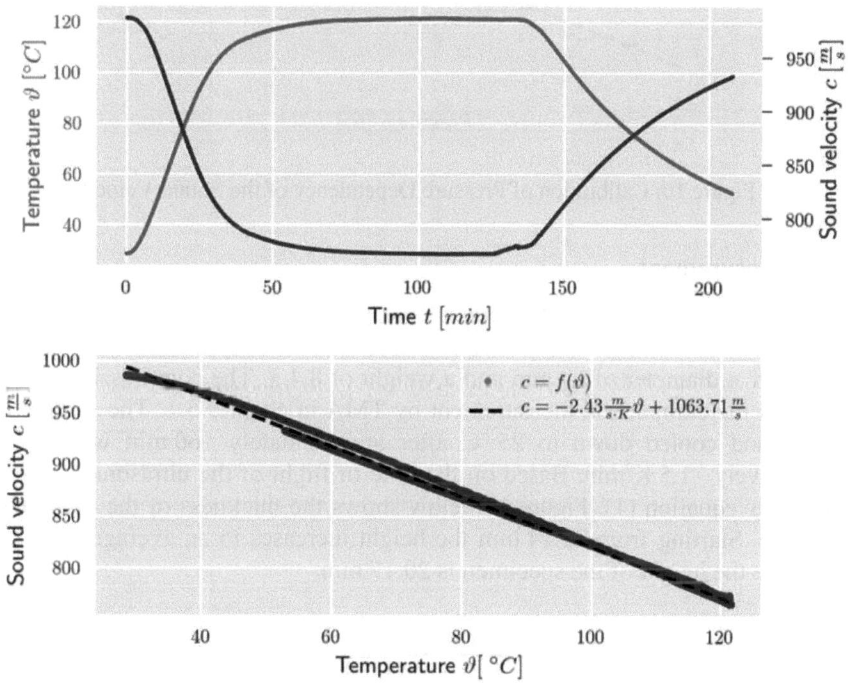

Figure 9: Calibration of Temperature Dependency of the Sound Velocity

Figure 10 shows the relationship of the fluid pressure and the sound velocity, which is less important. A pressure decrease of 1 bar has approximately the same effect as a temperature increase of 3.5 K.

In the test chamber both influences are compensated in a dedicated calibration setup, where the fluid's current sound velocity is measured over a fixed distance. Alternatively compensation can be derived from the temperature and pressure measurement and the presented calibrations.

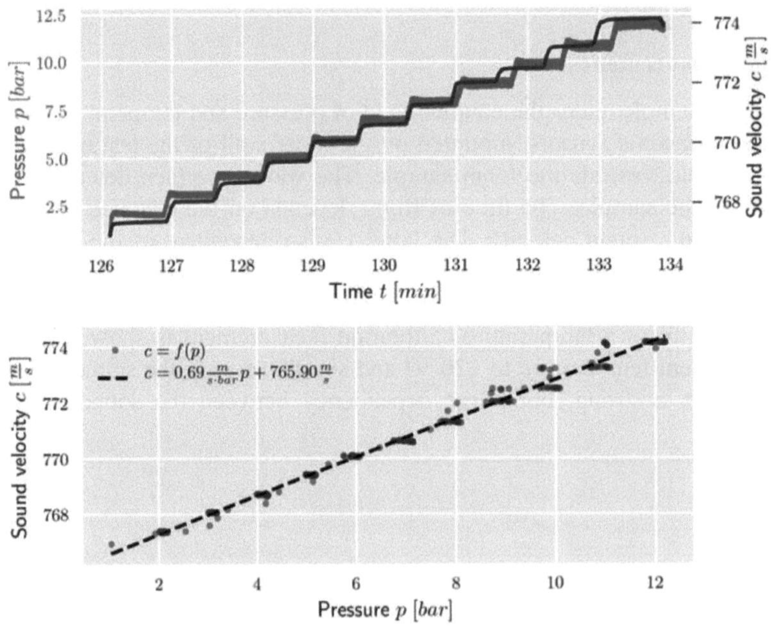

Figure 10: Calibration of Pressure Dependency of the Sound Velocity

8.2 PET foam measurement

A measurement of AIREX T92.100 PET foam core was performed by the procedure described in chapter 7.4 to proof the measurement principle introduced in chapter 7.1. The test specimen has a height of 20.14 mm, a diameter of 69 mm and a weight of 8.4 g. The compression load is set to 1 bar taking into account the comparison measurement by TMA in chapter 5.1. The sample is heated from 25 °C to 120 °C and cooled down to 25 °C after approximately 160 min with a heating rate of 1.5 K/min, respectively –1.5 K/min. Based on the time of flight of the ultrasonic signal the change in size is calculated by equation (1). Figure 11 below shows the thickness of the foam according to the temperature profile. Starting from 20.14 mm the height increases to an average of 23.55 mm. At the end of the test cycle the height of the specimen is 20.17 mm.

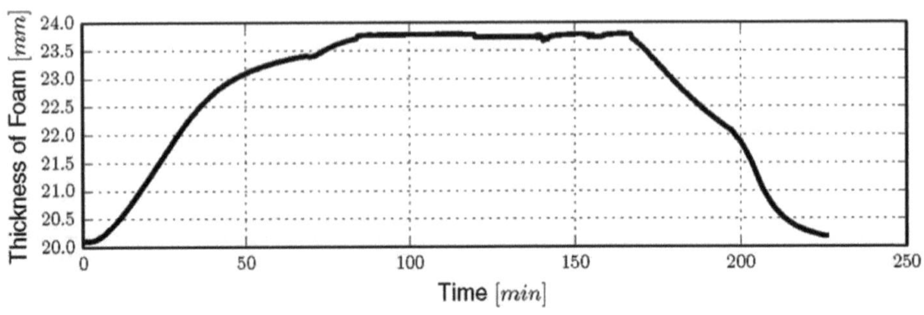

Figure 11: Proof of principle by PET foam thickness measurement

In terms of the standard ISO 11359-2, the linear coefficient of thermal expansion is calculated from the change in thickness at about 178.2 * 10^{-6} K^{-1}. This is approximately equivalent to the second run CTE measurement in Table 1. Nevertheless, the value is about 2.5 times the data sheet value. In addition, the sample showed an elliptical shape after removal. The smallest diameter is 62.36 mm and the largest diameter is 68.98 mm, hence a ratio of 0.9. The height increased slightly to 20.17 mm (increase of 0.1 %), as well as the weight to 12.3 g (increase of 46 %).

8.3 Results

In the first series of tests, the measuring principle as well as the functionality of the measuring system was tested. The verification of the measuring principle was carried out successfully by determination of CTEs. Ultrasonic sensors only consisting of bare piezoelectric ceramics can be used to measure the distance between foam core and sensor in a closed tooling. The temperature, pressure and fluid dependence of the ultrasound signal can be compensated by calibration. The measured deformation and therefore the coefficient of thermal expansion are of the same order of magnitude as the reference method. However, the final dimensions of the sample show uneven strain behavior, indicating directional material properties. In contrast to the given data the permanent deformation of the specimen shows an anisotropic behavior. Also a significant resin uptake of about 46 % compared to sample body weight was observed.

9 CONCLUSIONS

An efficient design of the processes requires not only process knowledge but also a profound and comprehensive understanding of the used materials. For this reason, a simple but highly efficient method based on ultrasonic sensors has been developed. Embedded into an innovative measuring system presented in this paper a quantified property description of foam materials under realistic manufacturing conditions is done for the first time. The main advantage of ultrasonic sensors is that the measurement is carried out without contact to the test specimen. In addition, the measurement speed is relatively high enabling the detection of fast or small state changes. Whereas temperature control system and hydraulic system can simulate the production conditions in an optimal way. And also the size of the sample takes into account material-equivalent influences as well as manufacturing and component-relevant influences. Due to 24 ultrasonic sensors positioned in the chamber, a multi-axis measurement and thus directional determination of the foam characteristics under temperature and pressure will be possible.

Beside the material characterization, the use of experimentally obtained data is primarily seen in the development of efficient, material-adapted impregnation strategies. The identified process parameter can help to find an ideal processing window and to establish a quality assurance. Furthermore, the data can be used for numerical simulation methods in the early development process.

ACKNOWLEDGEMENTS

The research leading to these results has received funding from the project Next Generation Car (NGC) by the German Aerospace Center (DLR)

REFERENCES

[1] S. Vohrer, C. David, M. Ruff, and H. E. Friedrich, “Funktionsintegrierte Faserverbundstrukturen im Fahrzeugbau”, in *Tagungsband LLC 2017*, 2017, pp. 198–208.

[2] G. Kopp, M. Münster, M. Kriescher, M. Ruff, S. Vohrer, and G. Kopp, “Next Generation Car: vehicle concepts, vehicle architectures and structural design for future road mobility”, in *7. Tagung für neue Fahrzeug- und Werkstoffkonzepte*, 2017.

[3] S. Vohrer, C. David, and M. Ruff, “Fiber reinforced composite structures in the Next-Generation-Car: Interurban Vehicle (NGC-IUV)”, *Carbon Composites Magazin*, no. 4/2016, p. 42, Nov-2016.

[4] S. Torstrick-v.d.Lieth, M. Wiedemann, and J. Stüve, “Is Carbon Already Obsolete for Non-Super Sports Cars?”, Dec. 2017.

[5] R. Kötte, "Der Resin-Transfer-Molding-Prozess: Analyse eines Harzinjektionsverfahrens", RWTH Aachen, 1991.

[6] S. Laurenzi and M. Marchetti, "Advanced Composite Materials by Resin Transfer Molding for Aerospace Applications", *Compos. Their Prop.*, pp. 197–226, 2012.

[7] S. Schmidhuber and S. Fenske, "Die ganze Palette des Leichtbaus", *K Magazin*, 2016.

[8] C. D. Rudd, A. C. Long, K. . Kendall, and C. G. . Magin, *Liquid moulding technologies*, vol. 1. 1997.

[9] T. M. Kruckenberg and R. Paton, *Resin Transfer Moulding for Aerospace Structures*. Dordrecht: Springer Netherlands, 1998.

[10] D. Bertling, R. Kaps, and E. Mulugeta, "Analysis of dry-spot behavior in the pressure field of a liquid composite molding process", *CEAS Aeronaut. J.*, vol. 7, no. 4, pp. 577–585, Dec. 2016.

[11] L. Herbeck, M. Kleineberg, and C. Schöppinger, "Foam Cores in RTM Structures, Manufacturing Aid of High-Performance Sandwich?", in *Go For the Best Through Advanced Materials & Processes*, Page Bros., Norwich, UK, 2002, pp. 515–525.

[12] DIAB, "DIAB Guide to Core and Sandwich", p. 48, 2012.

[13] A. C. Garay, J. A. Souza, and S. C. Amico, "Evaluation of mechanical properties of sandwich structures with polyethylene terephthalate and polyvinyl chloride core", *J. Sandw. Struct. Mater.*, vol. 18, no. 2, pp. 229–241, 2016.

[14] M. Schön *et al.*, "Leichtbau durch funktionsintegrierende Strukturen in Multi Material Design – hybride Sandwichbauweisen für Schienenfahrzeuge : PURtrain ; Abschlussbericht des Forschungsvorhabens", 2012.

[15] F. C. Campbell, "The Case Against Honeycomb Core", in *49th International SAMPE symposium and exhibition*, 2004, p. 301.

[16] J. Grünewald, P. Parlevliet, and V. Altstädt, "Manufacturing of thermoplastic composite sandwich structures", *J. Thermoplast. Compos. Mater.*, vol. 30, no. 4, pp. 437–464, Apr. 2017.

[17] M. Flemming, G. Ziegmann, and S. Roth, *Faserverbundbauweisen – Fertigungsverfahren mit duroplastischer Matrix*. Berlin, Heidelberg: Springer Berlin Heidelberg, 1999.

[18] J. Pflug, "Sandwich Materials Selection Charts", *J. Sandw. Struct. Mater.*, vol. 8, no. 5, pp. 407–421, Sep. 2006.

[19] 3A Composites, "Data Sheet: Airex T92", p. 3, 2017.

[20] G. Menges, E. Haberstroh, W. Michaeli, and E. Schmachtenberg, *Menges Werkstoffkunde Kunststoffe*. München: Carl Hanser Verlag GmbH & Co. KG, 2011.

[21] W. Grellmann and S. Seidler, *Kunststoffprüfung*. München: Carl Hanser Verlag GmbH & Co. KG, 2015.

[22] K. Bungardt *et al.*, Eds., *Die Prüfung der metallischen Werkstoffe*. Berlin, Heidelberg: Springer Berlin Heidelberg, 1939.

[23] C. Ferreira, "Measurement of the Nonuniform Thermal Expansion Coefficient of a PVC Foam Core by Speckle Interferometry – Influence on the Mechanical Behavior of Sandwich Structures", *J. Cell. Plast.*, vol. 42, no. 5, pp. 393–404, 2006.

[24] J. Scherble and T. Jahn, "New Rohacell Development For Resin Infusion Processes", in *Sandwich Structures 7: Advancing with Sandwich Structures and Materials*, 2005, pp. 753–761.

[25] P. Viot, "Hydrostatic compression on polypropylene foam", *Int. J. Impact Eng.*, vol. 36, no. 7, pp. 975–989, Jul. 2009.

[26] O. Almanza, Y. Masso-Moreu, N. J. Mills, and M. A. Rodrícuez-PÉREZ, "Thermal expansion coefficient and bulk modulus of polyethylene closed-cell foams", *J. Polym. Sci. Part B Polym. Phys.*, vol. 42, no. 20, pp. 3741–3749, 2004.

[27] U. Lang, “Rohacell – High performance foam cores for aircraft application”. Evonik, p. 91, 2011.

[28] E. Kappel, “Spring-in of curved CFRP/foam-core sandwich structures”, *Compos. Struct.*, vol. 128, pp. 155–164, Sep. 2015.

[29] D. Q. Deng and L. Xu, “Measurements of thermal expansion coefficient of phenolic foam at low temperatures”, *Cryogenics (Guildf).*, vol. 43, no. 8, pp. 465–468, Aug. 2003.

[30] J. Krautkrämer and H. Krautkrämer, *Werkstoffprüfung mit Ultraschall.* Berlin, Heidelberg: Springer Berlin Heidelberg, 1986.

[31] N. Liebers, F. Raddatz, and F. Schadow, “EFFECTIVE AND FLEXIBLE ULTRASOUND SENSORS FOR CURE MONITORING FOR INDUSTRIAL COMPOSITE PRODUCTION”, in *Deutscher Luft- und Raumfahrtkongress (DGLR)*, 2012.

[32] N. Liebers, F. Raddatz, and F. Schadow, “Überwachung eines Herstellungsprozesses”, 2013.

[33] M. Kleineberg *et al.*, “Fully controlled production environment for autoclave injection processes”, in *Jec Europe Composites Show and Conference*, 2015, pp. 1–32.

[34] D. Bertling, N. Liebers, and M. Opitz, “Verfahren und Vorrichtung zur Analyse von Formkörpern als Komponenten für Faserverbundbauteile”, 2016.

[35] T. Nörtershäuser and M. Opitz, “Entwicklung eines Messsystems zur Analyse von Formkörpern hinsichtlich der Verwendung in Resin Transfer Moulding Prozessen”, Hochschule Bonn-Rhein-Sieg, 2016.